Discrete Mathematics and Theoretical Computer Science

Springer
London
Berlin
Heidelberg
New York
Barcelona
Hong Kong
Milan
Paris
Singapore
Tokyo

We can see this also by substituting R into S, or vice versa:

$$\begin{aligned} R(S,S^*) &= SS^* = (y/x)(y/x)^* = (y/x)(x^2/y) = x, \\ S(R,R^*) &= R^*/R = (x^2y)/(xy) = x. \end{aligned}$$

A more interesting example is the self-inverse

$$R(x,y) = \frac{x^2+y+1}{xy}.$$

In fact we have

$$R(x,y)^* = R(y,x^2) = \frac{x^2+y^2+1}{x^2y}$$

and

$$\begin{aligned} R(R,R^*) &= \frac{R^2+R^*+1}{RR^*} \\ &= \frac{x\left(x^4+y^2+1+(x^2+y^2+1)y+x^2y^2\right)}{(x^2+y+1)(x^2+y^2+1)} \\ &= \frac{x\left(x^4+(y^2+y)x^2+(y+1)^3\right)}{(x^2+y+1)(x^2+y^2+1)} \\ &= x \end{aligned}$$

As already mentioned we can compute the inverse for global $\mathcal{P}$-forms explicitly. We demonstrate this at our example. The fundamental equational system is

$$\begin{aligned} \frac{x^2+y+1}{xy} &= a, \\ \frac{x^2+y^2+1}{x^2y} &= b, \end{aligned}$$

The first equation means $(ax+1)y = x^2+1$, and substituting y in the second equation gives $a^2+1 = b(ax+1)$ after division by $x^2(x^2+1)$, that is

$$x = R^{-1}(a,b) = \frac{a^2+b+1}{ab},$$

which confirms again that $R^{-1} = R$.

This example is the special case $m = 2$ of the following theorem (cf. [9]):

Theorem 1. *For arbitrary m set $\epsilon = 0$ if m is odd and $\epsilon = 1$ if m is even. Then*

$$Q_m(x_0,\ldots,x_{m-1}) = \frac{x_0^2+\sum_{i=1}^{m-1} x_i + \epsilon}{x_0x_1} = \frac{\left(\sum_{i<m} x_i\right)^* + \epsilon}{x_0x_1}$$

is a global $\mathcal{P}$-form with inverse

$$Q'_m(x_0, ..., x_{m-1}) = \sum_{i=1}^{m} \sum_{e} x_0^{e_0} x_1^{e_1} \cdots x_{i-1}^{e_{i-1}},$$

where for each i, in the right hand sum e runs through all vectors of $\{1, -1\}^i$ such that

$e_{i-1} = e_0 = -1$, *or* $e_0 = 1$ *if* $i = m$,
$(e_{j-1}, e_j) \neq (1, 1)$ *for all* $j < i$.

Proof. In [9] it has been shown, expressed in our terminology, that the fundamental equational system, i.e. $Q_m(x_0, ..., x_{m-1}) = a_0$ and its conjugates, implies that

$$x_0 = Q'_m(a_0, ..., a_{m-1}).$$

This implies that $(Q_m)_L$ is a permutation with inverse $(Q'_m)_L$:

$$(Q_m)_L\left((Q'_m)_L(\xi)\right) = \left((Q_m)_L \circ (Q'_m)_L\right)(\xi) = (Q_m \circ Q'_m)_L(\xi) = \xi$$

for all L. In other words all local interpretations of $(Q_m \circ Q'_m) + \mathbf{1}$ vanish, and the same argument applies to $(Q'_m \circ Q_m) + \mathbf{1}$. Therefore by Lemma 1 we conclude that $Q_m^{-1} = Q'_m$. □

The Q'_m can also be described by a recursion (see [9]). Define A_i and B_i as follows:

$$\begin{aligned} A_1(x_0) &= 0, \\ A_2(x_0, x_1) &= x_0 x_1^{-1}, \\ A_{i+2}(x_0, ..., x_{i+1}) &= A_{i+1}(x_0, ..., x_i)\, x_{i+1}^{-1} + A_i(x_0, ..., x_{i-1})\, x_i x_{i+1}^{-1}, \; i \geq 1, \end{aligned}$$

$$\begin{aligned} B_1(x_0) &= x_0^{-1}, \\ B_2(x_0, x_1) &= x_0^{-1} x_1^{-1}, \\ B_{i+2}(x_0, ..., x_{i+1}) &= B_{i+1}(x_0, ..., x_i)\, x_{i+1}^{-1} + B_i(x_0, ..., x_{i-1})\, x_i x_{i+1}^{-1}, \; i \geq 1. \end{aligned}$$

Then we have

$$Q'_m(x_0, ..., x_{m-1}) = A_m(x_0, ..., x_{m-1}) + {\sum_{i=1}^{m}}' B_i(x_0, ..., x_{i-1}).$$

In particular

$$Q'_2(x_0, x_1) = x_0 x_1^{-1} + x_0^{-1} + x_0^{-1} x_1^{-1} = \frac{x_0^2 + x_1 + 1}{x_0 x_1} = Q_2(x_0, x_1),$$

which confirms that for $m = 2$ Theorem 1 gives the example considered above.

T. Helleseth, P.V. Kumar and K. Yang (Eds)

Sequences and their Applications

Proceedings of SETA '01

Springer

T. Helleseth
Department of Informatics, University of Bergen, N-5020 Bergen, Norway

P.V. Kumar
Communication Sciences Institute, Department of Electrical Engineering Systems, University of Southern California, Los Angeles, CA 90089-2565, USA

K. Yang
Department of Electronic and Electrical Engineering, Pohang University of Science and Technology (POSTECH), Pohang, Kyungbuk 790-784, Korea

British Library Cataloguing in Publication Data
Sequences and their applications : proceedings of SETA '01.
- (Discrete mathematics and theoretical computer science)
1.Sequences (Mathematics) - Congresses 2.Sequences (Mathematics) - Data processing - Congresses
I.Helleseth, Tor II.Kumar, P.V. III.Yang, K. IV.SETA '01 (Conference : 2nd : 2001 : Bergen, Norway)
515.2'4
ISBN-13:978-1-85233-529-8 e-ISBN-13:978-1-4471-0673-9
DOI: 10.1007/978-1-4471-0673-9

Library of Congress Cataloging-in-Publication Data
A catalog record for this book is available from the Library of Congress

Discrete Mathematics and Theoretical Computer Science Series ISSN 1439-9911

ISBN-13:978-1-85233-529-8 e-ISBN-13:978-1-4471-0673-9
Springer-Verlag London Berlin Heidelberg
a member of BertelsmannSpringer Science+Business Media GmbH
http://www.springer.co.uk

Softcover reprint of the hardcover 1st edition 2002

Typesetting: Camera ready by editors

34/3830-543210 Printed on acid-free paper SPIN 10842797

For odd m, the permutation polynomials associated to $1/Q_m$ are linearized modifications of the MCM polynomials, found 1994 by Müller [14], Cohen and Matthews [4]. The main motivation to introduce the Q_m in [7] was the fact that they can be used to derive the so-called APN property of Kasami power functions. The Q'_m were introduced in [9] to give a trace expansion for a conjectured (see Section 3 of [9]) and recently confirmed (see [6]) new class of difference sets with Singer parameters.

Theorem 1 states that $Q_m, Q'_m \in \mathcal{G}_m$. Up to trivial examples like

$$\eta(x_0, ..., x_{m-1}) = x_{m-1}x_{m-2}\cdots x_0, \quad \eta^{-1}(x_0, ..., x_{m-1}) = x_{m-1}/x_0$$

and the building of compositions, no other elements in $\mathcal{G}_m$ are known at present. We propose a further investigation of $\mathcal{G}_m$, for instance the following topics could be studied:

- find more examples of global $\mathcal{P}$-forms (i.e. elements of $\mathcal{G}_m$),
- find a small subset of $\mathcal{G}_m$ generating it (as a group),
- what is the structure of the (infinite countable) group $\mathcal{G}_m$?
- are there members of $\mathcal{G}_m$ with finite order for $m > 2$, different from trivial examples like $1/x_0$?
- does Q_m have finite order for $m > 2$?

Recall that for $m = 2$ we have seen above that $Q_2 \circ Q_2 = \mathbf{1}$,

4 Non-global $\mathcal{P}$-forms

In this section we shall demonstrate how the $\mathcal{L}$-method works for non-global $\mathcal{P}$-forms (resp. $\mathcal{O}$-forms) and that it applies to all of the presently known five (non-trivial) examples, as we have checked with computer support.

Three of these examples (see Subsection 4.2, 4.3 and 4.5) are well-known results of Glynn [11] and Cherowitzo [5] on o-polynomials, which where originally proven by ad-hoc arguments for the different cases.

The other two example (see Subsections 4.1 and 4.4) are results of the author, found in connection with the recent confirmation of long-standing conjectures of Welch and Niho on binary m-sequences (see [1], [2] and [12]). The first example in Subsection 4.1, which requires the least computational effort, is used to explain our technique. The original proof [8] for the example in Subsections 4.4 was the first application of the $\mathcal{L}$-method in the sense described below, and it has motived this work.

We cannot write down the complete computations for the examples in Subsections 4.2 – 4.5, where we shall restrict ourselves to give the fundamental equational system and the fundamental polynomial ϕ, respectively.

4.1 First Example ($m = 2$)

In [7] it has been shown, using the terminology introduced in Section 2:

Theorem 2. *The function*

$$R(x,y) = x(y + x^2 + 1)$$

is a non-global $\mathcal{P}$-form.

We shall describe the $\mathcal{L}$-method for non-global $\mathcal{P}$-forms in detail at this example. (The proof given in [7] is in the last part slightly different and uses particular properties of this special case. The general $\mathcal{L}$-method, which will be explained below, is applied first in [8].) We consider the fundamental equational system:

$$R(x,y) + a = 0,$$
$$S(x,y) + b = 0,$$

where $S(x,y) = R(x,y)^* = R(y,x^2)$. We have to show that if we interprete $R(x,y) + a = 0$ in L, setting $\alpha = R_L(\xi_0)$, then there is no other preimage of α in L than ξ_0:

$$R_L(\xi) = \alpha \text{ implies } \xi = \xi_0 \text{ for } \xi \in L.$$

Step 1. We eliminate the variable y from the fundamental equational system. To this end multiply $R(x,y) + a$ resp. $S(x,y) + b$ with suitable polynomials in x and y such that $R(x,y) + a$ resp. $S(x,y) + b$ become polynomials, say P resp. Q. Compute the resultant of P and Q with respect to y. This gives a polynomial ρ in x with coefficients in $\mathbf{F}_2[a,b]$, and we conclude

$$\rho(x) = 0.$$

In the present case we have

$$\rho(x) = x^9 + ax^6 + x^5 + ax^4 + (a^2 + b)x^3 + a^2x + a^3$$

For $a \in \mathbf{F}_2$ has to be shown that $x \in \mathbf{F}_2$. (This is a routine matter.) Then we can assume $x, y, a, b \notin \mathbf{F}_2$. Factorize ρ, divide it by trivial factors (like for instance powers of x or $x + 1$) and replace x by the symbol t to get the *fundamental polynomial* ϕ. Here we have no trivial factors and therefore

$$\phi(t) = t^9 + at^6 + t^5 + at^4 + (a^2 + b)t^3 + a^2t + a^3.$$

The conjugated polynomial is

$$\phi^*(u) = u^9 + bu^6 + u^5 + bu^4 + (a^2 + b^2)u^3 + b^2u + b^3.$$

Step 2. Compute the discriminant of ϕ and decompose it into irreducible factors to show that it is non-zero. (Possibly discuss exceptional cases.) For our example

$$\mathrm{disc}(\phi) = a^{12}b^6.$$

Step 3. Define the polynomial $\tilde{\phi}$ by substituting a by $R(x,y)$ and b by $S(x,y)$ and by mulitiplying with suitable polynomial to avoid the denominators. (The latter is not necessary here.) Hence the coefficients of $\tilde{\phi}$ are polynomials in x and y. We know that x is a zero of ϕ and therefore of $\tilde{\phi}$. Hence we can factor $\tilde{\phi}$ by $t+x$. We define

$$\psi(t) = \frac{\tilde{\phi(t)}}{t+x}.$$

Here we have

$$\begin{aligned}\psi(t) = {} & t^8 + xt^7 + x^2t^6 + x(y+1)t^5 + (x^2+x^2y+1)t^4 + xy(x^2+1)t^3 + \\ & (x^6+x^4y+x^2y^2+x^2+y^3+y)t^2 + \\ & x(x^6+x^4y+x^2y^2+x^2+y^3+y)t + x^2(x^2+x+y)^3.\end{aligned}$$

Step 4. Decompose ψ into irreducible factors. (As demonstrated in [8] there is a very efficient algorithm to compute decompositions of multivariate polynomials over $\mathbf{F}_2$.) Concretely we have here

$$\psi(t) = \psi^{(1)}(t)\,\psi^{(2)}(t)$$

with

$$\begin{aligned}\psi^{(1)}(t) &= t^4 + xt^3 + yt^2 + x(y+x^2+1)t + (y+x^2+1)^2,\\ \psi^{(2)}(t) &= t^4 + (x^2+y)t^2 + y^2t + x^2(y+x^2+1).\end{aligned}$$

Thus for the conjugate $\psi^*(u) = (\psi^{(1)})^*(u)\,(\psi^{(2)})^*(u)$ with

$$\begin{aligned}(\psi^{(1)})^*(u) &= u^4 + yu^3 + x^2u^2 + y(x^2+y^2+1)u + (x^2+y^2+1)^2,\\ (\psi^{(2)})^*(u) &= u^4 + (x^2+y^2)u^2 + x^4u + y^2(x^2+y^2+1).\end{aligned}$$

Step 5. Via the computation of the resultant of $R(x,y)+a$ and $S(x,y)+b$ with respect to y (see Step 1) we get an equation of the form $g(x)y = f(x)$. In our example we have

$$\begin{aligned}f(x) &= x^3 + x + a,\\ g(x) &= x.\end{aligned}$$

Show that $g(x)$ is non-zero if $x \notin \mathbf{F}_2$. (More precisely discuss possibly occurring exceptional cases where g_L has a zero in $L \setminus \mathbf{F}_2$.) We call

$$\Gamma(t) = \sqrt{\left(\frac{f(t)}{g(t)}\right)^*}$$

the *fundamental mapping*. (Consider the square root in the definition of Γ as a formal symbol. It can be evaluated only if a substitution for t has the property that $f(t)/g(t)$ is a square. However, Γ_L is a mapping in view of $\sqrt{\xi} = \xi^{2^{n-1}}$.) Note that x is a fixpoint of Γ:

$$\Gamma(x) = \sqrt{\left(\frac{f(x)}{g(x)}\right)^*} = \sqrt{y^*} = \sqrt{x^2} = x.$$

We verify two statements:

$$\psi^{(1)}(t) \textit{ is a divisor of the numerator of } (\psi^{(2)})^*(\tilde{f}(t)/\tilde{g}(t)). \tag{1}$$

$$\psi^{(2)}(t) \textit{ is a divisor of the numerator of } (\psi^{(1)})^*(\tilde{f}(t)/\tilde{g}(t)). \tag{2}$$

Here $\tilde{f}(t)$, resp. $\tilde{g}(t)$, is obtained from $f(t)$, resp. $g(t)$, by substituting $a = R(x,y)$, $b = S(x,y)$ and by multiplication with a suitable polynomial such that both, $\tilde{f}(t)$ and $\tilde{g}(t)$ become a polynomial. Checking these divisibilities is a straightforward computation. In our example we have

$$\tilde{f}(t) = t^3 + t + xy + x^3 + x$$
$$\tilde{g}(t) = t$$

The numerators of $(\psi^{(1)})^*(\tilde{f}(t)/\tilde{g}(t))$ and $(\psi^{(2)})^*(\tilde{f}(t)/\tilde{g}(t))$ are respectively

$$N_1(t) = \tilde{g}(t)^4 \ (\psi^{(1)})^*(\tilde{f}(t)/\tilde{g}(t)),$$
$$N_2(t) = \tilde{g}(t)^4 \ (\psi^{(2)})^*(\tilde{f}(t)/\tilde{g}(t)),$$

since $(\psi^{(1)})^*$ and $(\psi^{(2)})^*$ have degree 4, and in our example we have

$$\begin{aligned} N_2(t)/\psi^{(1)}(t) = {} & t^8 + xt^7 + (x^2+y)t^6 + x(y+1)t^5 + (y+1)(x^2+y+1)t^4 + \\ & x^3yt^3 + x^2(y^2+x^2+x^2y+x^4)t^2 + x^5(x^2+y+1)t + \\ & x^4(x^2+y+1)^2, \\ N_1(t)/\psi^{(2)}(t) = {} & t^8 + x^2t^6 + xyt^5 + yt^4 + xy(x^2+1)t^3 + (x^6+x^4+x^4y + \\ & x^2y^2+y^3+y^2)t^2 + x^2(x^2+x+y)^3, \end{aligned}$$

confirming (1) and (2).

Basic Lemma. *From Steps 1 – 5 it follows that R is a $\mathcal{P}$-form.*

Proof. Suppose $R_L(\xi_0) = \alpha$ for fixed $\xi_0, \alpha \in L$. Then ϕ_L, Γ_L etc. are fixed polynomials over L if we define them by interpreting their coefficients according to $x_L = \xi_0$, $a_L = \alpha$.

From Step 2 we conclude, after discussion of the trivial case $\alpha \in \mathbf{F}_2$, that

- *ϕ_L has only simple zeros, and one of them is ξ_0.*

We claim:

$$\textit{If } \xi \textit{ is a zero of } \psi_L^{(1)} \textit{ in } L \textit{ then } \Gamma_L(\xi) \in L \textit{ is a zero of } \psi_L^{(2)}. \tag{3}$$

$$\textit{If } \xi \textit{ is a zero of } \psi_L^{(2)} \textit{ in } L \textit{ then } \Gamma_L(\xi) \in L \textit{ is a zero of } \psi_L^{(1)}. \tag{4}$$

In fact for instance from $\psi_L^{(1)}(\xi) = 0$ we derive using (1) above that

$$(\psi_L^{(2)})^*(\tilde{f}_L(\xi)/\tilde{g}_L(\xi)) = 0,$$

where the exceptional case $\tilde{g}_L(\xi) = g_L(\xi) = 0$ needs extra considerations. We conclude

$$\begin{aligned}\left(\psi_L^{(2)}(\Gamma_L(\xi))\right)^* &= (\psi_L^{(2)})^*(\Gamma_L(\xi)^*)\\ &= (\psi_L^{(2)})^*(f_L(\xi)/g_L(\xi))\\ &= (\psi_L^{(2)})^*(\tilde{f}_L(\xi)/\tilde{g}_L(\xi))\\ &= 0,\end{aligned}$$

where * has to be interpreted as φ_L. Therefore

$$\psi_L^{(2)}(\Gamma_L(\xi)) = 0,$$

we conclude that

- *Γ_L has precisely one fixpoint among the zeros of ϕ_L in L, namely ξ_0,*

since according to the definition of the ψ_i $(i < 2)$ we have

$$\phi_L(t) = \gamma\,(t + \xi_0)\;\psi_L^{(1)}(t)\;\psi_L^{(2)}(t)$$

for some $\gamma \in L$, and a zero of $\psi_L^{(1)}$ or $\psi_L^{(2)}$ cannot be a fixpoint of Γ_L by (3) and (4) above and the fact that ϕ has only simple zeros. It is now important to note that ϕ and Γ only depend of a and b (and not on x or y). Hence

- *ϕ_L and Γ_L depend only on α.*

This means that another R_L-preimage of α, say ξ_1, leads to the same ϕ_L and Γ_L as ξ_0, and therefore also ξ_1 is a zero of ϕ_L and a fixpoint of Γ_L. But as shown above, these properties characterize ξ_0, that is $\xi_1 = \xi_0$, and it follows that R_L is a permutation polynomial, since L is finite. This completes the proof of the basic lemma. □

We list again which computations had to be done here mainly in order to show that R is a $\mathcal{P}$-form.

Step 1:
Compute the polynomial ϕ over $\mathbf{F}_2[a, b]$ (eliminate y from the equations $R(x, y) = a$ and $Q(x, y) = b$). The degree of ϕ is 9 (as in all other examples we know).

Step 2:

Compute the discriminant of ϕ, factorize it to comfirm that it does not vanish. (Possibly discuss exceptional cases.)

Step 3:

Compute the polynomial ψ over $\mathbf{F}_2[x,y]$ with degree 8 by first substituting $R(x,y)$ for a and $Q(x,y) = R(x,y)^*$ for b in ϕ and then dividing by the linear factor corresponding to the zero x of ϕ.

Step 4:

Compute the decomposition $\psi = \psi^{(1)}\,\psi^{(2)}$, where $\psi^{(1)}$ and $\psi^{(2)}$ both have degree 4. (For all examples, which we know, ψ splits this way.)

Step 5:

Compute the equation $g(x)y = f(x)$ with polynomials f and g of minimal degree; $f, g \in \mathbf{F}_2[a,b]$. (Discuss the exceptional case $g(x) = 0$.)

Compute the polynomials $\tilde{f}(t)$ and $\tilde{g}(t)$ over $\mathbf{F}_2[x,y]$ by first substituting $R(x,y)$ for a and $S(x,y)$ for b in $f(t)$ resp. $g(t)$, and then multiplying with a suitable polynomial x and y to avoid the denominators.

Divide $\tilde{g}(t)^4\,(\psi^{(2)})^*(\tilde{f}(t)/\tilde{g}(t))$ by $\psi^{(1)}(t)$, resp. $\tilde{g}(t)^4\,(\psi^{(1)})^*(\tilde{f}(t)/\tilde{g}(t))$ by $\psi^{(2)}(t)$ in order to confirm that the remainder vanishes.

The $\mathcal{L}$-method works for $m = 4$ in precisely the same way as shown in [8] (see Subsections 4.4 and 4.5). Up to trivial cases we have no non-global examples for $m \neq 2, 4$. The original proof [7] for the preceding example in this subsection is simpler than the proof given here to illustrate the $\mathcal{L}$-method in the general case. Acutally Step 5 can be avoided here, since one can show easily that the interpretations of ψ_1 and ψ_2 have no zeros at all in L. We mention that Steve Cohen [3] has found a short proof for this case in the spirit of Dickson polynomials.

4.2 Second Example: Glynn's o-Polynomials of Type II ($m = 2$)

In this subsection we assume $m = 2$, that is $n = 2m' - 1$. Glynn [11] showed that the power function

$$\xi^{3\times 2^{m'}+4}$$

is an o-polynomial on $L = \mathbf{F}_{2^n}$; expressed in our terminology:

Theorem 3. *The function*

$$G(x,y) = x^4 y^3$$

is an $\mathcal{O}$-form.

It is trivial that G is a $\mathcal{P}$-form. (We have $G = \eta \circ \eta$ for $\eta(x,y) = x^2y = (xy)^*$.) In the monomial case, in order to confirm the $\mathcal{O}$-form condition (see Section 1) it suffices to consider the cases $c = 0$ and $c = 1$ for the constant. For $c = 0$ we have to verify that $G_0(x,y) = G(x,y)/x = (xy)^3$ is a $\mathcal{P}$-form, which is obvious. Thus the main statement of the preceding theorem is that

$$G_1(x,y) = \frac{G((x,y)+1)+G(1)}{x} = \frac{(x+1)^4(y+1)^3+1}{x}.$$

Here the fundamental equational system is

$$\begin{aligned} y^3x^4 + y^3 + y^2x^4 + y^2 + yx^4 + y + x^4 + ax &= 0,\\ x^6y^4 + x^6 + x^4y^4 + x^4 + x^2y^4 + x^2 + y^4 + by &= 0. \end{aligned}$$

The fundamental polynomials is:

$$\begin{aligned} \phi(t) = {} & a^4t^9 + a^4t^7 + a^3bt^6 + (a^4 + a^2b + 1)t^5 + abt^4 + (b^3 + a^2b^2 + b^2 + a^4)t^3 + \\ & a^3bt^2 + (a^2b^2 + a^2b) + b^2 + 1)t + ab^3 + ab \end{aligned}$$

4.3 Third Example: Cherowitzo's o-Polynomials ($m = 2$)

In this subsection we assume $m = 2$, that is $n = 2m' - 1$. Cherowitzo [5] showed that

$$\xi^{2^{m'}} + \xi^{2^{m'}+2} + \xi^{3\times 2^{m'}+4}$$

is an o-polynomial on $L = \mathbf{F}_{2^n}$; expressed in our terminology:

Theorem 4. *The function*

$$C(x,y) = x^4y^3 + x^2y + y,$$

is an $\mathcal{O}$-form.

We have $C = Q_2 \circ \eta$ for

$$\eta(x,y) = \frac{1}{x^2y}.$$

Thus C is a $\mathcal{P}$-form by Theorem 1, and the main statements of the preceding theorem are that

$$\begin{aligned} C_0(x,y) &= x^3y^3 + xy + y/x,\\ C_{c,d}(x,y) &= \frac{C(x+c,y+d)+C(c,d)}{x} \end{aligned}$$

are $\mathcal{P}$-forms (with $c^* = d$ and $d^* = c^2$). Although $C_{c,d}$ covers C_0 for $c = d = 0$, we have to consider the (degenerated) case $c = d = 0$ separately. For C_0 the fundamental equational system is

$$\begin{aligned} x^4y^3 + x^2y + y &= ax,\\ x^6y^4 + x^2y^2 + x^2 &= by, \end{aligned}$$

and we have

$$\phi(t) = a^4t^9 + a^3bt^6 + t^5 + abt^4 + b^2t^3 + a^2b^2t^3 + b^2t + ab^3.$$

For $C_{c,d}$ the fundamental equational system is

$$\begin{aligned}(y+d)^3x^4 + (y+d)x^2 + c^4y^3 + c^4dy^2 + (c^2 + c^4d^2 + 1)y &= ax,\\ (x+c)^6y^4 + (x+c)^2y^2 + d^4x^6 + d^4c^2x^4 + (d^2 + d^4c^4 + 1)x^2 &= by,\end{aligned}$$

and we have

$$\begin{aligned}\phi(t) = {} & a^4t^9 + a^4c^2t^7 + a^3bt^6 + \\ & (a^2bc^2d + a^2bc^4d^3 + a^2c^6d^4 + a^2c^2d^2 + a^4c^4 + a^2c^2 + c^{12}d^8 + c^8d^4 \\ & \quad b^2d^2 + c^{16}d^{12} + c^4d^4 + c^4 + d^4 + 1)t^5 + \\ & (abc^2d^2 + abc^4d^2 + abc^6d^4 + abc^2 + abc^8d^6 + ab^2d + abd^2 + ab)t^4 + \\ & (a^2b^2 + b^2c^2d^2 + b^3d^3 + a^4c^6 + b^2d^2 + b^2c^6d^6 + b^2c^2d^4 + b^2 + b^2c^4d^4)t^3 + \\ & a^3bc^4t^2 + (b^2c^6d^4 + a^2c^6d^2 + b^2c^8d^6 + c^{20}d^{12} + b^2 + a^2c^{10}d^4 + b^2d^2 + \\ & \quad a^2bc^8d^3 + b^2c^2d^2 + c^{12}d^4 + c^8d^4 + b^2c^2 + a^2b^2c^2 + \\ & \quad b^3d + a^2c^6 + c^{16}d^8 + d^4c^4 + c^4 + a^2bc^6d + c^8)t + abc^6d^2 + abc^8d^2 + \\ & abc^4 + abc^{10}d^4 + ab^3 + ab^2c^4d + abc^4d^2 + abc^6 + abc^{12}d^6\end{aligned}$$

4.4 Fourth Example ($m = 4$)

In this subsection we assume $m = 4$. The author proved in [8]:

Theorem 5. *The function*

$$D(x, y, z, u) = \frac{(x^2+1)z(u+1)}{x} + x$$

is a (non-global) $\mathcal{P}$-form.

The fundamental equational system is:

$$\begin{aligned}(x^2+1)z(u+1) + x^2 + ax &= 0,\\ (y^2+1)u(x^2+1) + y^2 + by &= 0,\\ (z^2+1)x^2(y^2+1) + z^2 + cz &= 0,\\ (u^2+1)y^2(z^2+1) + u^2 + du &= 0.\end{aligned}$$

The fundamental polynomial is

$$\begin{aligned}\phi(t) = {} & d^2(a^2b^2d^2 + (c^2 + d^2 + 1)(a^2 + c + 1))t^9 + \\ & acd^2(c^2 + d^2 + 1)t^8 + \\ & d^2(a^2cb^2d + b^2 + a^2b^2c + a^2c^2 + b^2d^2 + a^2b^2 + a^4c^2 + b^2c^2 + \\ & \quad a^2b^2d^2 + a^2c^3)t^7 +\end{aligned}$$

$$
\begin{aligned}
&acd^2(bd+b^2+c^2)t^6+\\
&(1+a^2+c+b^2+d^2+a^2c^2+a^4+a^6+c^2+c^3+b^2c+a^2b^2+a^4b^2+\\
&\ a^4d^2+cd^2+a^2d^2+b^2d^4+a^6b^2+a^6d^2+a^4c^2+a^6c^2+c^3d^2+c^2d^2+\\
&\ a^4c^3+b^2c^3+b^2c^2+b^2cd+a^2b^2d^2+cb^2d^2+a^4c+b^2cd^3+a^4b^2c+\\
&\ a^4cd^2+a^4b^2d^2+b^2c^3d+a^2b^2d^4+b^2c^2d^2+a^2c^2d^2+a^2b^2c^2+a^4b^2cd)t^5+\\
&ac(1+b^2d^2+c^2+a^4+a^4b^2d+d^2+b^2d^3+b^2d+b^2c^2+c^2d^2+b^2+\\
&\ a^4b^2+b^2c^2d+a^4c^2+a^4d^2)t^4+\\
&(a^2c^3d^2+a^6c^2+a^2c^2d^2+a^4b^2c^2d^2+b^2d^2+a^2b^2d^4+a^4c^2+a^2b^2cd^2+\\
&\ a^2c^2+a^2c^3+a^2b^2d^2+a^4c^2d^2+a^2b^2cd^3+b^2d^4+a^6c^3+a^8c^2)t^3+\\
&a^3c(a^4c^2+c^2+c^2d^2+b^2d^3+b^2d^2)t^2+\\
&(1+a^2+c+b^2+a^4+d^4+a^6+b^2c+a^2b^2+a^4b^2+a^4d^2+cd^4+\\
&\ a^2d^4+b^2d^4+a^6b^2+a^6d^2+b^2cd+a^2b^2d^2+b^2cd^2+a^4c+b^2cd^3+\\
&\ a^4b^2cd+a^4b^2c+a^4cd^2+a^4b^2d^2+a^4b^2c^2+a^4b^2c^3+a^6b^2c^2+\\
&\ a^4b^2c^2d^2+a^4b^2c^3d)t+\\
&ac(1+d)(a^4b^2c^2+(a^4+d^2+1)(b^2+d+1)).
\end{aligned}
$$

4.5 Fifth Example: Glynn's o-Polynomials of Type I ($m = 4$)

In this subsection we assume $m = 4$, that is $4m' \equiv 1 \pmod{n}$. Glynn [11] showed that the power function

$$\xi^{2^{m'}+2^{2m'}}$$

is a monomial o-polymnomial on $L = \mathbf{F}_{2^n}$; expressed in our terminology:

Theorem 6. *The function*

$$G(x, y, z, u) = yz$$

is an $\mathcal{O}$-form.

It is trivial that G and $G_0(x, y, z, u) = G(x, y, z, u)/x$ are $\mathcal{P}$-forms. The essential part of the preceding theorem is the statement that

$$G_1(x,y,z,u) = \frac{G((x,y,z,u)+1)+G(1)}{x} = \frac{(y+1)(z+1)+1}{x} = \frac{yz+y+z}{x}$$

is a $\mathcal{P}$-form. Here the fundamental equational system is

$$
\begin{aligned}
yz+y+z+ax &= 0,\\
zu+z+u+by &= 0,\\
x^2u+u+x^2+cz &= 0,\\
x^2y^2+x^2+y^2+du &= 0.
\end{aligned}
$$

The fundamental polynomial is

$$\begin{aligned}\phi(t) = {} & a^4c^2t^9 + a^4c^2t^7 + a^3bc^2dt^6 + \\ & (a^2 + a^2d + a^2cd + a^4c^2 + b^2d^2 + d^2 + b^2 + c^2 + a^2c^2 + a^2bcd + 1)t^5 + \\ & (ab^2cd^2 + abcd^2 + abc^2d + acb^2d + abcd)t^4 + \\ & (b^2d^3 + a^2b^2c^2d^2 + b^2d^2 + a^4c^2 + b^2cd^2 + b^3cd^2)t^3 + \\ & a^3bc^2dt^2 + \\ & (a^2 + b^3cd^3 + b^2cd^2 + a^2c^2 + a^2cd + b^2d^2 + d^2 + b^2c^2d^2 + \\ & \quad b^2cd^3 + b^3cd^2 + b^2 + a^2b^2c^2d^2 + a^2d + a^2bcd + c^2 + 1)t + \\ & abcd + abc^2d + ab^3c^2d^3 + ab^2cd^2 + abcd^2 + ab^2cd\end{aligned}$$

5 Local Inverses of Non-Global $\mathcal{P}$-forms

We consider again the first example, see Subsection 4.1:

$$R(x,y) = xy + x^3 + x,$$

where we have

$$\phi(t) = t^9 + at^6 + t^5 + at^4 + (a^2 + b)t^3 + a^2t + a^3.$$

The (global) variant of the $\mathcal{L}$-method, which we have developed in the preceding section, can be characterized as *implicit* in the sense that we showed that there is not more than one pre-image, respectively, but we do not get a formula for it. Nevertheless we can vary the $\mathcal{L}$-method such that we can compute the inversion locally. The key observation in this connection is that ϕ divides a linearized polynomial ℓ of degree 16:

$$\begin{aligned}\ell(t) &= t^{16} + (a^2+1)\,t^8 + (a^4 + b^2 + a^2b + a^2)\,t^4 + (a^4 + a^2b)\,t^2 + a^3b\,t \\ &= \phi(t)\,\left(t^7 + a\,t^4 + t^3 + a\,t^2 + b\,t\right).\end{aligned}$$

Starting with

$$\ell^{(0)} = \ell$$

we derive by induction a sequence of new linearized polynomials

$$\begin{aligned}\ell^{(i)}(t) &= t^{2^{i+4}} + A_it^8 + B_it^4 + C_it^2 + D_it, \\ \ell^{(i+1)}(t) &= \ell^{(i)}(t)^2 + A_i^2\ell^{(0)}(t),\end{aligned}$$

which leads to the recursions

$$\begin{aligned}A_0 &= a^2 + 1, \\ B_0 &= a^4 + b^2 + a^2b + a^2, \\ C_0 &= a^4 + a^2b, \\ D_0 &= a^3b,\end{aligned}$$

$$\begin{aligned}
A_{i+1} &= B_i^2 + A_i^2 A_0, \\
B_{i+1} &= C_i^2 + A_i^2 B_0, \\
C_{i+1} &= D_i^2 + A_i^2 C_0, \\
D_{i+1} &= A_i^2 D_0.
\end{aligned}$$

Note that $\phi(x) = 0$ implies $\ell^{(i)}(x) = 0$ for all i. We can now localize the situation if the combine $\ell^{(i)}(x) = 0$ for $i + 4 = n$ with the equation $x^{2^n} = x$ defining $L = \mathbf{F}_{2^n}$ and get the local equation:

$$A_{n-4}x^8 + B_{n-4}x^4 + C_{n-4}x^2 + (D_{n-4} + 1)x = 0.$$

Thus if we use

$$\ell_L^{(n-4)}(\xi) = 0,$$

the interpretation of the latter equation for $L = \mathbf{F}_{2^n}$ in addition to the (interpretations) of the fundamental equational system $R_L(\xi) = \alpha$ and $\phi_L(\xi) = 0$, then we are able to derive a polynomial representation of ξ depending on α.

This method also works for the other examples studied in Section 4. We shall not go into the details here. Instead we consider a similar situation, where the R_L are not inducing permutations but two-to-one mappings. With the same method as described above we can derive a trace expansion of the image of R_L and thereby the 2-rank. We consider an example in the next section.

6 Trace Expansion of Glynn-Maschietti Difference Sets

Suppose that ξ^k is a monomial o-polynomial on $L = \mathbf{F}_{2^n}$, i.e.

$$(k, 2^n - 1) = (k - 1, 2^n - 1) = 1,$$

and

$$\pi_k(\xi) = \frac{(\xi + 1)^k + 1}{\xi}$$

is a permutation polynomial on L. Then the mapping

$$\tau_k(\xi) = \xi^k + \xi$$

from L to L is two-to-one, as shown by Maschietti [13] recently. It is also well-known that with ξ^k also the exponents

$$1 - k, \quad \frac{1}{k}, \quad \frac{1}{1-k}, \quad \frac{k}{k-1}, \quad \frac{k-1}{k} \quad (\text{modulo } 2^n - 1)$$

give monomial o-polynomials. Maschietti showed that

$$D_k = L \setminus \tau_k[L]$$

is a difference set in the cyclic multiplicative group $L \setminus \{0\}$.

Recently Evans et alii [10] determined the 2-rank of D_k for the case $k = 6$, these o-polynomial is due to Segre, and the two Glynn o-polynomials. In the Segre case, but not in the two Glynn cases, Evans et alii [10] could even derive a trace expansion for D_k. By such a *trace expansion* one understands a characterization of D_k in the form

$$\xi \in D_k \quad \text{iff} \quad \mathrm{Tr}_L(P(\xi)) = 1 \quad \text{for all} \quad \xi \in L,$$

where P is a polynomial over $\mathbf{F}_2$ and Tr_L denotes the absolute trace on L. W.l.o.g. on can assume that P is in *reduced normalform,* which means that all occuring exponents in P are smaller than $2^n - 1$ and minimal in their cyclotomic class. The 2-rank of D_k is then n times the weight of P. In this sense a trace expansion also gives the 2-rank.

In [9] we derived also a trace expansion (and therefore the 2-rank) of D_k for the case $k = 6$ as a particular conclusion from Theorem 1 in Section 3. But we mentioned that we can, with similar methods, derive a result stronger than the result of Evans et alii [10], namely a trace expansion for the difference sets associated by Maschietti to Glynn o-polynomials.

In the sequel we shall outline a variant of the $\mathcal{L}$-method explained in the preceding section to derive such a trace expansion for Glynn o-polynomials of Type II (see Section 4.2).

From the global point of view, in the case of Glynn's o-polynomial of Type II, we have to study

$$R(x, y) = x^4y^3 + x.$$

(Of course R is *not* a $\mathcal{P}$-form, since we know that R_L maps two-to-one. Nevertheless the $\mathcal{L}$-method can be applied also here.) The fundamental equational system is

$$\begin{aligned} &x^4y^3 + x + a = 0, \\ &x^6y^4 + y + b = 0, \end{aligned}$$

which leads to the fundamental polynomial

$$\phi(t) = t^{10} + t^7 + (a^4 + a)t^6 + a^2t^5 + (a^3 + b^3 + 1)t^4 + a^2t^2 + t + a.$$

The linearized polynomial is here

$$\begin{aligned} \ell(t) &= t^{16} + a^8t^8 + (b^6 + b^3 + 1)t^4 + a^2b^3t^2 + b^3t + a^4 + ab^3 \\ &= \phi(t)\ \left(t^6 + t^3 + (a^4 + a)t^2 + a^2t + a^3 + b^3\right). \end{aligned}$$

Consequently the coefficients of the

$$\ell^{(i)}(t) = t^{2^{i+4}} + A_it^8 + B_it^4 + C_it^2 + D_it + E_i$$

are defined by the recursion

$$\begin{aligned}
A_0 &= a^8,\\
B_0 &= b^6 + b^3 + 1,\\
C_0 &= a^2 b^3,\\
D_0 &= b^3,\\
E_0 &= a(a^3 + b^3),\\
A_{i+1} &= B_i^2 + A_i^2 A_0,\\
B_{i+1} &= C_i^2 + A_i^2 B_0,\\
C_{i+1} &= D_i^2 + A_i^2 C_0,\\
D_{i+1} &= A_i^2 D_0,\\
E_{i+1} &= E_i^2 + A_i^2 E_0.
\end{aligned}$$

We have $\ell^{(i)}(x) = 0$, and for $i = n - 4$ with the L defining (local) equation $x^{2^n} = x$ we get the local equation

$$A_{n-4}x^8 + B_{n-4}x^4 + C_{n-4}x^2 + (D_{n-4} + 1)x + E_{n-4} = 0,$$

which then leads to a quadratic equation. The latter, using the simple fact that a quadratic equation $\lambda x^2 + \mu x + \nu = 0$ has a solution in L iff $\mathrm{Tr}_L(\lambda\nu/\mu^2) = 0$, implies the desired trace expansion, where the details require rather extensive technical work.

Finally we remark that a trace expansion for the simplest case $k = 6$, which can be derived from Theorem 2 as shown in [9], also follows with the above approach. In this (degenerated) case

$$R(x, y) = x^6 + x + a,$$

we have

$$\phi(t) = t^6 + t + a$$

and the linearized polynomial is

$$\ell(t) = t^{16} + a^2 t^4 + t + a.$$

In fact ϕ divides ℓ; we have $\ell(t) = \phi(t)(t^{10} + t^5 + at^4 + 1)$.

References

1. A. Canteaut, P. Charpin and H. Dobbertin, *Couples der suites binaires de longueur maximale ayant une corrélation croisée à trois valeurs: conjecture de Welch,* C.R. Acad. Sci. Paris, Automatique théorique / Automation (theoretical), t. **328**, Serie I, (1999), pp. 173–178.

2. A. Canteaut, P. Charpin and H. Dobbertin, *A three-valued crosscorrelation function for binary m-sequences: a proof of Welch's conjecture,* IEEE Transactions on Information Theory, to appear.
3. S. D. Cohen, *private communication.*
4. S. D. Cohen and R. W. Matthews, *A class of exceptional polynomials,* Trans. Amer. Math. Soc. **345** (1994), pp. 897–909.
5. W. Cherowitzo, *α-flocks and hyperovals,* Geometriae Dedicata, **72** (1998), pp. 221–246.
6. J. Dillon and H. Dobbertin, *New difference sets with Singer parameters,* Finite Fields and their Applications, submitted.
7. H. Dobbertin, *Almost perfect nonlinear power functions on* $\mathrm{GF}(2^n)$*: The Welch case,* IEEE Transactions on Information Theory **45** (1999), pp. 1271–1275.
8. H. Dobbertin, *Almost perfect nonlinear power functions on* $\mathrm{GF}(2^n)$*: The Niho case,* Information and Computation **151** (1999), pp. 57–72.
9. H. Dobbertin, *Kasami power functions, permutation polynomials and cyclic difference sets,* Proceedings of the NATO-A.S.I. Workshop "Difference sets, sequences and their correlation properties", Bad Windsheim, August 3-14, 1998, Kluwer Verlag, to appear.
10. R. Evans, H. Hollmann, C. Krattenthaler and Q. Xiang, *Gauss sums, Jacobi sums, and p-ranks of cyclic difference sets,* J. Combin. Theory Ser. A, to appear.
11. D. Glynn, *Two new sequences of ovals in finite Desarguesian planes of even order,* Lecture Notes in Mathematics **1036**, Springer, 1983, pp. 217–229.
12. H. Hollmann and Q. Xiang, *A proof of the Welch and Niho conjectures on crosscorrelation on binary m-sequences,* Finite Fields and their Applications, submitted
13. A. Maschietti, *Difference sets and hyperovals,* Designs, Codes and Cryptography **14** (1998), pp. 89–98.
14. P. Müller, *New examples of exceptional polynomials,* Finite Fields: Theory, Applications and Algorithms, (G. L. Mullen and P. J. Shiue, eds.), Contemp. Math. **168** (1994), pp. 245–249.

New p-ary Perfect Sequences and Difference Sets with Singer Parameters

J. F. Dillon

National Security Agency, Fort George G. Meade, MD 20755, USA

Summary. For each odd prime p, each odd integer $m > 1$ and each even integer k, $0 \leq k \leq m-1$, the quadric Ω_k defined by the quadratic form $\mathrm{Tr}(x^{p^k+1})$ on $\mathrm{GF(p^m)}$ gives a difference set in $G = \mathrm{GF(p^m)}^\times/\mathrm{GF(p)}^\times$ which is equivalent to the classical one given by a hyperplane. We show that the equivalent quadrics $D_k = L_k(\Omega_k)$, where L_k is the invertible linear map $x \mapsto x + x^{p^{2k}}$, are also difference sets; and we modulate the functions g_k defined by $g_k(L_k(x)) = \mathrm{Tr}(x^{p^k+1})$ to obtain p-ary functions h_k whose corresponding sequences $\{h_k(w^t)\}$, w primitive in $\mathrm{GF(p^m)}^\times$, have ideal autocorrelation.

1 Introduction

For any prime p, a p-ary sequence $\{a_t\}$ of period $p^m - 1$ has *ideal autocorrelation* if

$$c(s) := \sum_{t=0}^{p^m-2} \zeta_p^{\,a_{t+s}-a_t} = \begin{cases} p^m - 1 & \text{if } p^m - 1 | s \\ -1 & \text{otherwise} \end{cases}$$

where ζ_p is the complex p^{th} root of unity $e^{\frac{2\pi i}{p}}$. We may identify such a sequence with the function h on the multiplicative group $< w >$ of $GF(p^m)$ given by $h(\omega^t) = a_t$ and extend it to all of $GF(p^m)$ by defining $h(0) = 0$. The corresponding complex-valued function $H = \zeta_p^h$ then has the property

$$\sum_{x \in GF(p^m)} H(\alpha x)\overline{H(x)} = p^m \delta_{1,\alpha} \quad \forall \alpha \in GF(p^m). \tag{1}$$

Moreover, if the function h should have the homogeneity property [9]

$$h(cx) = c^d h(x) \quad \forall x \in GF(p^m) \quad \forall c \in GF(p)$$

for some d relatively prime to $p-1$, then the support of h is a difference set with Singer parameters $(v, k, \lambda) = ((p^m - 1)/(p-1), p^{m-1}, (p-1)p^{m-2})$ in the cyclic group $G = GF(p^m)^\times/GF(p)^\times$. We shall call such sequences and functions *perfect*. The best known examples of such perfect functions are the trace maps $\mathrm{Tr} : \mathrm{GF(p^m)} \to \mathrm{GF(p)}$ which give the trace with respect to the extension $\mathrm{GF}(\mathrm{p^m})/\mathrm{GF}(\mathrm{p})$ and which are, of course, *p-linear* maps. They give rise to the *m-sequences* $\{a_t\} = \{\mathrm{Tr}(\omega^t)\}$ and the Singer difference sets

$$D = \{x \in \mathrm{GF}(\mathrm{p^m})^\times/\mathrm{GF}(\mathrm{p})^\times : \mathrm{Tr}(x) \neq 0\}.$$

In this paper we shall obtain perfect *p*-ary functions from quadratic forms on $\mathrm{GF}(\mathrm{p^m})$ which are equivalent to $\mathrm{Tr}(x^2)$. The resulting difference sets will therefore have the interesting property that they are equivalent quadrics *additively* ...i.e. under the action of the general linear group... but inequivalent multiplicatively.

For a general background in coding theory, geometry and difference sets the reader is referred to the book of E. F. Assmus, Jr. and J. D. Key, *Designs and their Codes* [1]. For more information on cyclic difference sets, Marshall Hall's textbook *Combinatorial Theory* [8] and the encyclopedic treatise of L. D. Baumert [2] are highly recommended. A comprehensive treatment of general difference sets can be found in T. Beth, D. Jungnickel and H. Lenz, *Design Theory* [4].

2 Quadratic forms and Gauss sums.

If V is a finite dimensional vector space over the field F, a map $q : V \longrightarrow F$ is a *quadratic form* on V if it satisfies the two conditions:

1. $q(\lambda x) = \lambda^2 q(x)$ for all λ in F;
2. $B_q(x,y) = q(x+y) - q(x) - q(y)$ is bilinear on V;

If we choose a basis $b_1, b_2, \cdots, b_m$ for V over F, then the quadratic form q is uniquely expressed in terms of the coordinate functionals as the more familiar

$$q(X) = \sum_{i \leq j} q_{ij} X_i X_j, \quad q_{ij} \in F.$$

q is *nondegenerate* if $V^\perp := \{y : B_q(x,y) = 0 \quad \forall x \in V\} = \{0\}$. Two such forms q_1 and q_2 are *equivalent* if $q_2(x) = q_1(L(x))$ for some $L \in GL(V)$. In this paper we shall be concerned only with nondegenerate quadratic forms on $\mathrm{GF}(\mathrm{p^m})$ regarded as an m-dimensional vector space over $\mathrm{GF}(\mathrm{p})$ where p and m are *odd*. In this setting $q(X)$ has the familiar symmetric matrix representation

$$q(X) = X^{\mathrm{T}} M_q X,$$

where

$$M_q(i,j) = \begin{cases} q_{ii} & \text{if } i = j \\ 1/2 q_{ij} & \text{if } i < j. \end{cases}$$

Dickson's Theorem [5] asserts that there is a choice of basis which puts q into one of the two canonical forms

$$q^+(X) = X_1^2 + X_2^2 + \cdots + X_m^2,$$

or

$$q^-(X) = \mu q^+(X),$$

where μ is any nonsquare in $\mathrm{GF(p)}^\times$. The equivalence type of such a nondegenerate form q is determined by the quadratic character of the determinant of its matrix; i.e.

$$q \sim \begin{cases} q^+ \text{ if } \mathtt{det} M_q \text{ is a square} \\ q^- \text{ otherwise.} \end{cases}$$

For our purposes here there is a more convenient invariant which distinguishes inequivalent forms, namely the value distribution as given by another theorem of Dickson [5]

Proposition 1. *The number of solutions in $GF(p)$, $p > 2$, of the equation*

$$\alpha_1\xi_1^2 + \alpha_2\xi_2^2 + \cdots + \alpha_m\xi_m^2 = c,$$

where each α_j is a nonzero element in the field and c is an element of the field, is $p^{m-1} + \omega p^{\frac{m-1}{2}}$, where ω is 1, -1 or 0 according as $(-1)^{\frac{m-1}{2}} c\alpha_1\alpha_2\cdots\alpha_m$ is a square, a nonsquare or zero in the field.

In particular, we see that every nondegenerate form takes the value 0 p^{m-1} times (in fact this is true for any form of odd rank) and the number of times it takes a nonzero value c depends only on whether or not c is a square. If we let $\mathtt{N}_q(c)$ denote the number of solutions of $q(x) = c$, then

$$\mathtt{N}_q(c) = \begin{cases} p^{m-1} \text{ if } c = 0 \\ \mathtt{N}_q(1) \text{ if } c \text{ is a square} \\ \mathtt{N}_q(\mu) \text{ if } c \text{ is a nonsquare} \end{cases}$$

where μ is any fixed nonsquare in $\mathrm{GF(p)}^\times$. If q_1 and q_2 are two nondegenerate forms on $\mathrm{GF(p^m)}$, then certainly $\mathtt{N}_{q_1}(c) = \mathtt{N}_{q_2}(c)$ for all c if $q_1 \sim q_2$. However, if q_1 and q_2 are of *opposite* types, then

$$\mathtt{N}_{q_2}(1) = \mathtt{N}_{q_1}(\mu) \quad \text{and} \quad \mathtt{N}_{q_2}(\mu) = \mathtt{N}_{q_1}(1).$$

Thus, the value distribution or, equivalently, the single multiplicity $\mathtt{N}_q(1)$ (i.e. the number of times that q takes the value 1) is a complete invariant for nondegenerate quadratic forms on the odd dimensional $\mathrm{GF(p)}$-space $\mathrm{GF(p^m)}$. This invariant information is captured by the sum

$$\Sigma_q := \sum_{x \in \mathrm{GF(p^m)}} {\zeta_p}^{q(x)} = \mathtt{N}_q(0) + \mathtt{N}_q(1)\mathcal{S} + \mathtt{N}_q(\mu)\mathcal{N},$$

where

$$\mathcal{S} := \sum_{t \in \mathrm{GF(p)}^{\times 2}} {\zeta_p}^t,$$

and $\mathcal{N}$ is the similar sum over the nonsquares in $\mathrm{GF(p)}^\times$. If q_1 and q_2 are nondegenerate forms of opposite type then

$$\begin{aligned}\Sigma_{q_1} + \Sigma_{q_2} &= \mathrm{N}_{q_1}(0) + \mathrm{N}_{q_1}(1)\mathcal{S} + \mathrm{N}_{q_1}(\mu)\mathcal{N} \\ &\quad + \mathrm{N}_{q_2}(0) + \mathrm{N}_{q_2}(1)\mathcal{S} + \mathrm{N}_{q_2}(\mu)\mathcal{N} \\ &= 2\mathrm{N}_{q_1}(0) + (\mathrm{N}_{q_1}(1) + \mathrm{N}_{q_1}(\mu))(\mathcal{S} + \mathcal{N}) \\ &= 2\mathrm{N}_{q_1}(0) - (p^m - \mathrm{N}_{q_1}(0))\,/((p-1)/2) \\ &= 2p(\mathrm{N}_{q_1}(0) - p^{m-1})/(p-1) \\ &= 0,\end{aligned}$$

and we see that $\Sigma_{q_2} = -\Sigma_{q_1}$.

We'll now consider a very simple class of nondegenerate quadratic forms. For any positive integer d, let s_d denote the function on $\mathrm{GF(p^m)}$ given by $s_d(x) := \mathrm{Tr}(x^d)$, where Tr is the trace with respect to the extension $\mathrm{GF(p^m)/GF(p)}$. Let S_d denote the corresponding complex-valued function $S_d = \zeta_p^{s_d}$. Note that for any k, $0 \le k \le m-1$, the even integers p^k+1 and p^m-1 have greatest common divisor exactly 2 since

$$(p^k+1, p^m-1) | (p^{2k}-1, p^m-1) = p^{(2k,m)} - 1 = p^{(k,m)} - 1 | p^k - 1,$$

and so $(p^k+1, p^m-1)|(p^k+1, p^k-1) = 2$. Now for any k, $0 \le k \le m-1$, s_{p^k+1} is a quadratic form on $V = \mathrm{GF(p^m)}$. Indeed, for all c in GF(p) and all x in $\mathrm{GF(p^m)}$ we have

$$s_{p^k+1}(cx) = Tr([cx]^{p^k+1}) = Tr(c^2 x^{p^k+1}) = c^2 Tr(x^{p^k+1}) = c^2 s_{p^k+1}(x);$$

and

$$B(x,y) := s_{p^k+1}(x+y) - s_{p^k+1}(x) - s_{p^k+1}(y) = Tr([y^{p^k} + y^{p^{-k}}]x)$$

is bilinear on $\mathrm{GF(p^m)}$.
Now $V^\perp := \{y \in \mathrm{GF(p^m)} : y^{p^k} + y^{p^{-k}} = 0\}$ is the nullspace of the GF(p)-linear operator $\mathcal{L}_k$ on $\mathrm{GF(p^m)}$ given by

$$\mathcal{L}_k(z) = z^{p^k} + z^{p^{-k}} \quad \forall z \in \mathrm{GF(p^m)};$$

and $\mathcal{L}_k$ has the same nullspace as the operator L_k given by

$$L_k(z) = (\mathcal{L}_k(z))^{p^k} = z + z^{p^{2k}} = z(1 + z^{p^{2k}-1}) \quad \forall z \in \mathrm{GF(p^m)}.$$

If $\xi \in V^\perp$ were nonzero, then $1 = (-1)^{p^{2k}+1} = \xi^{p^{4k}-1}$, which implies that

$$\xi \in \mathrm{GF(p^{4k})} \cap \mathrm{GF(p^m)} = \mathrm{GF(p^{(4k,m)})} = \mathrm{GF(p^{(k,m)})} \subseteq \mathrm{GF(p^k)}.$$

In particular, $1 = \xi^{p^{2k}-1} = -1$, which is absurd since p is odd.
Thus, $V^\perp = \{0\}$ and s_{p^k+1} is nondegenerate. Now for each k, $0 \le k \le m-1$, and for all x in $\mathrm{GF(p^m)}$ we have

$$\mathrm{Tr}(x^{p^k+1}) = (\mathrm{Tr}(x^{p^k+1}))^{p^{m-k}} = \mathrm{Tr}((x^{p^k+1})^{p^{m-k}}) = \mathrm{Tr}(x^{p^{m-k}+1}).$$

Thus, $s_{p^{m-k}+1} = s_{p^k+1}$ for all k, $0 \le k \le m-1$, and restricting k to be *even* gives us $(m+1)/2$ distinct nondegenerate quadratic forms which are all equivalent to s_2 since

$$s_{p^k+1}(x) = s_2(\pi_k(x)),$$

where, because $((p^k+1)/2, p^m-1) = 1$, $\pi_k(z) := z^{(p^k+1)/2}$ is a permutation on $\mathrm{GF(p^m)}$ and so s_{p^k+1} and s_2 have the same value distribution; i.e. $\Sigma_{s_{p^k+1}} = \Sigma_{s_2}$ for all such k. In the sequel we shall therefore restrict k to be even; and we shall freely exploit the fact that $x \mapsto x^{(p^k+1)/2}$ is a permutation on $\mathrm{GF(p^m)}$. Moreover, since $(p^k-1)/(p-1)$ is even, $(p^k+1)/2 = 1 + (1/2)((p^k-1)/(p-1))(p-1)$ is conguent to 1 mod $p-1$ and so this permutation fixes GF(p) pointwise.

Now note that

$$\Sigma_{s_2} := \sum_{x\in \mathrm{GF(p^m)}} S_2(x) = \sum_{x\in \mathrm{GF(p^m)}} \zeta_p^{\mathrm{Tr}(x^2)} \tag{2}$$

is precisely the quadratic Gauss sum $\mathcal{G}_m$ usually expressed as

$$\mathcal{G}_m := \sum_{x\in \mathrm{GF(p^m)}} \chi(x)\zeta_p^{\mathrm{Tr}(x)}, \tag{3}$$

where χ is the quadratic character on $\mathrm{GF(p^m)}$ defined by

$$\chi(x) = \begin{cases} 1 & \text{if } x \text{ is a nonzero square} \\ -1 & \text{if } x \text{ is a nonsquare} \\ 0 & \text{if } x \text{ is } 0. \end{cases}$$

Indeed, (2) is obtained from (3) by adding to it the sum

$$\sum_{x\in \mathrm{GF(p^m)}} \zeta_p^{\mathrm{Tr}(x)} = 0.$$

Since we are concerned here with sums such as (1), it shall be useful to employ Fourier analysis with respect to the additive group of the field $\mathrm{GF(p^m)}$. This technique was introduced in [6] and developed further in [7], both of these references treating fields of characteristic 2 in which case the Fourier transform was called the *Hadamard* transform. Here we shall call it simply the *additive Fourier transform.*

For any function $F : \mathrm{GF(p^m)} \longrightarrow C$, we denote by $\widehat{F}$ the additive Fourier transform of F defined by

$$\widehat{F}(\beta) := p^{-m/2} \sum_{y\in \mathrm{GF(p^m)}} F(y)\zeta_p{}^{-\mathrm{Tr}(\beta y)} \quad \forall \beta \in \mathrm{GF(p^m)}.$$

The Fourier transform $F \mapsto \widehat{F}$ is a unitary operator on the inner product space $C^{\mathrm{GF(p^m)}}$ of all complex-valued functions on $\mathrm{GF(p^m)}$ with inner product

$$< F_1, F_2 >= \sum_{x\in \mathrm{GF(p^m)}} F_1(x)\overline{F_2(x)}.$$

In this notation the Fourier coefficient $\widehat{F}(\beta)$ is simply the inner product $< F, \psi_\beta >$ of F with the normalized additive character ψ_β given by

$$\psi_\beta(x) = p^{-m/2}\zeta_p^{\mathrm{Tr}(\beta x)} \quad \forall x \in \mathrm{GF(p^m)};$$

i.e.

$$F = \sum_{\beta\in \mathrm{GF(p^m)}} < F, \psi_\beta > \psi_\beta.$$

For the convenience of the reader we list a few standard properties of the Fourier transform which we shall use in the sequel.

Proposition 2. (Properties of the Fourier Transform on C^V, $V = \mathrm{GF(p^m)}$.)
(Plancherel's Theorem.) $< \widehat{F}_1, \widehat{F}_2 >=< F_1, F_2 >$;
(GL(V) action.) $F_2(x) = F_1(L(x)) \Rightarrow \widehat{F}_2(x) = \widehat{F}_1(L^{*-1}(x)), \quad \forall L \in GL(V)$, *where* L^* *is the adjoint of* L *given by* $\mathrm{Tr}(L(x)y) = \mathrm{Tr}(xL^*(y))$;
(Galois action.)$F_2(x) = F_1(x^p) \Rightarrow \widehat{F}_2(x) = \widehat{F}_1(x^p)$;
(Cyclotomic action.)*If* F_1 *takes values in* $Q(\zeta_p)$, *and* $F_2(x) = F_1(x)^\sigma$, *where* $\sigma \in Gal(Q(\zeta_p)/Q)$ *is induced by* $\zeta_p \mapsto \zeta_p^t$, $(t,p) = 1$, *then* $\widehat{F}_2(x) = \widehat{F}_1(t^{-1}x)^\sigma$.

Now the Gauss sum $\mathcal{G}_m$ is just the unnormalized 0th Fourier coefficient $\widehat{S}_2(0)$; i.e.

$$\widehat{S}_2(0) = p^{-m/2}\mathcal{G}_m.$$

More generally, for any $\beta \in \mathrm{GF(p^m)}$ we have

$$\begin{aligned}\widehat{S}_2(\beta) &= p^{-m/2} \sum_{x\in\mathrm{GF(p^m)}} \zeta_p{}^{\mathrm{Tr}(x^2-\beta x)} \\ &= p^{-m/2} \sum_{x\in\mathrm{GF(p^m)}} \zeta_p{}^{\mathrm{Tr}([x-\frac{\beta}{2}]^2-[\frac{\beta}{2}]^2)} \\ &= \widehat{S_2}(0)\overline{S_2(\beta/2)},\end{aligned}$$

which we see follows simply by completing the square. Since all the quadratic forms s_{p^k+1} are equivalent to s_2, a similar relationship should exist between S_{p^k+1} and $\widehat{S}_{p^k+1}$. Indeed, replacing x by $x+\gamma$ in the summation

$$\widehat{S}_{p^k+1}(\beta) = p^{-m/2} \sum_{x\in\mathrm{GF(p^m)}} , \zeta_p{}^{\mathrm{Tr}(x^{p^k+1})-\mathrm{Tr}(\beta x)}$$

gives

$$\begin{aligned}\widehat{S}_{p^k+1}(\beta) &= p^{-m/2} \sum_{x\in\mathrm{GF(p^m)}} \zeta_p{}^{\mathrm{Tr}(x^{p^k+1}+[\gamma^{p^k}+\gamma^{p^{-k}}]x+\gamma^{p^k+1})-\mathrm{Tr}(\beta[x+\gamma])} \\ &= p^{-m/2} \sum_{x\in\mathrm{GF(p^m)}} \zeta_p{}^{\mathrm{Tr}(x^{p^k+1})+\mathrm{Tr}([\mathcal{L}_k(\gamma)-\beta]x)+\mathrm{Tr}(\gamma^{p^k+1}-\beta\gamma)}.\end{aligned}$$

But we showed above that $\mathcal{L}_k$ is invertible, so there is a unique γ for which $\gamma^{p^k} + \gamma^{p^{-k}} = \mathcal{L}_k(\gamma) = \beta$. In that case

$$\mathrm{Tr}(\gamma^{p^k+1} - \beta\gamma) = -\mathrm{Tr}(\gamma^{p^{-k}+1}) = -\mathrm{Tr}(\gamma^{p^k+1});$$

and it follows that

$$\widehat{S}_{p^k+1}(\beta) = \widehat{S}_{p^k+1}(0)\overline{S_{p^k+1}(\gamma)} = \widehat{S}_{p^k+1}(0)\overline{S_2(\gamma^{(p^k+1)/2})} = \widehat{S}_2([2\gamma]^{(p^k+1)/2}); i.e.$$

$$\widehat{S}_{p^k+1}(\mathcal{L}_k(\gamma)) = \widehat{S}_2([2\gamma]^{(p^k+1)/2}) \quad \forall\gamma \in \mathrm{GF(p^m)}.$$

These results imply

Proposition 3. $\widehat{S}_{p^k+1}(1/2\mathcal{L}_k(x)) = \widehat{S}_2(x^{(p^k+1)/2})$.

Finally, for completeness, we record the precise value of the Gauss sum [3].

Proposition 4.

$$\mathcal{G}_m := \sum_{x\in \mathrm{GF(p^m)}} \zeta_p^{Tr(x^2)} = \begin{cases} (-1)^{m-1}\sqrt{p^m} & \textit{if } p = 1 \bmod 4 \\ (-1)^{m-1}i^m\sqrt{p^m} & \textit{if } p = 3 \bmod 4. \end{cases}$$

We summarize the above as

Proposition 5. *Let p be an odd prime and let* GF(p^m) *be regarded as an m-dimensional vector space over* GF(p) *where m is odd. Let μ be any nonsquare in* GF(p)$^\times$. *Let q be any nondegenerate quadratic form on* GF(p^m). *Then either*

$$\Sigma_q := p^{m/2}\widehat{Q}(0) = \mathcal{G}_m \quad \textit{and} \quad q(x) \sim s_2(x) := Tr(x^2)$$

or

$$\Sigma_q := p^{m/2}\widehat{Q}(0) = -\mathcal{G}_m \quad \textit{and} \quad q(x) \sim \mu s_2(x) := \mu Tr(x^2).$$

3 A New Construction.

Theorem 1. *Let p be any odd prime and let $L = GF(p^m)$, $m > 1$ odd. For any even integer k, $0 \le k \le m-1$, let $g_k : GF(p^m) \to GF(p)$ be the quadratic form given by $g_k(x + x^{p^{2k}}) = Tr(x^{p^k+1})$ and let h_k be the related function given by $h_k(x) = x^f g_k(x)$, where f is the odd part of $p^m - 1$.*
a) *$\{a_t\} = \{h_k(\omega^t)\}$ has ideal autocorrelation;*
b) *$D_k :=$ support h_k = support g_k is a difference set with Singer parameters $(v, k, \lambda) = ((p^m-1)/(p-1), p^{m-1}, (p-1)p^{m-2})$ in $G = GF(p^m)^\times/GF(p)^\times$.*

Proof. We shall prove the theorem for the equivalent functions g_k given by $g_k(1/2\mathcal{L}_k(x)) = Tr(x^{p^k+1})$, where $\mathcal{L}_k(x) = x^{p^k} + x^{p^{-k}}$; these g_k and corresponding h_k are those of the statement of the theorem simply multiplied by the nonzero scalar 4. By the basic results on quadratic forms reviewed in the previous section, for each k, $0 \le k \le m-1$, g_k is a quadratic form on the GF(p)-space GF($\mathrm{p^m}$) which is equivalent to $\mathrm{Tr}(x^{p^k+1})$ and hence also to $\mathrm{Tr}(x^2)$. Thus, for all λ in GF(p), $g_k(\lambda x) = \lambda^2 g_k(x)$ for all x in GF($\mathrm{p^m}$). In particular, $g_k(-x) = g_k(x)$ $\forall x \in$ GF($\mathrm{p^m}$), so that the sequence $\{g_k(\omega^t)\}$ has period dividing $(p^m-1)/2$. We shall modulate the function g_k to remove this deficiency.

Define $h_k : GF(p^m) \to GF(p)$ by $h_k(x) = x^f g_k(x)$ for all x in GF($\mathrm{p^m}$), where f is the odd part of $p^m - 1$. Thus, if $e = 2^t$ is the greatest power of 2 dividing $p^m - 1$, then $p^m - 1 = ef$; and, since $(p^m-1)/(p-1)$ is odd, e is actually a divisor of $p-1$. Now for all λ in GF(p) we have

$$h_k(\lambda x) = (\lambda x)^f g_k(\lambda x) = \lambda^{f+2} x^f g_k(x) = \lambda^{f+2} h_k(x).$$

Put $\delta = f+2$ and note that, since f is the odd part of $p^m - 1$ and δ is odd, we must have $(\delta, p^m-1) = (\delta, f) = (f+2, f) = (2, f) = 1$. We note for future reference that a similar argument shows that $f + p^k + 1$ is prime to $p^m - 1$ also; and therefore the maps $x \mapsto x^{f+2}$ and $x \mapsto x^{f+p^k+1}$ are permutations on GF($\mathrm{p^m}$). Now, for every element $\xi \in$ GF($\mathrm{p^m}$)$^\times$ for which $h_k(\xi) \neq 0$, h_k maps GF(p)ξ onto GF(p). In particular, h_k takes every nonzero value in GF(p) equally often; and, since g_k is a nondegenerate quadratic form on the *odd*-dimensional GF(p)-space GF($\mathrm{p^m}$), g_k and hence also h_k takes the value 0 exactly p^{m-1} times. Thus, h_k takes every value in GF(p) exactly p^{m-1} times; i.e. h_k is balanced. Let H_k be the complex-valued function $\zeta_p^{h_k}$. We shall now show that h_k has perfect autocorrelation; i.e.

$$\sum_{x \in GF(p^m)} H_k(\alpha x)\overline{H_k(x)} = p^m \delta_{1,\alpha} \quad \forall \alpha \in GF(p^m).$$

Note that for $\alpha = 0$ this is true because h_k is balanced. So suppose that α is any nonzero element in GF($\mathrm{p^m}$). The map $\sigma : x \mapsto x^f$ is a homomorphism from GF($\mathrm{p^m}$)$^\times$ onto the 2-subgroup E of GF(p)$^\times$ which we shall identify with the Galois group of the cyclotomic field $Q(\zeta_p)/Q$ in the usual way. Note that σ induces an automorphism on E and that $\sigma(x)$ is a square if and only if x is a square. Letting $G_k = \zeta_p^{g_k}$, and breaking up the sum over GF($\mathrm{p^m}$)$^\times$ into two sums, one over squares y^2 and the other over nonsquares μy^2, μ a nonsquare in GF($\mathrm{p^m}$)$^\times$ which we may take to be in GF(p)$^\times$, we have

$$\sum_{x \in \mathrm{GF(p^m)}} H_k(\alpha x)\overline{H_k(x)} = 1 + \sum_{x \in \mathrm{GF(p^m)^\times}} G_k^{\sigma(\alpha x)}(\alpha x) G_k^{-\sigma(x)}(x)$$

$$
\begin{aligned}
&= 1 + 1/2 \sum_{y \in \mathrm{GF}(p^m)^\times} G_k^{\sigma(\alpha)\sigma(y^2)}(\alpha y^2) G_k^{-\sigma(y^2)}(y^2) \\
&\quad + 1/2 \sum_{y \in \mathrm{GF}(p^m)^\times} G_k^{\sigma(\alpha)\sigma(\mu)\sigma(y^2)}(\alpha\mu y^2) G_k^{-\sigma(\mu y^2)}(\mu y^2) \\
&= 1 + 1/2 \sum_{y \in \mathrm{GF}(p^m)^\times} G_k^{\sigma(\alpha)}(\alpha\sigma(y)y^2) G_k^{-1}(\sigma(y)y^2) \\
&\quad + 1/2 \sum_{y \in \mathrm{GF}(p^m)^\times} G_k^{\sigma(\alpha)\sigma(\mu)}(\alpha\mu\sigma(y)y^2) G_k^{-\sigma(\mu)}(\mu\sigma(y)y^2) \\
&= 1 + 1/2 \sum_{y \in \mathrm{GF}(p^m)^\times} G_k^{\sigma(\alpha)}(\alpha y) G_k^{-1}(y) \\
&\quad + 1/2 \sum_{y \in \mathrm{GF}(p^m)^\times} G_k^{\sigma(\alpha)\sigma(\mu)}(\alpha y) G_k^{-\sigma(\mu)}(y) \\
&= 1/2 \left[\sum_{y \in \mathrm{GF}(p^m)} G_k^{\sigma(\alpha)}(\alpha y) G_k^{-1}(y) + \sum_{y \in \mathrm{GF}(p^m)} G_k^{\sigma(\alpha)\sigma(\mu)}(\alpha y) G_k^{-\sigma(\mu)}(y) \right] \\
&= 1/2 \left[\mathcal{S}_\alpha + \mathcal{S}_\alpha^{\sigma(\mu)} \right],
\end{aligned}
$$

where

$$
\mathcal{S}_\alpha = \sum_{y \in \mathrm{GF}(p^m)} G_k^{\sigma(\alpha)}(\alpha y) \overline{G_k(y)}.
$$

But $G_k(1/2\mathcal{L}_k(x)) = S_{p^k+1}(x)$ implies that $\widehat{G_k}(y) = \widehat{S}_{p^k+1}(1/2\mathcal{L}_k(y)) = \widehat{S}_2(y^{(p^k+1)/2})$, the last equality by Proposition 3. Using this result and properties of the Fourier transform given in Proposition 2, we derive

$$
\begin{aligned}
\mathcal{S}_\alpha &= \sum_{\beta \in \mathrm{GF}(p^m)} \widehat{G_k^{\sigma(\alpha)}}(\alpha^{-1}\beta) \overline{\widehat{G_k}(\beta)} \\
&= \sum_{\beta \in \mathrm{GF}(p^m)} \widehat{G_k}^{\sigma(\alpha)}(\sigma(\alpha)^{-1}\alpha^{-1}\beta) \overline{\widehat{G_k}(\beta)} \\
&= \sum_{\beta \in \mathrm{GF}(p^m)} \widehat{S_2}^{\sigma(\alpha)}([\sigma(\alpha)^{-1}\alpha^{-1}\beta]^{(p^k+1)/2}) \overline{\widehat{S_2}(\beta^{(p^k+1)/2})} \\
&= \sum_{\beta \in \mathrm{GF}(p^m)} \widehat{S_2}^{\sigma(\alpha)}(\sigma(\alpha)^{-1}\alpha^{-(p^k+1)/2}\beta) \overline{\widehat{S_2}(\beta)} \\
&= \sum_{\beta \in \mathrm{GF}(p^m)} \widehat{S_2^{\sigma(\alpha)}}(\alpha^{-(p^k+1)/2}\beta) \overline{\widehat{S_2}(\beta)} \\
&= \sum_{y \in \mathrm{GF}(p^m)} S_2^{\sigma(\alpha)}(\alpha^{(p^k+1)/2}y) \overline{S_2(y)} \\
&= \sum_{y \in \mathrm{GF}(p^m)} \zeta_p^{\sigma(\alpha)\mathrm{Tr}(\alpha^{p^k+1}y^2) - \mathrm{Tr}(y^2)} \\
&= \sum_{y \in \mathrm{GF}(p^m)} \zeta_p^{\mathrm{Tr}([\sigma(\alpha)\alpha^{p^k+1}-1]y^2)}.
\end{aligned}
$$

Since the map $\alpha \mapsto \sigma(\alpha)\alpha^{p^k+1} = \alpha^{f+p^k+1}$ is an automorphism of $\mathrm{GF(p^m)}^\times$, $q_\alpha(x) := Tr([\sigma(\alpha)\alpha^{p^k+1} - 1]x^2)$ is a nondegenerate quadratic form unless $\alpha = 1$ when it is identically 0. Furthermore, when $\alpha \neq 1$ $\sigma(\mu)q_\alpha(x)$ is a nondegenerate quadratic form of the opposite type since $\sigma(\mu)$ is not a square. Hence we have

$$\sum_{x\in \mathrm{GF(p^m)}} H_k(\alpha x)\overline{H_k(x)} = \frac{1}{2}(\Sigma_{q_\alpha} + \Sigma_{\sigma(\mu)q_\alpha}) = \begin{cases} \frac{1}{2}\,[p^m + p^m] = p^m & \text{if } \alpha = 1 \\ \frac{1}{2}\,[\mathcal{G}_m + (-\mathcal{G}_m)] = 0 & \text{if } \alpha \neq 1. \end{cases}$$

This completes the proof of **a)**; and, since h_k is δ-homogeneous with $(\delta, p-1) = 1$, **b)** follows immediately. **Q.E.D.**

We note here that computer results indicate that the $(m+1)/2$ difference sets constructed here are pairwise inequivalent if $m > 3$. Indeed, for all choices of p and *prime* m examined in detail, the $(m+1)/2$ translate designs have distinct mod p ranks. Notice that for all $x \in \mathrm{GF(p^m)}$, we have

$$g_k(1/2\mathcal{L}_k(x)) = \mathrm{Tr}(x^{p^k+1}) = \mathrm{Tr}(1/2\mathcal{L}_k(x)x).$$

Since the linear operator $1/2\mathcal{L}_k$ has inverse $\mathcal{T}_k$ given by

$$\mathcal{T}_k(x) = \sum_{i=0}^{m-1} (-1)^i x^{p^{(2i+1)k}},$$

it follows that

$$\begin{aligned} g_k(x) = \mathrm{Tr}(x\mathcal{T}_k(x)) &= \mathrm{Tr}(\sum_{i=0}^{m-1} (-1)^i x^{p^{(2i+1)k}+1}) \\ &= (-1)^{\frac{m-1}{2}}(\mathrm{Tr}(x^2) + 2\sum_{i=1}^{\frac{m-1}{2}} (-1)^i \mathrm{Tr}(x^{p^{2ik}+1})); \end{aligned}$$

and so

$$h_k(x) = x^f g_k(x) = (-1)^{\frac{m-1}{2}}(\mathrm{Tr}(x^{f+2}) + 2\sum_{i=1}^{\frac{m-1}{2}} (-1)^i \mathrm{Tr}(x^{f+p^{2ik}+1})).$$

The p-rank of D_k is the p-rank of the circulant (0,1)-matrix $\left[g_k^{p-1}(\omega^{j-i})\right]$, or, equivalently, the dimension of the p-ary cyclic code generated by the (0,1)-valued characteristic sequence $\{g_k^{p-1}(\omega^t)\}$. This is [1] the number of terms in the polynomial $g_k^{p-1}(Z) \in \mathrm{GF(p)}[Z]$, where

$$g_k(Z) = \sum_{0\leq i,j\leq m-1} (-1)^i Z^{(p^{(2i+1)k}+1)p^j},$$

and exponents are interpreted mod $p^m - 1$. Thus, the rank is bounded above by $m^{2(p-1)}$ and is in the cases alluded to above a polynomial in m of degree $2(p-1)$ if $k > 0$. In the case $k = 0$ $g_k(x)$ and $h_k(x)$ degenerate to $g_0(x) = \mathrm{Tr}(x^2)$ and $h_0(x) = \mathrm{Tr}(x^{f+2})$; and the sequence $\{a_t\} = \{h_0(\omega^t)\}$ is an m-sequence. In this case the polynomial $g_0(Z)$ degenerates to $\sum_{j=0}^{m-1} Z^{2p^j}$ and the number of terms in the polynomial $g_0^{p-1}(Z)$ is easily seen to be $\binom{m+p-2}{p-1}$, the well-known rank of the Singer set [1]. For $k > 0$ the polynomial $g_k^{p-1}(Z)$ may be expanded as

$$\sum_{u=0}^{p^m-2} d_u Z^u,$$

where

$$d_u = \sum (-1)^{i_1+i_2+\cdots+i_{p-1}} \in \mathrm{GF(p)},$$

and this last sum is over all $i_1, i_2, \ldots, i_{p-1}, j_1, j_2, \ldots, j_{p-1}$ in $\{0, 1, \ldots, m-1\}$ satisfying

$$\sum_{t=1}^{p-1} (p^{(2i_t+1)k} + 1)p^{j_t} \equiv u \pmod{p^m - 1}.$$

We shall return to this interesting equivalence question in another paper.

References

1. E. F. Assmus, Jr., and J. D. Key, *Designs and their codes*, Cambridge Tracts in Mathematics **103**, Cambridge University Press, Cambridge, 1992.
2. L. D. Baumert, *Cyclic Difference Sets*, Lecture Notes in Math. **182**, Springer-Verlag, Berlin-Heidelberg-New York, 1971.
3. Bruce C. Berndt, Ronald J. Evans and Kenneth S. Williams, *Gauss and Jacobi Sums*, Can. Math. Soc. Monographs Volume **21**, Wiley, New York, 1998.
4. T. Beth, D. Jungnickel and H. Lenz, *Design Theory, 2nd Edition*, Cambridge University Press, Cambridge, 1999.
5. Leonard Eugene Dickson, *Linear Groups with an Exposition of the Galois Field Theory*, Dover, New York, 1958.
6. J. F. Dillon, Multiplicative Difference Sets via Additive Characters, *Designs, Codes and Cryptography, Assmus Memorial Issue* **17** (1999), pp. 225-235.
7. J. F. Dillon and H. Dobbertin, New Cyclic Difference Sets with Singer Parameters, submitted to *Finite Fields and Their Applications.*
8. Marshall Hall, Jr., *Combinatorial Theory, 2nd ed.*, Wiley, New York, 1986.
9. Jong-Seon No, New cyclic difference sets with Singer parameters constructed from d-homogeneous functions, preprint.

On the Crosscorrelation of m-Sequences and Related Sequences with Ideal Autocorrelation

Tor Helleseth

Department of Informatics
University of Bergen
høyteknologisenteret
N-5020 Bergen
NORWAY

Summary. We present a survey of the crosscorrelation between m-sequences and point out some of the most interesting and challenging remaining open problems in this area. Some families of nonbinary sequences with ideal two-level autocorrelation functions are constructed. Methods for designing sequences of even length with good autocorrelation properties using cyclotomy are given. In particular, binary sequences of even length with optimal autocorrelation functions are presented.

1 Introduction

Sequences with good correlation properties are important for many practical applications. We will give a survey of some methods for constructing sequences with good auto and crosscorrelation and discuss some open problems for further research.

A well studied problem in sequence designs is to study sequences of length $n = p^m - 1$, p a prime, with optimal autocorrelation. The prime examples of such sequences are the m-sequences which have a two-valued autocorrelation with a constant out-of-phase autocorrelation of -1. It is known that the crosscorrelation between two different m-sequences has at least three different values. During the last 35 years the crosscorrelation of m-sequences has been thouroughly studied by many researchers.

In the second section we give a brief survey of the crosscorrelation function between m-sequences and an overview of some of the remaining open problems in this area. Some of these problems have been open for more than 25 years and this survey hopes to give some renewed interest in these problems that may contribute to their solutions. In this section we will also describe some related sequences with ideal two-level autocorrelation

In contrast to m-sequences of length $n = 2^m - 1$ with their two-valued ideal autocorrelation function, it is known that balanced binary sequences of other lengths n, where $n \not\equiv 3 \pmod 4$, always have at least three values in their autocorrelation function. In the third section we give an overview of results of sequences of even length and optimal autocorrelation. Some constructions of such sequences based on cyclotomy are given.

2 Crosscorrelation of m-sequences

The crosscorrelation of two sequences $\{u(t)\}$ and $\{v(t)\}$ of period n over an alphabet of size p is defined as

$$\theta_{u,v}(\tau) = \sum_{t=0}^{n-1} \omega^{u(t+\tau)-v(t)},$$

where ω is a complex, primitive pth root of unity. If the two sequences are the same we call it the autocorrelation function.

Let F_{p^m} denote the finite field with p^m elements and let the *trace function* be the mapping from the finite field F_{p^m} to the subfield F_p given by

$$Tr(x) = \sum_{i=0}^{m-1} x^{p^i}.$$

A sequence $\{s(t)\}$ of period $n = p^m - 1$, p a prime, generated by a primitive polynomial of degree m is called a maximum-length sequence (denoted an m-sequence for short). An m-sequence can be written, (after a cyclic shift) in terms of the trace function from F_{p^m} to F_p, as $s(t) = Tr(\alpha^t)$ where α is an element of order $p^m - 1$. This is often called natural orientation for $\{s(t)\}$ and it follows from the trace representation that $\{s(t)\} = \{s(pt)\}$.

The m-sequence is balanced in the sense that all nonzero elements occur equally often p^{m-1} times and the zero element occurs $p^{m-1} - 1$ times during a period of the sequence. Since $\{s(t - \tau) - s(t)\}$ is also an m-sequence for all $\tau \neq 0$, and thus balanced, it follows by definition that the autocorrelation function of the m-sequence has an ideal two-level autocorrelation function with an out-of-phase correlation value of -1.

Any m-sequence of length $n = p^m - 1$ can (after a cyclic shift) be obtained by decimating $\{s(t)\}$ by a d relatively prime to n. We use $C_d(\tau)$ to denote the crosscorrelation function between the m-sequence $\{s(t)\}$ and its decimation $\{s(dt)\}$. By definition, we have

$$C_d(\tau) = \sum_{i=0}^{p^m-2} \omega^{s(t+\tau)-s(dt)}.$$

Using well known properties of the trace function, this can be reformulated as follows:

$$C_d(\tau) = \sum_{x \in F_{p^m}^*} \omega^{Tr(cx - x^d)}, \quad c = \alpha^\tau.$$

Hence, finding the crosscorrelation function between m-sequences is equivalent to evaluating the exponential sum above. ¿From the properties of the trace function and the results in Trachtenberg [19] and Helleseth [9], we obtain the following:

Lemma 1. *(1) $C_d(\tau)$ is a real number.*

(2) Let $d \cdot d^{-1} = 1 \pmod{p^m - 1}$, then the values and the number of occurrences of each value of $C_d(\tau)$ and $C_{d^{-1}}(\tau)$ are the same.

(3) $C_{dp^j}(\tau) = C_d(\tau)$.

(4) $C_d(p^j\tau) = C_d(\tau)$.

(5) $\sum_{\tau=0}^{p^m-2} C_d(\tau) = 1$.

(6) $\sum_{\tau=0}^{p^m-2} C_d^2(\tau) = p^{2m} - p^m - 1$.

(7) $\sum_{\tau=0}^{p^m-2} C_d^l(\tau) = -(p^m - 1)^{l-1} + 2(-1)^{l-1} + a_l p^{2m}$ where a_l is the number of solutions of the equations

$$x_1 + x_2 + \cdots + x_{l-1} + 1 = 0$$
$$x_1^d + x_2^d + \cdots + x_{l-1}^d + 1 = 0$$

where $x_i \in F_{p^m}^*$ for $i = 1, 2, \cdots, l-1$.

In the case when $\{s(t)\} = \{s(dt)\}$, i.e., when $d \equiv p^i \pmod{p^m - 1}$, then we have the two-valued autocorrelation function of the m-sequence $\{s(t)\}$. When the two sequences are distinct, i.e., when $d \not\equiv p^i \pmod{p^m - 1}$, then the crosscorrelation function $C_d(\tau)$ is known to take on at least three different values when $\tau = 0, 1, \cdots, p^m - 2$.

Theorem 1. *If $d \notin \{1, p, \cdots, p^{m-1}\}$, (i.e., when the two m-sequences are cyclically distinct), then $C_d(\tau)$ takes on at least three values as τ varies over the set $\{0, 1, \cdots, p^m - 2\}$.*

Proof: This follows from Lemma 1 (5) and (6) by showing that these two relations can not be satisfied for a two-valued crosscorrelation function, except when $d \in \{1, p, \cdots, p^{m-1}\}$. The full proof is given in Helleseth [9].

It is therefore of special interest to study the cases when exactly three values occur. In the binary case when $p = 2$ the following six decimations give three-valued crosscorrelation. These decimations cover all the known cases found by computer search.

(1) $d = 2^k + 1$, $m/gcd(m, k)$ odd.
(2) $d = 2^{2k} - 2^k + 1$, $m/gcd(m, k)$ odd.
(3) $d = 2^{\frac{m}{2}} + 2^{\frac{m+2}{4}} + 1$, $m \equiv 2 \pmod 4$.
(4) $d = 2^{\frac{m+2}{2}} + 3$, $m \equiv 2 \pmod 4$.
(5) $d = 2^{\frac{m-1}{2}} + 3$, m odd.
(6)

$$d = \begin{cases} 2^{\frac{m-1}{2}} + 2^{\frac{m-1}{4}} - 1 & \text{if } m \equiv 1 \pmod 4 \\ 2^{\frac{m-1}{2}} + 2^{\frac{3m-1}{4}} - 1 & \text{if } m \equiv 3 \pmod 4. \end{cases}$$

Case (1) is the oldest result due to Gold [8] in 1968. Case (2) is proved by Kasami but he mentions that this result has also been proved by Welch. The cases (3) and (4) were proved by Cusick and Dobbertin [2]. The case (5) was proved by Canteaut, Charpin and Dobbertin [1] and case (6) by Hollmann and Xiang [12].

In the nonbinary case when $p > 2$ there are fewer cases which give three-valued crosscorrelation. The following two cases have a three-valued cross-correlation. These were proved by Trachtenberg [19] for m odd and extended to the case $m/gcd(m,k)$ odd in Helleseth [9].

(1) $d = (p^{2k}+1)/2$, $m/gcd(m,k)$ odd.
(2) $d = p^{2k} - p^k + 1$, $m/gcd(m,k)$ odd.

These decimation are the natural analogues of the Gold and Kasami-Welch decimations respectively. For $p > 3$ there are no other decimations known by computer search that lead to three-valued crosscorrelation.

In the case of ternary sequences when $p = 3$, a recent result by Dobbertin, Helleseth, Kumar and Martinsen [7] provides a new decimation for ternary m-sequences that gives a three-valued crosscorrelation function.

Theorem 2. *Let $d = 2 \cdot 3^{(m-1)/2} + 1$, where m is odd, then the crosscorrelation function $C_d(\tau)$ takes on the following three values:*

$$\begin{array}{llll} -1+3^{(m+1)/2} & \textit{occurs} & \frac{1}{2}(3^{m-1}+3^{(m-1)/2}) & \textit{times} \\ -1 & \textit{occurs} & 3^m - 3^{m-1} - 1 & \textit{times} \\ -1-3^{(m+1)/2} & \textit{occurs} & \frac{1}{2}(3^{m-1}-3^{(m-1)/2}) & \textit{times.} \end{array}$$

Numerical results revealed also some other decimations with the same crosscorrelation properties. These decimations seem to be natural generalizations of the so-called binary Niho decimations in case (6) above.

Conjecture 1 *Let $d = 2 \cdot 3^r + 1$, where m is odd, and*

$$r = \begin{cases} \frac{m-1}{4} & \text{if } m \equiv 1 \pmod 4, \\ \frac{3m-1}{4} & \text{if } m \equiv 3 \pmod 4, \end{cases}$$

then the crosscorrelation function $C_d(\tau)$ is as in Theorem 2.

Even though some ideas from the proof of Theorem 2 can be applied, it seems that this conjecture is somewhat harder to prove than Theorem 2, and thus the "ternary Niho conjecture" is still open. At present this is the only known open case of three-valued crosscorrelation of m-sequences.

It turns out from cases (1) and (2) in the binary as well as in the nonbinary case that it is always possible, *except* when $m = 2^i$, to select $m/gcd(m,k)$ to be odd and thus find sequences with three-valued crosscorrelation. In the case $m = 2^i$ no decimations with three-valued crosscorrelation are known and it is conjectured [9] that none exist in this case.

Conjecture 2 *If $m = 2^i$ then the crosscorrelation function always takes on at least four different values.*

Another open problem is due to the observation that the value -1 always occurs in all the known examples in the binary case. In the nonbinary case the values of the crosscorrelation function are always real but not always integers. It has been shown that when $d \equiv 1 \pmod{p-1}$ the crosscorrelation function has only integer values. In this case numerically -1 always occurs in the known examples and this is conjectured in [9].

Conjecture 3 *Let $d \equiv 1 \pmod{p-1}$, then -1 always occurs as a value of the crosscorrelation function $C_d(\tau)$.*

It is possible to prove that the conjecture is related to the following system of equations:

$$\begin{aligned} x_0 + \alpha x_1 + \alpha^2 x_2 + \cdots + \alpha^{q-2} x_{q-2} &= 0 \\ x_0^d + x_1^d + x_2^d + \cdots + x_{q-2}^d &= 0 \end{aligned}$$

where α is a primitive element of F_q, where $q = p^m$. The conjecture holds if and only if this system has exactly q^{q-3} solutions where $x_i \in F_q$ for $i = 0, 1, \ldots, q-2$. Loosely speaking the number of solutions is the same as if this was a linear system of two equations, which it almost is in the sense that the first equation is linear and the other is linear if one substitues $y_i = x_i^d$.

Recently, there has also been a considerable interest in binary sequences with period $n = 2^m - 1$ with constant out-of-phase autcorrelation of -1 due to their close connections to difference sets. For recent work on binary sequences with two-level ideal autocorrelation function the reader is referred to [17], [4] and [3].

It is perhaps interesting to observe that the decimation in Theorem 2 occurs in a different connection in a conjecture by Lin [14] leading to some ternary sequences with a -1 out-of-phase autocorrelation.

Conjecture 4 *(Lin's conjecture) Let $d = 2 \cdot 3^{(m-1)/2} + 1$, where m is odd, and let $\{u(t)\}$ be the ternary sequence given by*

$$u(t) = Tr(\alpha^t + \alpha^{dt})$$

where α is a primitive element of F_{3^m}. Then the sequence $\{u(t)\}$ has a constant out-of-phase autocorrelation of -1.

Using a different decimation corresponding to the Kasami-Welch exponent, Helleseth, Kumar and Martinsen [11] found a new family of ternary sequences with ideal autocorrelation which also in a natural way leads to difference sets.

Theorem 3. *Let $d = 3^{2k} - 3^k + 1$, $m = 3k$ and let $\{v(t)\}$ be the ternary sequence given by*

$$v(t) = Tr(\alpha^t + \alpha^{dt})$$

where α is a primitive element of F_{3^m}. Then the sequence $\{v(t)\}$ has a constant out-of-phase autocorrelation of -1.

In this case when the sequence $\{v(t)\}$ is a sum a two m-sequences that differ by a decimation d it follows directly from the definitions that the autocorrelation of such a sequence equals the crosscorrelation between two m-sequences that differ by a decimation d. Therefore, it follows straightforward by Theorem 2 that the autocorrelation function is at most three-valued. A closer investigation of the autocorrelation function in this case, using techniques from quadratic forms, reveals that the out-of-phase autocorrelation function always takes on the value -1.

Recently, Helleseth and Gong [10] generalized this result and constructed a new familie of nonbinary sequences of period $p^m - 1$ for any prime number $p \geq 3$.

Theorem 4. *Let α be a primitive element of F_{p^m}. Let $m = (2r+1)k$ and $q = p^k$. Define*

$$f(x) = \sum_{i=0}^{r} u_i x^{(q^{2i}+1)/2}.$$

and let $b_0 = 2u_0$, $u_i = b_{2i} = b_{2r+1-2i}$ for $i = 1, 2, ..., r$. Suppose $b_0 = \pm 1$ and $b_i = (-1)^i$ for $i = 1, 2, ..., r$, then the sequence over F_p defined by

$$s(t) = Tr(f(\alpha^t))$$

has an ideal two-level autocorrelation, where indices of b_i are taken mod $2r+1$.

The ternary sequences constructed by Helleseth, Kumar and Martinsen [11] can be obtained as a special case, when $p = 3$ and $r = 1$, by a suitable decimation of some of the sequences obtained from Theorem 4.

3 Sequences with even lengths and optimal autocorrelation

In this section we will study the problem to find balanced binary sequences of even length n with a low maximum absolute value of the autocorrelation. It is known the best one can do is to have an autocorrelation of -2 and 2 when $n \equiv 2 \pmod 4$ and 0 and -4 when $n \equiv 0 \pmod 4$.

One well known construction of sequences of even length $n = p^m - 1$, for an odd prime p, with the autocorrelation properties above is given by Lempel, Cohn and Eastman [13] in 1977. Their construction is as follows: Let α be a primitive element of F_{p^m}, and define

$$S = \{\alpha^{2i+1} - 1 \mid i = 0, 1, \ldots, (p^m - 3)/2\}$$

and let the characteristic sequence $\{s(t)\}$ of the set S be defined as:

$$s(t) = \begin{cases} 1 \text{ if } \alpha^t \in S \\ 0 \text{ otherwise.} \end{cases}$$

The autocorrelation of this sequence of period $n = p^m - 1$ is optimal, i.e., takes on the values -2 and 2 when $n \equiv 2 \pmod 4$ and 0 and -4 when $n \equiv 0 \pmod 4$. It was later pointed out that this construction was a special case of a construction due to Sidelnikov [18].

Recently, in [16] sequences with the same length and the same optimal autocorrelation functions were constructed by replacing the set S in the construction above by the set

$$N = \{(z+1)^d + az^d + b \mid z \in F_{p^m}\}.$$

For several choices of a, b and d this leads to balanced (and almost balanced) binary sequences with optimal autocorrelation. Some values of d that work with suitable a and b can be found in [16] and include,

1) $d = (p^m + 1)/2$.
2) $d = p^m - p^{m-1} - 1$.
3) $p \geq 3$, $d = 2$.
4) $p = 3$, $d = 4$.
5) $p \geq 5$, $d = 3$.

Two sequences are said to be equivalent if one can be obtained from the other by a combination of cyclic shifting and decimation by a number relatively prime to the period. It is clear that equivalent sequences have the same autocorrelation function. In most cases the sequences constructed in [16] can be shown to be essentially equivalent to the ones obtained from Lempel, Cohn and Eastman [13] and Sidelnikov [18].

A computer search found all inequivalent balanced sequences of even lengths ≤ 34. It turned out that there are several occurrences of sequences with optimal autocorrelation for lengths n not of the form $p^m - 1$ where p is a prime. In addition there are examples of lengths n such that optimal sequences do not exist at all. Therefore, investigating this problem further seems to be quite interesting and challenging.

When $n \equiv 2 \pmod 4$ the number of equivalence classes with optimal autocorrelation for each length ≤ 34 are (where $n(\lambda)$ means λ equivalence classes of length n): 6(1), 10(1), 14(0), 18(3), 22(2), 26(5), 30(6), 34(2).

Table 1 shows the nonequivalent sequences with optimal autocorrelation function in the case $n \equiv 2 \pmod 4$ and $n \leq 34$, and where * is used, in the case of several sequences, to denote the construction due to Lempel, Cohn and Eastman. One observes that there are no sequences with autocorrelation ± 2 of length $n = 14$.

When $n \equiv 0 \pmod 4$ the number of equivalence classes with optimal autocorrelation for each length ≤ 32 are : 4(1), 8(2), 12(2), 16(2), 20(3), 24(8), 28(4), 32(0). Table 2 shows the nonequivalent sequences with optimal autocorrelation function in the case $n \equiv 0 \pmod 4$ and $n \leq 32$, and where * is used, in the case of several sequences, to denote the construction due to

Table 1. Optimal balanced sequences with $n \equiv 2 \pmod 4$

n	Sequences
6	011010
10	0000110111
14	None
18	000001011011100111
	000001100110101111*
	000010011010101111
22	0000010010111001101111*
	0000010101110110100111
26	00000010110110101011100111
	00000011011001010111100111
	00000100101100111010101111
	00000100111000101111011011*
	00000100111101011100011011
30	000000010110101100110111010111
	000000100110110001110101101111
	000000101011111000111011010011
	000000110001101101101010101111
	000000110001110110110010101111*
	000010001110111010011101001011
34	1000000010110010111001011100111011
	1000000010110101011111000111001101

Lempel, Cohn and Eastman. We observe that there are no sequences with autocorrelation $0, -4$ of length $n = 32$.

Recently, a new family of even length sequences were constructed using cyclotomy. These sequences are different from all the previous families since the period is in general not of the form $p^m - 1$ for an odd prime p.

Let $p = ef + 1$ where p is an odd prime. The cyclotomic classes C_i in F_p, $i = 0, 1, \ldots, e-1$, are defined by

$$C_i = \{\alpha^{i+te} \mid t = 0, 1, \ldots, f-1\}.$$

where α is a primitive element of F_p. The cyclotomic numbers of order e are defined by

$$(i, j) = |(C_i + 1) \cap C_j|.$$

Table 2. Optimal balanced sequences with $n \equiv 0 \pmod 4$

N	Sequences
4	0011
8	00010111
	00011011*
12	000010111011
	000010110111*
16	0000011010110111*
	0000011011010111
20	00000101011011001111
	00000101101110011101
	00000110011010110111
24	000000110011110101011011
	000001001110011011010111
	000001100101011001111011
	000001001110101101100111*
	000000110110100111010111
	000001010110001110110111
	000001100111101100101011
	000010011110100111010011
28	1110001101011011011101000000
	0100011010111011001011110000
	0000111010011001111011010100*
	0000001110101011011001111001
32	None

The cyclotomic numbers are known for almost all values of e less than 24 and in a few other cases.

Let $p = 4f+1$ be a prime and consider cyclotomic classes with respect to p and cyclotomic numbers of order 4. Let $p = x^2+4y^2$, $x \equiv 1 \pmod 4$. There are at most five possible distinct cyclotomic numbers when f is odd. In fact, we have $(0,0) = (2,2) = (2,0)$, $(0,1) = (1,3) = (3,2)$, $(1,2) = (0,3) = (3,1)$, $(3,3) = (0,1)$ and

$$(0,0) = (p-7+2x)/16,$$
$$(0,1) = (p+1+2x-8y)/16,$$

$$
\begin{aligned}
(1,2) &= (p+1+2x+8y)/16,\\
(0,2) &= (p+1-6x)/16,\\
\text{the rest} &= (p-3-2x)/16.
\end{aligned}
$$

where $p = x^2 + 4y^2$ and $x \equiv 1 \pmod 4$. Here y is two-valued depending on the choice of the primitive element α employed to define the cyclotomic classes.

Let

$$
\begin{aligned}
D_0 &= C_i \cup C_j \cup \{0\}\\
D_1 &= C_l \cup C_j
\end{aligned}
$$

where i, j and l are pairwise distinct integers between 0 and 3.

By the Chinese Remainder Theorem $Z_N \cong Z_2 \times Z_p$ under the isomorphism given by $\phi : w \mapsto (w \pmod 2, w \pmod p)$. Let $D = (0, D_0) \cup (1, D_1)$, then $|D| = p$ and therefore the characteristic sequence of length $2p$ constructed from the set $T = \phi^{-1}(D)$ is balanced. Furthermore, by construction it is straightforward to see that the autocorrelation function of the sequence can be expressed in terms of the cyclotomic numbers. A careful analysis of these expressions show that in some cases the sequence will have an optimal autocorrelation. Some examples are the following.

Theorem 5. *Let $p = 4f+1 = x^2+4y^2$, where $x \equiv 1 \pmod 4$ and f is odd. Define $T = \phi^{-1}(D)$, then the characteristic sequence $\{s(t)\}$ of T has period $2p$, is balanced and has an optimal autocorrelation if $(i,j,l) = (0,1,3)$ and $y = 1$.*

Theorem 6. *Let $p = 4f+1 = x^2+4y^2$, where $x = 1$ and f is odd. Define $T = \phi^{-1}(D)$, then the characteristic sequence $\{s(t)\}$ of T has period $2p$, is balanced and has an optimal autocorrelation if $(i,j,l) = (1,2,3)$.*

It natural problem that needs further investigation is to construct other families of balanced sequences of optimal autocorrelation of even period.

4 Nonbinary cyclotomic sequences

Finally, we discuss some nonbinary sequences with good autocorrelation properties that can be calculated explicitly based on the cyclotomic numbers. The sequences have length an odd prime p over an alphabet of size e dividing $p-1$ and can be constructed as follows:

The cyclotomic sequence $\{s(t)\}$ of length p (of order e where $e|(p-1)$) is defined as $s(t) = i$ for $t \in C_i$ and $s(0) = 0$. The sequence is balanced in the sense that all elements occur equally often, except for an extra 0.

Let $e = 6$ and $q = 31$, then 3 is a primitive element mod 31 and the cyclotomic classes are : $C_0 = \{1, 16, 8, 4, 2\}$, $C_1 = \{3, 17, 24, 12, 6\}$, $C_2 =$

$\{9, 20, 10, 5, 18\}$, $C_3 = \{27, 29, 30, 15, 23\}$, $C_4 = \{19, 25, 28, 14, 7\}$ and $C_5 = \{26, 13, 22, 11, 21\}$. This leads to a cyclotomic sequence of order 5 given by

$$s(t) = 0001021402251543012425531453433.$$

The maximum absolute value of the autocorrelation function of the sequence turns out to be equal to 2.

Let $A(\tau)$ denote the autocorrelation of a sequence $\{s(t)\}$ of period p over an alphabet of size e, i.e.,

$$A(\tau) = \sum_{t=0}^{p-1} \omega^{s(t+\tau)-s(t)}$$

where ω is a complex, primitive eth root of unity.

The well known Legendre sequences are special cases of cyclotomic sequences of order $e = 2$ and period being an odd prime p. These sequences have maximum absolute value of the autocorrelation function of 1 or 3 depending on whether $(p-1)/2$ is odd or even respectively.

It is rather straightforward to see that the autocorrelation function can be determined if the cyclotomic numbers are known. Recently, in [6] it has been demonstrated that even in the cases when the cyclotomic numbers are not known it is possible to find the autocorrelation of these sequences.

Theorem 7. *The autocorrelation of the cyclotomic sequence of period p and order e dividing $p-1$ is given by:*

Case 1: *f even,*

$$A(\tau) = \begin{cases} p & \text{if } \tau = 0, \\ -1 + \omega^k + \omega^{-k} & \text{if } \tau \in C_k. \end{cases}$$

Case 2: *f odd,*

$$A(\tau) = \begin{cases} p & \text{if } \tau = 0, \\ -1 + \omega^k - \omega^{-k} & \text{if } \tau \in C_k. \end{cases}$$

5 Conclusions

An overview of the crosscorrelation of m-sequences has been given along with some of the open remaining problems in this area. Constructions of nonbinary sequences of period $p^m - 1$ with ideal autocorrelation have been constructed. Binary sequences with even length and good autocorrelation properties based on cyclotomic numbers have been constructed.

Acknowledgment

This research work is supported in part by the Norwegian Research Council. The author thanks Halvard Martinsen for computer assistance and many useful discussions during this work.

References

1. Canteaut, A., Charpin, P., Dobbertin, H. (2000) Binary m-sequences with three-valued crosscorrelation: a proof of Welch's conjecture. IEEE Trans. Inform. Theory, **46**, 4–8
2. Cusick, T. W., Dobbertin, H. (1996) Some new three-valued crosscorrelation functions for binary m-sequences. IEEE Trans. Inform. Theory, **42**, 1238–1240
3. Dillon, J. F. (1999) Multiplicative difference sets via additive characters. Designs, Codes and Cryptography, **17**, 225–235
4. Dillon, J. F., H. Dobbertin, H. (2000) New cyclic difference sets with Singer parameters. Preprint.
5. Ding, C., Helleseth, T., Martinsen, H. (2001) New classes of binary sequences with optimal three-valued autocorrelation. IEEE Trans. Inform. Theory, **47**, 428–433
6. Ding, C., Helleseth, T., and Parker, M. G. (2000) Autocorrelation of cyclotomic sequences. Submitted for publication.
7. Dobbertin, H., Helleseth, T., Kumar, P. V., Martinsen, H. (2001) Ternary m-sequences with three-valued crosscorrelation function: Two new decimations. To appear in IEEE Trans. Inform. Theory.
8. Gold, R. (1968) Maximal recursive sequences with 3-valued recursive cross-correlation functions. IEEE Trans. Inform. Theory, **14**, 154–156
9. Helleseth, T. (1976) Some results about the cross-correlation function between two maximal linear sequences. Discrete Math. **16**, 209–232
10. Helleseth, T., Gong, G. (2001) New nonbinary sequences with ideal two-level autocorrelation function. Submitted for publication.
11. Helleseth, T., Kumar, P. V., Martinsen, H. (2001) A new family of ternary sequences with ideal autocorrelation function. Designs, Codes and Cryptography **23**, 157–166
12. Hollmann, H.D.L., Xiang, Q. (1999) A proof of the Welch and Niho conjectures on crosscorrelation of binary m-sequences. Submitted to Finite Fields and Their Applications.
13. Lempel, A., Cohn, M., Eastman, W. (1977) A class of balanced binary sequences with optimal autocorrelation properties. IEEE Trans. Inform. Theory, **23**, 38–44
14. Lin, H. A. (1998) From cyclic Hadamard difference sets to perfectly balanced sequences. Ph.D. Thesis, University of Southern California, Los Angeles, USA.
15. Niho, Y. (1972) Multi-valued cross-correlation functions between two maximal linear recursive sequences. Ph.D. Thesis, University of Southern California, 1972.
16. No, J. S., Chung, H., Song, H. Y., Yang, K., Lee, J. D., Helleseth, T. (2001) New construction for binary sequences of period $p^m - 1$ with optimal autocorrelation using $(z+1)^d + az^d + b$. IEEE Trans. Inform. Theory, **47**, 1638–1644
17. No, J. S., Chung, H., and Yun, M.S., (1998) Binary pseudorandom sequences of period $2^m - 1$ with ideal autocorrelation generated by the polynomial $z^d + (z+1)^d$. IEEE Trans. Inform. Theory, **44**, 1278–1282
18. V. M. Sidelnikov, V. M. (1969) Some k-valued pseudo-random sequences ane nearly equidistant codes. Problem. Peredach. Inform., **5**, 16–22
19. Trachtenberg, H. M.(1970) On the cross-correlation functions of maximal linear sequences. Ph. D. Thesis, University of Southern California, Los Angeles.

Sequences for OFDM and Multi-Code CDMA: Two Problems in Algebraic Coding Theory

Kenneth G. Paterson

Hewlett-Packard Laboratories,
Filton Road, Stoke Gifford,
Bristol BS34 8QZ, UK.

Summary. We study the peak-to-average power ratio (PAPR) problem for two different kinds of communications systems, Orthogonal Frequency Division Multiplexing (OFDM) and Multi-Code Code-Division Multiple Access (MC-CDMA). We describe a common coding theoretic approach to reducing the PAPR of both kinds of transmissions. In both cases, the classical Reed-Muller codes turn out to play a critical role. There is an intimate connection between Reed-Muller codes and Golay complementary sequences which can be exploited to produce codes suitable for OFDM. For MC-CDMA, it turns out that bent functions lead to transmissions with ideal power characteristics. In this way, the problem of finding good codes for OFDM and MC-CDMA can be closely related to some old and new problems in algebraic coding theory and sequence design.

1 Introduction

In this paper, we study the peak-to-average power ratio (PAPR) problem for two different kinds of communications systems, Orthogonal Frequency Division Multiplexing (OFDM) and Multi-Code Code-Division Multiple Access (MC-CDMA).

OFDM is a method of transmitting data simultaneously over multiple equally-spaced carrier frequencies, using Fourier transform processing for modulation and demodulation [2,8]. The method has been proposed for many types of radio systems such as wireless local area networks [1] and digital audio and digital video broadcasting [44]. OFDM offers many well-documented advantages for multicarrier transmission at high data rates, particularly in mobile applications.

Code-Division Multiple-Access (CDMA) dominates amongst proposals for 3rd Generation cellular communications systems [56]. Multi-code CDMA is a very simple, backwards-compatible technique for supporting users who demand widely varying data rates for different applications. In MC-CDMA, a user who wishes to transmit at a higher data rate is simply assigned additional orthogonal transmission channels and appears to the base station as multiple users [21,22]. (We note that the abbreviation MC-CDMA has been widely used for both multi-carrier CDMA, where characteristics of OFDM and CDMA systems are combined [20], and multi-code CDMA. Here we use it to abbreviate the latter.)

Both OFDM and MC-CDMA involve signals that are the sums of some number of basic signals from an orthogonal set. In the former case, these are continuous-time sinusoidal signals and in the latter, discrete-time Walsh-Hadamard sequences. In both cases, the signal is an orthogonal transform of the data to be transmitted: in OFDM, the signal is related to a Fourier transform and in MC-CDMA, it is simply a Walsh-Hadamard transform. Because of 'constructive interference' in the summation of basic signals, both systems can suffer from high PAPR, a severe handicap in low-cost mobile applications [4,22,23,26,34].

Coding, selecting for transmission only those sequences with low PAPR, is a possible solution to the PAPR problem. For OFDM, it turns out that Golay complementary sequences have excellent PAPR properties. Recently, Davis and Jedwab gave an explicit description of a large class (possibly all) of Golay complementary sequences in terms of certain cosets of the first order Reed-Muller codes [10,11]. Because of this intimate connection to algebraic coding theory, the codes for OFDM in Davis and Jedwab's work have not only low PAPR but also simple encoding algorithms and good error-correcting properties. We will review this work and some generalisations [35,42,45,46] and present some new problems in algebraic coding and sequence design which arise from it.

Then we'll turn our attention to the PAPR problem for MC-CDMA. Here, the sequences with optimal PAPR are those whose Walsh-Hadamard transforms are uniformly small. These correspond to the class of Boolean functions known in the literature as bent functions. Based on this connection, we will describe some classes of codes for MC-CDMA. These codes include earlier coding schemes of Ottosson [33,34] and Wada *et al* [52,53] as special cases. In view of the relationship between the Walsh-Hadamard transform and Reed-Muller codes, it is perhaps not surprising that the Reed-Muller codes will once again play a crucial role. We will develop links to old and new problems about bent functions, Kerdock codes and their relatives, the Delsarte-Goethals codes.

We will close by speculating on the connections between coding for OFDM and coding for MC-CDMA.

This paper draws heavily on material contained in [11,35,36]. These references contain full details and proofs as well as a full account of independent work on Golay complementary sequences [30,32], and on power control in OFDM and MC-CDMA. Further work adopting different approaches to coding for OFDM can be found in [38,47].

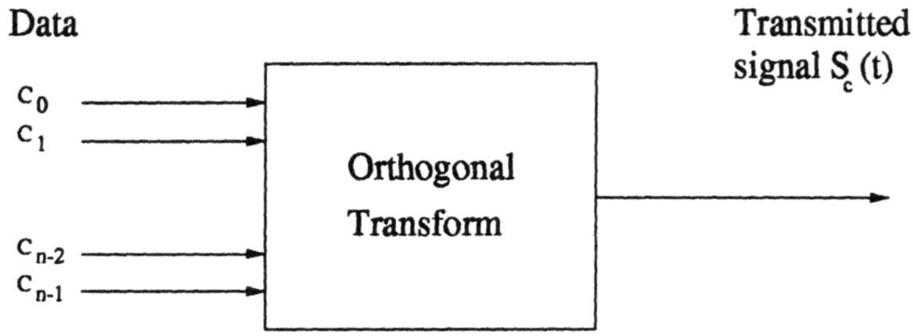

Fig. 1. Model for communication by orthogonal transforms

2 Communication by Orthogonal Transforms

2.1 Model of Communications Systems

In this section, we outline a general model for communication which includes OFDM and MC-CDMA. This model allows us to examine the common features shared by OFDM and MC-CDMA.

In our model (Fig. 1), a binary data vector $c = (c_0, c_1, \ldots, c_{n-1})$ is input to an *orthogonal transform*. Each data bit c_i modifies the sign of one of n orthogonal functions $f_i(t)$ of time t, and the output is the sum of these n modulated functions, the transmitted signal $S_c(t)$. So

$$S_c(t) = \sum_{i=0}^{n-1} (-1)^{c_i} f_i(t)$$

At the receiver (not shown in the figure), a noise-corrupted version of the signal is received. By virtue of orthogonality, computing the inner product of each of the orthogonal functions with the received signal recovers an estimate for each data bit c_i. In practice, the inner products are computed simultaneously via computation of the *inverse transform*.

For OFDM, the orthogonal transform is actually a kind of *Fourier* transform. The resulting signal is resistant to multi-path fading, which makes OFDM an attractive transmission technique in certain wireless environments. More specifically, we have $f_i(t) = e^{2\pi j(f+i\Delta f)t}$ where $j = \sqrt{-1}$, and given $c = (c_0, c_1, \ldots, c_{n-1})$, the OFDM signal is:

$$S_c(t) = \sum_{i=0}^{n-1} (-1)^{c_i} f_i(t) = \sum_{i=0}^{n-1} (-1)^{c_i} e^{2\pi j(f+i\Delta f)t}.$$

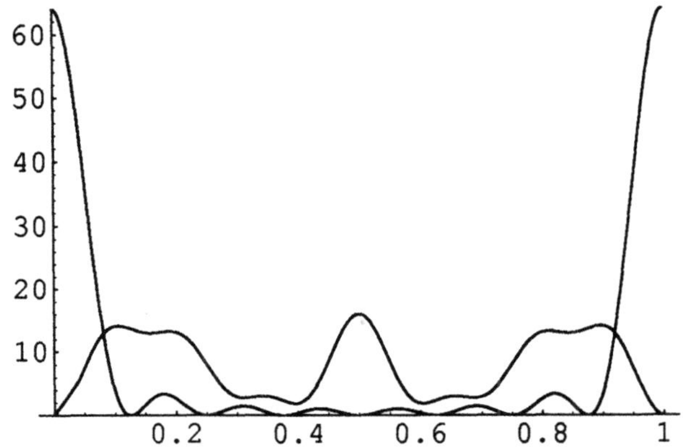

Fig. 2. (Envelope) powers of OFDM signals for $c = 00000000$ and $c = 00011101$, with t normalised to $[0, 1]$ by taking $\Delta f = 1$

Here, the n functions f_i are called *carriers* and are orthogonal on the interval $t \in [0, 1/\Delta f]$ with respect to the inner product

$$\langle f, g\rangle = \int_t f(t)g(t)^* dt.$$

In fact, the transmitted signal for OFDM is actually the real part of the complex function $S_c(t)$. However, it is more mathematically convenient (and a good approximation when considering power properties as we shall) to work with this complex signal, called the envelope signal. Fig. 2 shows the function $|S_c(t)|^2$ for two different OFDM signals. Typically $n = 2^m$ where m is small, say 4 up to 6 or so for mobile applications and as large as 10 or 11 for digital TV.

For multi-code CDMA, the transform is a *Walsh-Hadamard* transform. The system transmits at n times the rate of a basic CDMA system and thus caters for users demanding higher data rates. Given $c = (c_0, c_1, \ldots, c_{n-1})$ where $n = 2^m$, we model an MC-CDMA signal by:

$$S_c(t) = \sum_{i=0}^{n-1} (-1)^{c_i} \mathrm{WH}(n)_{it}, \quad t = 0, 1, \ldots, n-1,$$

where

$$\mathrm{WH}(2^m) = \begin{pmatrix} \mathrm{WH}(2^{m-1}) & \mathrm{WH}(2^{m-1}) \\ \mathrm{WH}(2^{m-1}) & -\mathrm{WH}(2^{m-1}) \end{pmatrix}, \qquad \mathrm{WH}(1) = (1)$$

is a $2^m \times 2^m$ Walsh-Hadamard matrix whose rows (and columns) are easily shown by induction to be orthogonal vectors. Notice that, in contrast

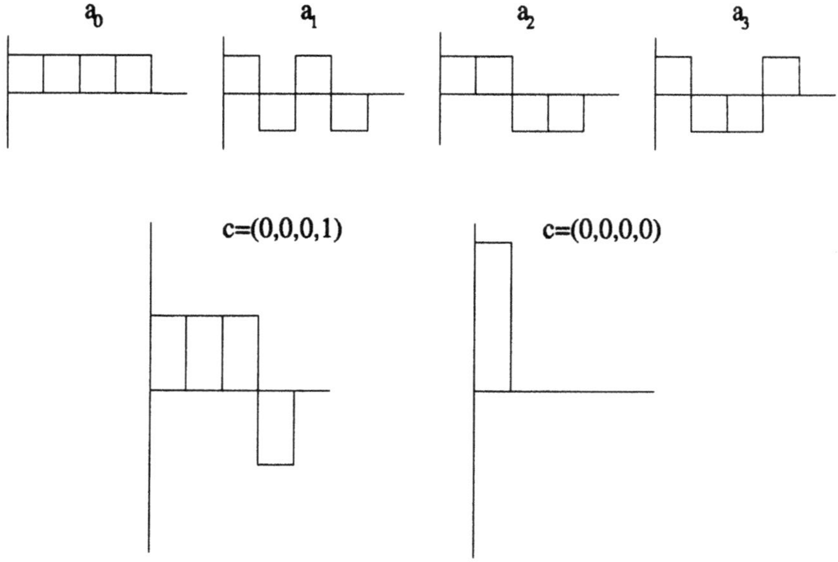

Fig. 3. MC-CDMA signal for $n = 4$

to OFDM, our MC-CDMA signal is modelled as a *discrete-time* signal and involves a *discrete* transform. We can write

$$S_c = (S_c(0), S_c(1), \ldots, S_c(n-1)) = \sum_{i=0}^{n-1} (-1)^{c_i} f_i$$

where f_i is the i-th row of WH(n). Thus we once again have that the signal is a sum of orthogonal functions (vectors in this case) and the transmitted signal is the sum of modulated versions of these signals. Fig. 3 shows the formation of two different MC-CDMA signals in the case $n = 4$. Typically in applications $n = 2^m$ with m between 2 and 6.

In both the OFDM and the MC-CDMA case, our model is only a rough approximation to what happens in a real system. Nevertheless, it captures the key parameter, power, that we want to study.

2.2 Peak-to-Average Power Ratio

Now consider the power of the transmitted signals in a system using an orthogonal transform.

We define the instantaneous power of the signal at time t for data c to be:

$$P_c(t) = |S_c(t)|^2.$$

In the context of OFDM, this real-valued function is usually called the *envelope power* of the signal. If we assume that each of the n orthogonal functions f_i is normalised so that it satisfies $\langle f_i, f_i \rangle = n$, then it is easy to show using orthogonality that the average value of the function $P_c(t)$ is equal to n. On the other hand, as is easily verified, for both OFDM and MC-CDMA, the signal $S_c(t)$ in the case $c = (0, 0, \ldots, 0)$ satisfies $S_c(0) = n$ so that $P_c(0) = n^2$. Informally, this arises because of a temporal alignment of peaks in the orthogonal functions, a kind of 'constructive interference' in the transform. This behaviour can be seen in both Figs. 3 and 2.

We define the *peak-to-average power ratio (PAPR)* of data c to be:

$$\mathrm{PAPR}(c) = \frac{1}{n} \cdot \max_t |S_c(t)|^2.$$

For OFDM, this function is more commonly referred to as the peak-to-mean envelope power ratio (PMEPR), with PAPR being reserved for the ratio $\frac{1}{n} \cdot \max_t \mathrm{Re}(S_c(t))^2$, representing the peak-to-average power ratio of the actual transmitted OFDM signal. Of course, this last function is bounded above by what we call here PAPR(c). Working with the complex signal $S_c(t)$ is somewhat easier than working with the actual transmitted signal.

The above discussion shows that PAPR(c) can be as large as n in a communication system using an orthogonal transform. If the peak power is subject to a design or regulatory limit then this has the effect of reducing the allowed average power relative to that which would be allowed with any constant power transmission scheme. This reduces the effective range of the transmissions and is particularly acute in mobile applications where battery power is a constraint. Moreover, to prevent signal distortions and spectral growth due to non-linearities inherent in electronic components, power amplifiers must be operated below their compression point where power is converted most efficiently. This results in more expensive and inefficiently used components. In summary, high PAPR is a serious drawback to both OFDM and MC-CDMA.

2.3 The Coding Solution

How then can the PAPR of transmissions be controlled? A very simple idea is to use coding: find a code $C \subset \{0,1\}^n$ in which every word has small PAPR and select for transmission only words $c \in C$. Thus we sacrifice transmission rate for PAPR reduction, inserting an encoder for C between data and orthogonal transform. At the receiver, the original data is recovered by performing the inverse transform and then decoding the resulting received word. This idea, perhaps obvious with the benefit of hindsight, appears to date back to the series of papers [23,24,55] for OFDM and to [33,52] for MC-CDMA. We illustrate the idea schematically in Fig. 4.

Several basic questions immediately arise from the idea of using coding to control PAPR:

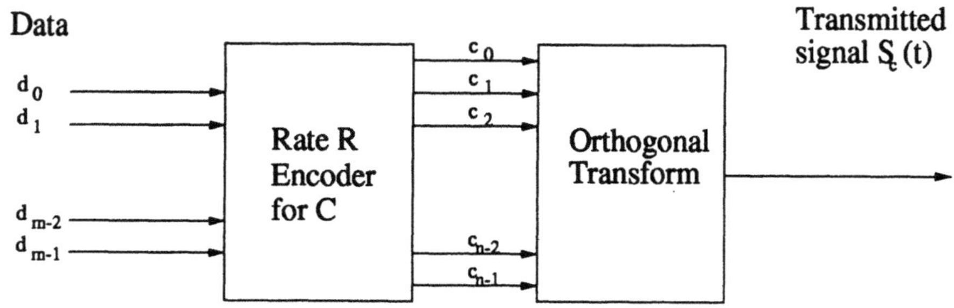

Fig. 4. Model for communication by orthogonal transforms with coding. Here R denotes the rate of the code C and $m = Rn$ is the number of data bits input to the encoder

- Given an orthogonal transform, which words can be used to form a code C with low PAPR?
- What is the maximum rate R that can be achieved while keeping the PAPR of C below a certain value?
- Can the redundancy introduced to reduce PAPR be exploited for error-correction?
- If so, for what triples $(R, d, \mathrm{PAPR}(C))$ can we construct codes? Here d denotes the minimum distance of C and $\mathrm{PAPR}(C)$ the peak-to-average power ratio of C.
- Are there efficient encoders (and decoders) for the code C?

In Secs. 3 and 4, we will describe at least partial answers to these questions for OFDM and MC-CDMA.

3 Coding for OFDM

3.1 Golay Complementary Pairs and Sets

A *Golay complementary pair* is a pair $\{c, d\}$ of binary sequences of length n such that:

$$A_c(u) + A_d(u) = 0, \quad u \neq 0$$

where

$$A_c(u) = \sum_i (-1)^{c_i - c_{i+u}}$$

is the *aperiodic auto-correlation* function of c (in which the summation is understood to be over only those integer values for which both i and $i + u$ lie within $\{0, 1, \ldots, n-1\}$).

Each member of a Golay complementary pair is called a *Golay complementary sequence.*

We are interested in using Golay complementary sequences as OFDM codewords because the resulting OFDM signals have PAPR of at most 2, a substantial and practically very useful reduction from the maximum value of n. This result in [39] generalises earlier work of [3]:

Theorem 1. *The PAPR of any Golay complementary sequence is at most 2.*

Proof. We have

$$\begin{aligned} P_c(t) &= S_c(t) \cdot S_c(t)^* \\ &= \sum_{i=0}^{n} (-1)^{c_i} e^{2\pi j(f+i\Delta f)t} \cdot \sum_{k=0}^{n} (-1)^{-c_k} e^{-2\pi j(f+k\Delta f)t} \\ &= \sum_{\substack{i-k=u \\ 0\le i,k<n}} (-1)^{c_i - c_k} \cdot e^{2\pi j u \Delta f t} \\ &= \sum_{u=1-n}^{n-1} A_c(u) e^{2\pi j u \Delta f t} \\ &= A_c(0) + 2 \cdot \mathrm{Re} \sum_{u=1}^{n-1} A_c(u) e^{2\pi j u \Delta f t}. \end{aligned}$$

Using the fact that $A_c(u) + A_d(u) = 0$ for every $u \neq 0$, we obtain

$$P_c(t) + P_d(t) = A_c(0) + A_d(0) = 2n.$$

Since the function $P_a(t)$ is real-valued and non-negative, we deduce that $P_a(t) \le 2n$ and the theorem follows.

An illustration of this theorem can be seen in Fig. 5. Notice how the sum of the two powers is exactly 16 for every time t, so that each of the sequences has instantaneous power at most 16 and hence PAPR at most 2.

Binary Golay complementary pairs were introduced by Golay [15,16] in connection with infrared multislit spectrometry and have since found application in a variety of fields [29,48]. They are known to exist for all lengths $n = 2^\alpha 10^\beta 26^\gamma$, where $\alpha, \beta, \gamma \ge 0$ [50], but do not exist for length n having any prime factor congruent to 3 modulo 4 [13]. A variety of recursive and direct constructions for Golay complementary pairs were given in [17]. For a survey of previous results on non-binary Golay complementary pairs, see [14, Chap. 13].

Golay complementary sets were introduced in [49] as a generalisation of Golay complementary pairs. For $1 \le j \le N$, let $c^j = (c_0^j, c_1^j, \dots, c_{n-1}^j)$ be a

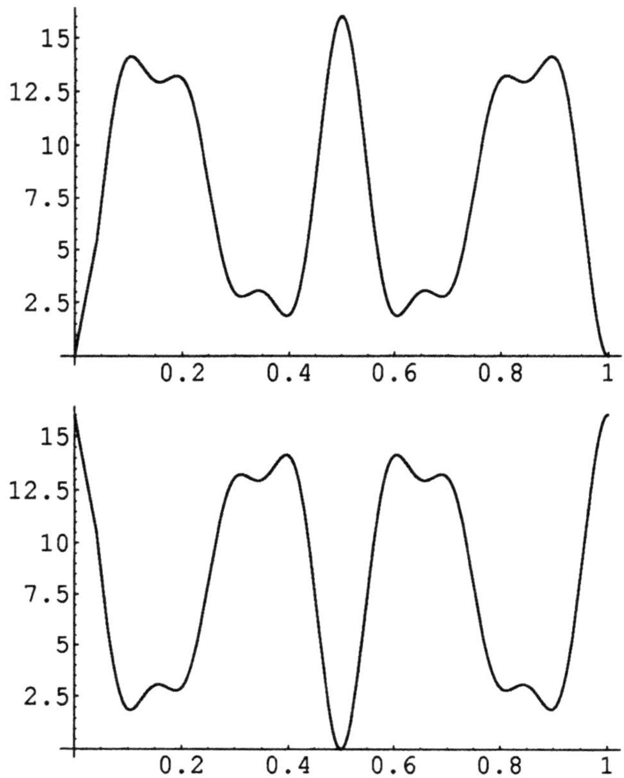

Fig. 5. OFDM Power for Golay Complementary Pair $\{00011101, 00010010\}$

binary sequence of length n. Let $C = \{c^1, c^2, \ldots, c^N\}$ The set C is called a *Golay complementary set of size N* if

$$\sum_{j=1}^{N} A_{c^j}(u) = 0 \quad \text{for each } u \neq 0.$$

Clearly, a Golay complementary set of size 2 is a Golay complementary pair. A survey of previous work on these sets and their applications can also be found in [14, Chap. 13].

As with Golay complementary sequences, our motivation for studying Golay complementary sets is that their sequences can have low PAPR. We have the following straightforward generalisation of Thm. 1.

Theorem 2. *The PAPR of any sequence from a Golay complementary set of size N is at most N.*

To make use of Golay complementary sequences (and more generally sequences from Golay complementary sets) to reduce PAPR in OFDM, we at least need to find large numbers of sequences. To build a practical system, we also need to answer the basic questions of how to encode and decode. Thus we need some structure in the set of sequences. But the received wisdom prior to the work of [10,11] was that *aperiodic* correlations are relatively difficult to understand and that sequences with controlled aperiodic correlations do not have any particular regularity. With this context, the results of [10,11] that we describe in Sect. 3.3 below are even more surprising.

3.2 Reed-Muller Codes

We recall the definition of the classical binary Reed-Muller codes RM(r,m), $0 \le r \le m$, from [28, Chap. 13]: the code RM(r,m) has

$$(n,k,d) = (2^m, \sum_{i=0}^{r} \binom{m}{i}, 2^{m-r})$$

and a generator matrix whose rows are vectors related to certain Boolean functions in m variables $x_0, x_1, \ldots, x_{m-1}$. Each row is obtained by evaluating a monomial function of non-linear order at most r over all 2^m possible inputs $(x_0, x_1, \ldots, x_{m-1})$ in their natural ordering. Notice that we label the codewords of the Reed-Muller codes somewhat unconventionally, c.f. [11,28]. The following example should clarify our notation:

Example 1. RM$(2,3)$ has length 8, dimension 7, minimum distance 2 and generator matrix:

$$\begin{bmatrix} 1111\,1111 \\ 0101\,0101 \\ 0011\,0011 \\ 0000\,1111 \\ 0001\,0001 \\ 0000\,0101 \\ 0000\,0011 \end{bmatrix} \quad \begin{matrix} 1 \\ x_0 \\ x_1 \\ x_2 \\ x_0x_1 \\ x_0x_2 \\ x_1x_2 \end{matrix}$$

Each codeword c of RM(r,m) is obtained from a unique Boolean function in m variables in which the maximum non-linear order is r. We use the two notions of codeword and Boolean function interchangeably in what follows, using c to denote both.

3.3 Golay Complementary Pairs and Sets from Reed-Muller Codes

Theorem 3. *[10,11] Let*

$$Q_\pi = x_{\pi(0)}x_{\pi(1)} + x_{\pi(1)}x_{\pi(2)} + \cdots + x_{\pi(m-2)}x_{\pi(m-1)}$$

where π is a permutation of $\{0, 1, \ldots, m-1\}$. *Then the coset* $Q_\pi + RM(1,m) \subset RM(2,m)$ *consists entirely of Golay complementary sequences. The Golay complementary pairs include:*

$$\{Q_\pi + w, Q_\pi + w + x_{\pi(0)}\}$$
$$\{Q_\pi + w, Q_\pi + w + x_{\pi(m-1)}\}$$

for each $w \in RM(1,m)$.

Thm. 3 explicitly determines $2^{m+1} \cdot m!/2$ binary Golay complementary sequences of length 2^m (using the factor $m!/2$ rather than $m!$ because the expression $\sum_{k=0}^{m-2} x_{\pi(k)} x_{\pi(k+1)}$ is invariant under the mapping $\pi \mapsto \pi'$, where $\pi'(k) = \pi(m-1-k)$). Exhaustive computations for $m \leq 6$ by T. Stinchcombe (personal communication) have shown that, for these parameters, Thm. 3 accounts for all binary Golay complementary sequences.

Open Problem 1 *Are there any more binary Golay complementary sequences of length* 2^m*?*

Each of the $2^{m+1} \cdot m!/2$ Golay complementary sequences of length 2^m identified in Thm. 3 has PAPR at most 2. They are neatly arranged in $m!/2$ second order cosets of RM$(1,m)$, i.e. they occur as a subcode of RM$(2,m)$. Each coset is identified with a quadratic form of the type Q_π. Therefore this set has all the properties that we need to build practical OFDM systems: a structure which can be exploited for encoding, a certain minimum distance (of at least 2^{m-2} determined by the minimum distance of RM$(2,m)$) and a relationship to well-understood error-correcting codes which can be used to build efficient decoders. For full details of the variety of coding options trading-off rate, minimum distance and PAPR that can be developed from this result, see [11]. For details on the important issue of efficient decoding, see [11,18,37]. We give just one example of the kind of trade-offs that can be made:

Example 2. In the case $m = 4$, Thm. 3 identifies 12 cosets of RM$(1,4)$ inside RM$(2,4)$ in which every word has PAPR at most 2. Selecting any 8 of these cosets (so that the number of encoded bits is an integer, easing implementation), we get an OFDM code with $(R, d, \text{PAPR}) = (1/2, 4, 2)$. We can increase the minimum distance at the expense of rate by considering just a single coset, giving a code with parameters $(5/16, 8, 2)$.

Extensive computations reported in [11] indicate that other second order cosets of RM$(1,m)$ have amazingly regular (but larger) PAPR. Where does this structure come from? The answer, as we shall see, is connected with Golay complementary sets. We need to introduce one further concept first. With each quadratic form Q in m variables $x_0, \ldots, x_{m-1}$, we can associate a labelled graph $G(Q)$ on m vertices, $0, 1, \ldots, m-1$. An example should make the correspondence clear:

Fig. 6. The graph corresponding to Q_π

Example 3. $Q_\pi = \sum_{i=0}^{m-2} x_{\pi(i)}x_{\pi(i+1)}$ has as its graph what we call a *path* on m vertices, as depicted in Fig. 6.

With the concept of the graph of a quadratic form in hand, we can now state:

Theorem 4. *[35] Suppose Q is a quadratic form in m variables. Suppose further that $G(Q)$ contains a set of $\ell \geq 0$ distinct vertices labelled $j_1, \ldots, j_\ell$ with the property that deleting those ℓ vertices and all their edges results in a path graph (necessarily on $m - \ell$ vertices). Let t be the label of either vertex of degree 1 in this path graph. Then for any choice of $c, c_k \in \mathbb{Z}_2$,*

$$\left\{ Q + \sum_{k=0}^{m-1} c_k x_k + c + \sum_{k=1}^{\ell} d_k x_{j_k} + d x_t \mid d_k, d \in \mathbb{Z}_2 \right\}$$

is a Golay complementary set of size $2^{\ell+1}$.

Thm. 4 provides a partial answer to an open problem posed in [49]:

Open Problem 2 *Obtain direct construction procedures for complementary sets with given parameters, namely, the number of sequences in the set and their lengths.*

Example 4. Let $m = 4$ and

$$Q = x_0x_1 + x_0x_2 + x_0x_3 + x_1x_2 + x_2x_3.$$

The graph $G(Q)$ is shown in Fig. 7. We see that deleting the vertex labelled 0 results in a path graph on vertices 1, 2 and 3. Applying Thm. 4 with $\ell = 1$, we get, for each choice of $c, c_0, c_1, c_2, c_3 \in \mathbb{Z}_2$, the following Golay complementary set of size 4:

$$\begin{aligned}
\{\ & Q + \textstyle\sum_{k=0}^{3} c_k x_k + c, \\
& Q + \textstyle\sum_{k=0}^{3} c_k x_k + c + x_0, \\
& Q + \textstyle\sum_{k=0}^{3} c_k x_k + c + x_1, \\
& Q + \textstyle\sum_{k=0}^{3} c_k x_k + c + x_0 + x_1\ \}.
\end{aligned}$$

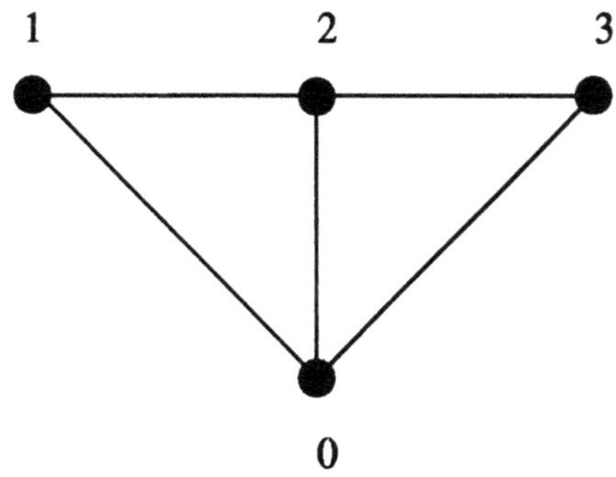

Fig. 7. The graph of the quadratic form $Q = x_0x_1 + x_0x_2 + x_0x_3 + x_1x_2 + x_2x_3$

Using Thm. 4, it is possible to give an explicit form (in terms of Boolean functions and graphs) for a large class of binary sequences that lie in Golay complementary sets of size $2^{\ell+1}$. These sequences all have PAPR at most $2^{\ell+1}$. The theorem also gives a bound for the PAPR of each second order coset $Q + \mathrm{RM}(1, m)$ in terms of a graph-theoretic parameter, the minimum number of vertices in $G(Q)$ whose deletion leads to a path. This bound goes some way to explaining the PAPR behaviour of second order cosets observed in [11]. We can also use the theorem and the bound it gives to construct codes for OFDM – see [35] for details.

Thm. 4 is also a generalisation of Thm. 3: the special case $\ell = 0$ of the latter theorem recovers the former. However, the methods that were used to prove the two results are rather different: [11] contains a direct proof of the result on Golay complementary pairs, while [35] uses a recursive approach containing as an intermediate step a result on a generalisation of binary Golay complementary pairs to alphabets $\{+1, -1, 0\}$. The direct approach has the benefit of being quick, but the recursive approach can be used to shed light on why the particular quadratic forms Q_π appear in the theorems.

3.4 Non-binary Sequences and Codes

So far, we have only considered binary sequences and binary modulation for OFDM. In practice, other kinds of modulation are also used, resulting in increased transmission rates. Here the OFDM signal becomes

$$S_c(t) = \sum_{i=0}^{n-1} z_i e^{2\pi j(f+i\Delta f)t}.$$

where $z_i \in \mathcal{Z}$ is some finite subset of the complex numbers, called a *constellation* or *signal set*. For the binary modulation considered so far, we have

$\mathcal{Z} = \{+1, -1\}$ and one bit of information is transported on each carrier. Another common form of modulation is ***phase-shift keying***. In q-PSK, we have $\mathcal{Z} = \{\omega^i, 0 \le i < q\}$ where q is some fixed integer and $\omega = e^{2\pi j/q}$ is a complex q-th root of unity. Typically $q = 2$ (which coincides with our usual binary modulation), 4 (2 bits per carrier, also called QPSK) or 8 (3 bits per carrier). Another important family are the QAM constellations, see [40, Chap. 4] for details.

There are nice (and practically valuable) generalisations of Thms. 3 and 4 to q-ary alphabets which lend themselves to q-PSK modulation. We sketch these generalisations next. In what follows q will be any even integer. We consider the generator matrix for RM(r,m), but now take linear combinations of rows modulo q. This defines a code that is linear over $\mathbb{Z}_q$, and which we denote RM$_q(r,m)$. The code RM$_q(r,m)$ is distinct from other generalisations of the binary Reed-Muller codes [25], but is closely connected to the quaternary code ZRM(r,m) introduced in [19]. It can be shown [11,35] that the code has the same minimum Hamming distance 2^{m-r} and 'dimension' as the classical binary code. The minimum Lee distance can also be calculated. Each codeword can be associated with a (generalised) Boolean function in m variables with coefficients from $\mathbb{Z}_q$.

We also need to define aperiodic correlation functions and Golay complementary sequences, pairs and sets for sequences over $\mathbb{Z}_q$. To do so, we simply replace -1 by $\omega = e^{2\pi j/q}$ in the definition of aperiodic correlation.

The motivation for the above definitions is the following pair of theorems:

Theorem 5. *[10,11] Let $q = 2^h$ where $h \ge 1$. Then the cosets $2^{h-1}Q_\pi + RM_q(1,m)$ are composed of q-ary Golay complementary sequences.*

Theorem 6. *[35] Let $q = 2h$ be any even integer. Suppose Q is a quadratic form in m variables. Associate with Q a labelled graph $G(Q)$ in which there is an edge between vertices i and j labelled q_{ij} whenever $q_{ij}x_ix_j$ appears as a non-zero term in Q. Suppose further that $G(Q)$ contains a set of $\ell \ge 0$ distinct vertices labelled $j_1, \dots, j_\ell$ with the property that deleting those ℓ vertices and all their edges results in a path graph in which each edge is labelled h. Then every word of the coset $Q + RM_q(1,m)$ is a sequence lying in a q-ary Golay complementary set of size $2^{\ell+1}$.*

Both theorems can be refined to explicitly identify the pair/set which contains any particular codeword. As well as being of theoretical interest, these theorems can be used to develop many OFDM coding options enjoying low PAPR for non-binary modulations. This is because Thms. 1 and 2 generalise straightforwardly to q-ary sequences. Details can be found in [11,35].

OFDM codes designed for 16-QAM alphabets were introduced in [42]: the codes there are constructed by cleverly decomposing the 16-QAM constellation into a sum of two 4-PSK constellations and using the two sequences from quaternary Golay complementary pairs to determine the modulation in the two signal sets. The resulting OFDM signals have PAPR at most 5.

Table 1. The weight distribution of $Q + \mathrm{RM}(1, m)$, $\mathrm{rank}(Q) = 2h$

Weight	Number of words
$2^{m-1} - 2^{m-h-1}$	2^{2h}
2^{m-1}	$2^{m+1} - 2^{2h+1}$
$2^{m-1} + 2^{m-h-1}$	2^{2h}

Yet another direction was explored in [46]: here, 4-PSK codes are obtained by using a binary code twice, once to encode bits onto a constellation $\{+1, -1\}$ and a second time onto $\{+j, -j\}$. The resulting OFDM signals have PAPR at most 4.

3.5 Ranks of Quadratic Forms and Lower Bounds on PAPR

Let Q be a (binary) quadratic form, $Q = \sum_{i<j} q_{ij} x_i x_j$. Let $A = (q_{ij})$ be an $m \times m$ matrix and $B = A + A^T$. Then the *rank* of the quadratic form Q, denoted $\mathrm{rank}(Q)$, is defined to be the rank of the matrix B (over $\mathbb{Z}_2$). The rank of a quadratic form is a fundamental invariant of the form. As one example of its importance we have:

Theorem 7. *[28, p. 441, Thm. 5] Let Q be a binary quadratic form in m variables. Then $\mathrm{rank}(Q) = 2h$ is even. The coset $Q + RM(1, m)$ has the weight distribution given in Table 1.*

We can use the notion of rank to obtain *lower* bounds on the PAPR of cosets. Recall that

$$S_c(t) = \sum_{i=0}^{n-1} (-1)^{c_i} e^{2\pi j(f + i\Delta f)t}.$$

Cammarano and Walker [5] first suggested considering power in the special case $t = 0$. Doing so, we get:

$$|S_c(0)| = \sum_{i=0}^{n-1} (-1)^{c_i} = n - 2\mathrm{wt}(c)$$

where $\mathrm{wt}(c)$ denotes the Hamming weight of c. So

$$\mathrm{PAPR}(c) \geq \frac{1}{n} (n - 2\mathrm{wt}(c))^2$$

and it immediately follows that

$$\mathrm{PAPR}(Q + \mathrm{RM}(1, m)) \geq 2^{m - \mathrm{rank}(Q)}.$$

This bound can be used to show that the PAPR bounds on cosets given by Thms. 3 and 4 are often tight. For example, consider the cosets Q_π+RM$(1, m)$ in Thm. 3 when m is odd: in this case, the maximum possible rank of Q_π is $m-1$ and hence PAPR$(Q_\pi + \text{RM}(1, m)) \geq 2$. This technique can be pushed further [35], but there are still many situations where the upper and lower bounds do not match.

The thesis [45] extends the graph theoretical methods of [35]. For example, it is shown that under certain conditions, up to two vertices of degree 0 (i.e. vertices with no incident edges) created in the process of deleting vertices in a graph $G(Q)$ can be ignored. The result is the identification of smaller Golay sets than are predicted by Thm. 6 and therefore, better bounds on PAPR of cosets. Improved lower bounds on the PAPR of binary cosets are also developed in [45], by examining more carefully the values $S_c(0)$. These can be used to demonstrate that degree 0 vertices cannot always be ignored and that certain sequences must lie in large Golay complementary sets. In this way, PAPR can be a useful tool in the study of the Golay set structure of cosets of RM$_q(1, m)$, itself a theoretically interesting sequence problem

Little more is known about lower bounds in the non-binary case. One result in [5] is that the quaternary cosets $2Q_\pi$+RM$_4(1, m)$ have PAPR exactly 2 when m is even. A result of [31] shows that certain cosets of RM$_8(1, m)$ have PAPR at least 3. Earlier computations in [11] suggest that these cosets should have PAPR exactly 3. But this PAPR behaviour cannot arise from Golay complementary sets of size 3 because such sets must be of even size over $\mathbb{Z}_8$.

Open Problem 3 *Obtain stronger lower and upper bounds on the PAPR of second order cosets of $RM_q(1, m)$.*

3.6 Kerdock Codes and Golay Pairs

Here we briefly outline another series of interesting open problems on binary Golay complementary pairs, motivated by the desire to generate more coding options.

We define an (m, h)-set to be a set of quadratic forms $\mathcal{Y}$ in m variables such that:

$$\text{rank}(Q + Q') \geq 2h \quad \forall Q, Q' \in \mathcal{Y}.$$

If such a set $\mathcal{Y}$ contains the all-zero form, then clearly every non-zero form in $\mathcal{Y}$ also has rank at least $2h$. The code

$$\bigcup_{Q \in \mathcal{Y}} Q + \text{RM}(1, m)$$

obtained from such a set contains $|\mathcal{Y}| \cdot 2^{m+1}$ codewords and has minimum distance $2^{m-1} - 2^{m-h-1}$ (because the distance between any two words in the same coset of RM$(1,m)$ is 2^{m-1} and the distance between any two words in different cosets is at least $2^{m-1} - 2^{m-h-1}$, being determined by the rank of the sum of the two forms). This can be substantially larger than the minimum distance of RM$(2,m)$. It is shown in [28, p. 667, Thm. 13] that for any (m,h)-set $\mathcal{Y}$, $|\mathcal{Y}| \leq c^{\lfloor m/2 \rfloor - h+1}$, where $c = 2^m$ for m odd and $c = 2^{m-1}$ for m even. In the even case, maximal (m,h)-sets give rise to the notorious Kerdock and Delsarte-Goethals codes.

How can these sets be used to produce better OFDM codes? It is shown in [11] that when $m = 4$, there is an $(m, m/2)$-set $\mathcal{Y}$ of size 6 in which every quadratic form is of the type Q_π for some π. Selecting any 4 of the forms and taking the union of cosets they determine leads to a binary OFDM code with parameters $(R, d, \text{PAPR}) = (7/16, 6, 2)$. This gives an attractive coding option which is midway between the codes of Example 2 in terms of rate and minimum distance. Actually the set of 6 forms $\mathcal{Y}$ in this case is a subset of the 'standard Kerdock set' pictured in [27, p. 55]. This example raises a series of open questions:

Open Problem 4 *Suppose m is even. What is the maximum size of an $(m, m/2)$-set containing only quadratic forms of the type Q_π?*

In fact, a simple argument proves an upper bound of $\binom{m}{2}$ on this intersection. Because of the special type of quadratic forms under consideration, the first rows of the corresponding matrices B have a 0 as their first entry and must have weight 1 or 2. A simple count shows that there are $\binom{m}{2}$ possible first rows. For the pair-wise sums of the forms in the set to have rank m, the matrices B must all have distinct first rows. Thus there are at most $\binom{m}{2}$ forms in the set. This bound is attained for $m = 4$. Is there a general construction? Does studying the problem in the $\mathbb{Z}_4$-domain help?

The problem can be doubly generalised to consider (m,h) sets as well as the more general quadratic forms considered in Thm. 4:

Open Problem 5 *What is the maximum size of an (m,h)-set containing only quadratic forms Q for which the deletion of some ℓ vertices in $G(Q)$ gives a path graph?*

Nothing is known about this more general formulation. Its solution may lead to interesting OFDM codes with high minimum distances and rates.

4 Coding for Multi-code CDMA

4.1 Reed-Muller Codes, Walsh-Hadamard Transforms and Bent Functions

We recall our discrete time model for an MC-CDMA system from Sect. 2.1. We have:

$$S_c = (S_c(0), S_c(1), \ldots, S_c(n-1)) = \sum_{i=0}^{n-1} (-1)^{c_i} f_i$$

where the f_i, $0 \le i < 2^m$, are the rows of the orthogonal matrix $\mathrm{WH}(2^m)$. The following key result depends on the observation that f_i is just the ± 1 version of the word $\sum_{k=0}^{m-1} i_k x_k \in \mathrm{RM}(1,m)$, where $i = \sum_{k=0}^{m-1} i_k x_k$ is the binary expansion of i.

Theorem 8. *[36] For any word c of length $n = 2^m$,*

$$PAPR(c) = n\left(1 - \frac{2d_*(c)}{n}\right)^2,$$

where $d_*(c) := \min\{d_H(c,w) : w \in RM(1,m)\}$ is the minimum Hamming distance between c and the first-order Reed-Muller code of length 2^m. In particular, $PAPR(c) = 1$ (the minimum possible value) if and only if $d_*(c) = 2^{m-1} - 2^{\frac{m}{2}-1}$.

If we write $d_*(C) = \min\{d_*(c) : c \in C\}$, then we have $\mathrm{PAPR}(C) = n(1 - \frac{2d_*(C)}{n})^2$. Thus codes which are *far* from $\mathrm{RM}(1,m)$ will have small PAPR for MC-CDMA. This idea is exploited in [36] to prove analogues of the Hamming and Gilbert-Varshamov sphere-packing bounds for MC-CDMA codes. Similar results for OFDM codes, involving more sophisticated technical machinery, can be found in [38].

The *Walsh-Hadamard transform* of the Boolean function c is defined to be the function $\hat{c}$ with

$$\hat{c}(u) = \sum_{v \in \mathbb{Z}_2^m} (-1)^{c(v)+L_u(v)}, \qquad u \in \mathbb{Z}_2^m$$

where

$$L_u = \sum_{k=0}^{m-1} u_k x_k \in \mathrm{RM}(1,m), \quad u = (u_0, u_1, \ldots, u_{m-1}).$$

Alternatively, working with vectors, the Walsh-Hadamard transform of c is just the vector $(-1)^c \cdot \mathrm{WH}(2^m)$. Thus the vector S_c containing the signal values for codeword c has components that are just the Walsh-Hadamard transform coefficients of c. We have:

Theorem 9. *Let c be a word of length $n = 2^m$. Then*

$$PAPR(c) = \frac{1}{n} \max_u |\hat{c}(u)|^2.$$

Moreover c has PAPR equal to 1 if and only if $|\hat{c}(u)| = \sqrt{n}$ for every $u \in \mathbb{Z}_2^m$.

A *bent function* is defined to be a Boolean function all of whose Walsh-Hadamard transform coefficients are equal in magnitude to $2^{m/2} = \sqrt{n}$. Clearly m must be even for such a function to exist. Equivalently, a bent function corresponds to a word satisfying $d_*(c) = 2^{m-1} - 2^{\frac{m}{2}-1}$, i.e. at maximum distance from $\mathrm{RM}(1,m)$. Thus:

Theorem 10. *[36,51] C is a constant amplitude MC-CDMA code, i.e. a code with PAPR equal to 1, if and only if every codeword of C is a bent function. In particular, constant amplitude codes of length $n = 2^m$ exist only for m even.*

Bent functions have received a good deal of attention, see for example [6,7,12,41,43,54] and the brief overview in [28, Chap. 14, Sect. 5]. It is known that any bent function has non-linear order at most $m/2$, that is, lies in the code $\text{RM}(m/2, m)$. So any code of bent functions will automatically have minimum distance at least $2^{m/2}$.

We are also interested in MC-CDMA codes with low PAPR in the case where m is odd. We know that $\text{PAPR}(C) = 1$ cannot be achieved in this case. We are therefore motivated to pose:

Open Problem 6 *When m is odd, how close to being uniform can the Walsh-Hadamard transform of a length 2^m word be? Additionally, how many words achieve this most uniform transform and what, if any, coding structure do they have?*

The first part of this problem is equivalent to determining the covering radius of $\text{RM}(1, m)$ for odd m. See [9] for further information on this old and difficult problem.

4.2 Codes from Bent Functions

In this section, we give several constructions of constant amplitude codes for MC-CDMA from bent functions. For the remainder of the paper, m will be even. It is easy to see that if c is bent, then so is every word of the coset $c + \text{RM}(1, m)$. So as with codes for OFDM, our codes will tend to be formed from unions of cosets of $\text{RM}(1, m)$. In this way, they are amenable to decoding techniques developed for OFDM codes, [11,18,37].

Example 5. The code $Q + \text{RM}(1, m)$ where Q is any bent function in m variables (for example $Q = x_0x_1 + x_2x_3 + \cdots + x_{m-2}x_{m-1}$) is a constant amplitude code of rate $(m+1)/2^m$ and minimum distance 2^{m-1}. When $m = 2$ this gives a code equivalent to that obtained in an *ad hoc* fashion in [52]. See also [53].

Example 6. A second family of constant amplitude codes is obtained by taking as the code at length $n = 2^m$ a union of many second-order cosets corresponding to quadratic forms Q of full rank m: combining Thms. 7 and 8, it is easy to see that such a code will have PAPR 1. Any code formed in this way has minimum distance at least 2^{m-2} as it is a subcode of $\text{RM}(2, m)$. For $m = 4$, the total number of full rank forms is equal to 28, and a pictorial list of them can be found in [28, p. 429]. Selecting any 16 of these forms gives a code with $(R, d, \text{PAPR}) = (9/16, 4, 1)$. This code has the same parameters as length 16 codes obtained by searching in [34] and an *ad hoc* construction in [52]. In the general case, the number of full rank forms is at least $2^{m(\frac{m}{2}-1)}$ and we obtain a code of rate at least $\frac{m(\frac{m}{2}+1)}{2^m}$, minimum distance 2^{m-2} and PAPR 1. In order to make these codes practical for larger values of m, an algorithm for encoding data bits directly onto full rank forms is needed.

Yet more families of codes can be obtained from what is known as the Maiorana-McFarland construction for bent functions:

Theorem 11. *Let π be a permutation on $\{0,1\}^t$ and let g be any Boolean function in t variables. Then*

$$f(x_0,\dots,x_{2t-1}) = \pi(x_0,\dots,x_{t-1})\cdot(x_t,\dots,x_{2t-1}) + g(x_0,\dots,x_{t-1})$$

is a bent function of $2t$ variables. (Note that we interpret π as a vector of t Boolean functions in t variables).

This construction produces a large number of bent functions with controllable non-linear order. It can be exploited to produce a variety of coding options. We refer to [36] for the details. To make these codes practical, efficient algorithms for encoding data bits into functions of the type appearing in Thm. 11 are required.

The above code families give new motivation to a longstanding research topic:

Open Problem 7 *Enumerate, construct and classify bent functions.*

As well as being interesting for its own sake, progress on this problem is likely to lead to better families of constant amplitude codes for MC-CDMA.

4.3 Codes from Kerdock and Delsarte-Goethals Codes

In this section we generate more coding options for MC-CDMA by considering subcodes of the Kerdock and Delsarte-Goethals codes.

We recall the definition of an (m,h)-set from Sect. 3.6. An $(m,m/2)$-set is called a Kerdock set. For each even m, a Kerdock set is constructed in [28, p. 457, eqn. (33)]. The set contains the zero quadratic form and $2^{m-1}-1$ quadratic forms of full rank. The resulting code $\mathcal{K}(m)$, known as the Kerdock code, contains RM$(1,m)$ as a subcode, has minimum distance $2^{m-1}-2^{(m/2)-1}$ and rate $2m/2^m$. Selecting any 2^{m-2} of the $2^{m-1}-1$ non-zero cosets of RM$(1,m)$ in the Kerdock code gives an MC-CDMA code with the same minimum distance, rate $(2m-1)/2^m$ and PAPR 1. For example, for $m=4$, we obtain a code with $(R,d,\text{PAPR}) = (7/16,6,1)$ which is a subcode of the Nordstrom-Robinson code.

It is unfortunate that we had to remove the zero coset from the Kerdock code here, since it forced us to reduce the rate from $2m/2^m$ to $(2m-1)/2^m$. It is not hard to show that any Kerdock set of quadratic forms must contain the zero form. However, moving to bent functions with higher non-linear order may help.

Open Problem 8 *Does there exist a code with the same parameters as the Kerdock code which consists entirely of bent functions? In particular, is there a Boolean function g (necessarily of non-linear order greater than 2) such that the set $g + \mathcal{K}(m)$ contains only bent functions?*

Next we consider subcodes of the Delsarte-Goethals codes [28, p. 461, Thm. 19]. The code $\mathcal{DG}(m,h)$, where $1 \leq h \leq m/2$, is constructed from a maximal (m,h)-set, has minimum distance $2^{m-1} - 2^{m-h-1}$ and contains $2^{(m-1)(m/2-h+1)+m+1}$ codewords arranged in cosets of $\mathrm{RM}(1,m)$.

The quadratic forms in the (m,h)-set include the zero form and so every non-zero form in the set has rank at least $2h$. But to construct a constant amplitude MC-CDMA subcode of $\mathcal{DG}(m,h)$, we must include only full rank quadratic forms. So to evaluate the rate of this subcode, we must find the number of full rank forms in the (m,h)-set used to construct the Delsarte-Goethals codes. Fortunately, this number can be calculated, using the results of [28, Chap. 21, Secs. 7 and 8]. We sketch this calculation next.

Given a set of quadratic forms $\mathcal{Y}$, we define the *inner distribution* of $\mathcal{Y}$ to be the $(m+1)$-tuple of real numbers $(B_0, B_1, \ldots, B_{m/2})$ where

$$B_i = \frac{1}{|\mathcal{Y}|}|\{(Q,Q') \in \mathcal{Y} \times \mathcal{Y} : \mathrm{rank}(Q+Q') = 2i\}|.$$

For $\mathcal{Y}$ equal to the (m,h)-set used to construct $\mathcal{DG}(m,h)$, we would like to know the numbers $(A_0, A_1, \ldots, A_{m/2})$, where

$$A_i = |\{Q \in \mathcal{Y} : \mathrm{rank}(Q) = 2i\}|,$$

in particular the number $A_{m/2}$. We have the following lemma:

Lemma 1. *Let $\mathcal{Y}$ be the (m,h)-set used to construct $\mathcal{DG}(m,h)$ and let A_i, B_i be defined as above. Then*

$$A_i = B_i, \qquad 0 \leq i \leq m/2.$$

Proof. The code $\mathcal{DG}(m,h)$ is the Gray image of a code that is linear over $\mathbb{Z}_4$ and so is distance invariant, i.e. the weight distribution and distance distribution of $\mathcal{DG}(m,h)$ are equal. (We refer to [19] for details of the Gray map and $\mathbb{Z}_4$-linearity.) But by virtue of the code's construction from second order cosets of $\mathrm{RM}(1,m)$, these two distributions are determined entirely by the numbers A_i and B_i respectively, with the number of words of weight $2^{m-1} \pm 2^{m-i-1}$ being determined by A_i and the number of times $2^{m-1} \pm 2^{m-i-1}$ appears in the distance distribution being determined by B_i. To obtain equality of these distributions we must then have $A_i = B_i$, $0 \leq i \leq m/2$.

The inner distribution $(B_0, B_1, \ldots, B_{m/2})$ of any maximal (m,h)-set is known exactly from [28, p. 668, Thm. 14]. We have:

$$B_{m/2-i} = \sum_{j=i}^{m/2-h} (-1)^{j-i} C_{i,j}$$

where

$$C_{i,j} = 4^{\binom{j-i}{2}} \begin{bmatrix} j \\ i \end{bmatrix} \begin{bmatrix} m/2 \\ j \end{bmatrix} (2^{(m-1)(m/2-h+1-j)} - 1).$$

Here, $\begin{bmatrix} x \\ y \end{bmatrix}$ denotes a 4-ary Gaussian binomial coefficient [28, p. 443].

The following result is proved in [36] by carefully examining these coefficients:

Theorem 12. *With notation as above, we have*

$$A_{m/2} \geq 2^{(m-1)(m/2-h+1)-2}.$$

Thm. 12 shows that considering only cosets of $\mathrm{RM}(1, m)$ corresponding to the full rank forms in the (m, h)-set used in constructing $\mathcal{DG}(m, h)$ results in a subcode which encodes 2 bits less than the entire code. Since $\mathcal{DG}(m, h)$ always contains the zero form, this is just one bit less than we would have obtained by considering all the non-zero cosets in the code. This full rank subcode has minimum distance $2^{m-1} - 2^{m-h-1}$, rate $(m-1)(m/2-h+2)/2^m$ and PAPR 1. For small values of m, the full rank quadratic forms in the (m, h)-set can be obtained by direct calculation. It would be convenient to find a simple method of selecting such forms directly for larger values of m.

4.4 Further Codes for MC-CDMA

Here we briefly mention two other areas worthy of further exploration.

We have concentrated exclusively on binary codes for MC-CDMA. But, as with OFDM, QPSK and other modulations may be used in place of BPSK in MC-CDMA.

Open Problem 9 *What can be said about quaternary (and larger alphabet) MC-CDMA codes with low PAPR?*

We have only looked at constant amplitude MC-CDMA codes, these having optimal PAPR and a nice connection to bent functions. But codes with approximately constant PAPR may also be useful in practice (especially for m odd where bent functions do not exist). Some ideas in this direction can be found in [36].

Open Problem 10 *Find constructions for large numbers of 'approximately bent' functions and study the trade-offs between rate, PAPR and minimum distance which can be made.*

5 Conclusions

We have seen how the practical problem of designing codes which reduce the PAPR of OFDM and MC-CDMA transmissions leads to new problems in sequence design and algebraic coding theory. We have also seen new motivation for attacking some well-known (and difficult) problems on Reed-Muller codes and bent functions.

Finally, we speculate on the similarities and differences between codes for OFDM and codes for MC-CDMA. In both cases, an orthogonal transform is used to transform data prior to transmission and the problem is to design codes which reduce the size of the transform values. The Walsh-Hadamard transform is a discrete analogue of the Fourier transform inherent in OFDM, so similar coding solutions might be expected. Indeed, in [11] it is shown that for m even, the $m!/2$ cosets of $\mathrm{RM}(1,m)$ which consist of binary Golay complementary sequences are bent cosets. Thus a code formed from these cosets will simultaneously enjoy low OFDM PAPR and ideal MC-CDMA PAPR. However the Reed-Muller code, and the constructions as unions of cosets of $\mathrm{RM}(1,m)$, appear to arise for different reasons in the two cases. For OFDM, an explanation relating the particular Boolean functions yielding Golay complementary sequences and the recursive constructions for those sequences was given in [35]. In MC-CDMA, the Reed-Muller code plays a role because of the connection between rows of the Walsh-Hadamard matrix and the codewords of $\mathrm{RM}(1,m)$ (though this link also has a recursive proof). A detailed explanation of the double appearance of the Reed-Muller codes may give greater insight into both practical and theoretical questions.

Finally, we ask:

Open Problem 11 *Are there more examples of orthogonal transforms in communications theory waiting to be identified? Are PAPR considerations important and if so, what structure do good codes have?*

Acknowledgements

I am grateful to Prof. J.A. Davis and the anonymous referee for comments which significantly improved the paper.

References

1. M. Aldinger, "Multicarrier COFDM scheme in high bitrate radio local area networks," in *5th IEEE Int. Symp. on Personal, Indoor and Mobile Radio Commun., The Hague*, pp. 969–973, 1994.
2. J.A.C. Bingham, "Multicarrier modulation for data transmission: an idea whose time has come," *IEEE Commun. Magazine*, vol. 28, pp. 5–14, May 1990.
3. S. Boyd, "Multitone signals with low crest factor," *IEEE Trans. Circuits and Systems* vol. CAS-33, pp. 1018–1022, 1986.

4. R.N. Braithwaite, "Using Walsh code selection to reduce the power variance of band-limited forward-link CDMA waveforms," *IEEE J. on Selected Areas in Communications*, vol. 18, no. 11, pp. 2260–2269, Nov. 2000.
5. M.W. Cammarano and M.L. Walker, "Integer maxima in power envelopes of Golay codewords," Technical Report TR-99-02, Dept. Math. Comp. Science, University of Richmond, 1997.
6. C. Carlet and P. Guillot, "A characterization of binary bent functions," *J. Combin. Theory Ser. A*, vol. 76, no. 2, pp. 328–335, 1996.
7. C. Carlet and P. Guillot, "An alternate characterization of the bentness of binary functions, with uniqueness," *Des. Codes Cryptogr.*, vol. 14, no. 2, pp. 133–140, 1998.
8. L.J. Cimini, Jr., "Analysis and simulation of a digital mobile channel using orthogonal frequency division multiplexing," *IEEE Trans. Commun.*, vol. 33, pp. 665–675, July 1985.
9. G. Cohen, I. Honkala, S. Litsyn and A. Lobstein, *Covering codes*, North-Holland, Amsterdam, 1997.
10. J.A. Davis and J. Jedwab, "Peak-to-mean power control and error correction for OFDM transmission using Golay sequences and Reed-Muller codes," *Elec. Lett.*, vol. 33, pp. 267–268, 1997.
11. J.A. Davis and J. Jedwab, "Peak-to-mean power control in OFDM, Golay complementary sequences and Reed-Muller codes," *IEEE Trans. Inform. Theory*, vol. 45, pp. 2397–2417, Nov. 1999.
12. H. Dobbertin, "Construction of bent functions and balanced Boolean functions with high nonlinearity," in *Lecture Notes in Computer Science*, vol. 1008, pp. 61–74, Springer, Berlin, 1995.
13. S. Eliahou, M. Kervaire and B. Saffari, "A new restriction on the lengths of Golay complementary sequences," *J. Combin. Theory (A)* vol. 55, pp. 49–59, 1990.
14. P. Fan and M. Darnell, *Sequence Design for Communications Applications*, Communications Systems, Techniques and Applications, Research Studies Press, Taunton, 1996.
15. M.J.E. Golay, "Multislit spectroscopy," *J. Opt. Soc. Amer.* vol. 39, pp. 437–444, 1949.
16. M.J.E. Golay, "Static multislit spectrometry and its application to the panoramic display of infrared spectra," *J. Opt. Soc. Amer.* vol. 41, pp. 468–472, 1951.
17. M.J.E. Golay, "Complementary series," *IRE Trans. Inform. Theory*, vol. IT-7, pp. 82–87, 1961.
18. A.J. Grant and R.D.J. van Nee, "Efficient maximum-likelihood decoding of *Q*-ary modulated Reed-M-uller codes," *IEEE Comm. Lett.*, vol. 2, pp. 134–136, 1998.
19. A.R. Hammons, Jr., P.V. Kumar, A.R. Calderbank, N.J.A. Sloane and P. Solé, "The $\mathbf{Z}_4$-linearity of Kerdock, Preparata, Goethals and related codes," *IEEE Trans. Inform. Theory*, vol. 40, pp. 301–319, 1994.
20. S. Hara and R. Prasad, "Overview of multicarrier CDMA,", *IEEE Communications Magazine*, vol. 35, no. 12, pp. 126–133, 1997.
21. C.-L. I and R.D. Gitlin, "Multi-code CDMA wireless personal communications networks" in *Proc. ICC'95*, Seattle, Wash., pp. 1060–1064, 1995.

22. C.-L. I, C.A. Webb III, H.C. Huang, S. ten Brink, S. Nanda and R.D. Gitlin, "IS-95 enhancements for multimedia services," *Bell Labs Tech. J.*, vol. 1, no. 2, pp. 60–87, 1996.
23. A.E. Jones, T.A. Wilkinson and S.K. Barton, "Block coding scheme for reduction of peak to mean envelope power ratio of multicarrier transmission schemes," *Elec. Lett.*, vol. 30, pp. 2098–2099, 1994.
24. A.E. Jones and T.A. Wilkinson, "Combined coding for error control and increased robustness to system nonlinearities in OFDM," in *IEEE 46th Vehicular Technology Conference*, Atlanta, USA, pp. 904–908, April–May 1996.
25. T. Kasami, S. Lin, and W. Peterson, "New generalizations of Reed-Muller codes, Part I: primitive codes," *IEEE Trans. Inform. Theory*, vol. IT-14, pp. 189–199, 1968.
26. V.K.N. Lau, "On the analysis of peak-to-average ratio (PAR) for IS95 and CDMA2000 systems," *IEEE Trans. Vehic. Tech.*, vol. 49, no. 6, pp. 2174–2188, Nov. 2000.
27. J.H. van Lint, *An introduction to coding theory*, 2nd edition, Springer, Berlin, 1992.
28. F.J. MacWilliams and N.J.A. Sloane, *The Theory of Error-Correcting Codes*, North-Holland, Amsterdam, 1977.
29. M. Nazarathy, S.A. Newton, R.P. Giffard, D.S. Moberly, F. Sischka and W.R. Trutna, Jr., "Real-time long range complementary correlation optical time domain reflectometer," *IEEE J. Lightwave Technology*, vol. 7, pp. 24–38, 1989.
30. R.D.J. van Nee, "OFDM codes for peak-to-average power reduction and error correction," in *Proc. IEEE Globecom 1996*, pp. 740–744, London, 1996.
31. K.M. Nieswand and K.N. Wagner, "Octary codewords with power envelopes of $3*2^m$," Technical Report TR-99-03, Dept. Math. Comp. Science, University of Richmond, 1998.
32. H. Ochiai and H. Imai, "Block coding scheme based on complementary sequences for multicarrier signals," *IEICE Trans. Fundamentals*, pp. 2136–2143, 1997.
33. T. Ottosson, "Precoding in multicode DS-CDMA Systems," in *Proc. 1997 IEEE Int. Symp. Info. Thy.*, Ulm, Germany, June 29 - July 4th, 1997, p. 351.
34. T. Ottosson, "Precoding for minimization of envelope variations in multicode DS-CDMA systems," *Wireless Personal Communications*, vol. 13, pp. 57–78, May 2000.
35. K.G. Paterson, "Generalised Reed-Muller codes and power control in OFDM modulation," *IEEE Trans. Inform. Theory*, vol. 46, pp. 104–120, Jan. 2000.
36. K.G. Paterson, "On codes with low peak-to-average power ratio for multicode CDMA," HP Laboratories Technical Report HPL-2001-115, May 2001. Submitted.
37. K.G. Paterson and A.E. Jones, "Efficient decoding algorithms for generalised Reed-Muller Codes, *IEEE Trans. Commun.*, vol. 48, no. 8, pp. 1272–1285, 2000.
38. K.G. Paterson and V. Tarokh, "On the existence and construction of good codes with low peak-to-average power ratios," *IEEE Trans. Inform. Theory*, vol. 46, no. 6, pp. 1974–1987, Sept. 2000.
39. B.M. Popović, "Synthesis of power efficient multitone signals with flat amplitude spectrum," *IEEE Trans. Commun.*, vol. 39, pp. 1031–1033, 1991.
40. J.G. Proakis, *Digital Communications (3rd Edition)*, McGraw-Hill, Inc., 1995.

41. B. Preneel, W. van Leekwijck, L. van Linden, R. Govaerts and J. Vandewalle, "Propagation characteristics of boolean functions," in *Proc. EuroCrypt90*, Lecture Notes in Computer Science, vol. 473, pp. 161–173, Springer, Berlin, 1991.
42. C. Rössing, "Golay complementary sequences for OFDM with 16-QAM," in *Proc. 2000 IEEE Int. Symp. Info. Thy.*, Sorrento, Italy, June 25 - 30, 2000, p. 331.
43. O.S. Rothaus, "On "bent" functions", *J. Comb. Thy. Ser. A*, vol. 20, pp. 300–305, 1976.
44. P. Shelswell, "The COFDM modulation system: the heart of digital audio broadcasting," *Elec. Commun. Eng. J.*, pp. 127–136, June 1995.
45. T.E. Stinchcombe, "Aperiodic correlations of length 2^m sequences, complementarity, and power control for OFDM," Ph.D. thesis, University of London, 2000.
46. V. Tarokh and C. Chong, "A simple encodable/decodable OFDM QPSK code with low peak envelope to average power ratio," *Proc. 2001 IEEE Int. Symp. Info. Thy.*, Washington D.C., USA, June 24 - 29, 2001.
47. V. Tarokh and H. Jafarkhani, "On Reducing the Peak to Average Power Ratio in Multicarrier Communications," *IEEE Trans. Comm.*, vol. 48, pp. 37–44, 2000.
48. C.-C. Tseng, "Signal multiplexing in surface-wave delay lines using orthogonal pairs of Golay's complementary sequences," *IEEE Trans. Sonics and Ultrasonics*, vol. SU-18, pp. 103–107, 1971.
49. C.-C. Tseng and C.L. Liu, "Complementary sets of sequences," *IEEE Trans. Inform. Theory*, vol. IT-18(5), pp. 644–652, Sept. 1972.
50. R.J. Turyn, "Hadamard matrices, Baumert-Hall units, four-symbol sequences, pulse compression, and surface wave encodings," *J. Combin. Theory (A)*, vol. 16, pp. 313–333, 1974.
51. T. Wada, "Characteristic of bit sequences applicable to constant amplitude orthogonal multicode systems," *IEICE Trans. Fundamentals*, vol. E83-A, no. 11, pp. 2160–2164, Nov. 2000.
52. T. Wada, T. Yamazato, M. Katayama and A. Ogawa, "A constant amplitude coding for orthogonal multi-code CDMA systems," *IEICE Trans. Fundamentals*, vol. E80-A, no. 12, pp. 2477–2484, Dec. 1997.
53. T. Wada, T. Yamazato, M. Katayama and A. Ogawa, "Error correcting capability of constant amplitude coding for orthogonal multi-code CDMA systems," *IEICE Trans. Fundamentals*, vol. E81-A, no. 10, pp. 2166–2169, Oct. 1998.
54. J. Wolfmann, "Bent functions and coding theory," in *Difference sets, sequences and their correlation properties (Bad Windsheim, 1998)*, NATO Adv. Sci. Inst. Ser. C Math. Phys. Sci., vol. 542, Kluwer Acad. Publ., Dordrecht, pp. 393–418, 1999
55. T.A. Wilkinson and A.E. Jones, "Minimisation of the peak to mean envelope power ratio of multicarrier transmission schemes by block coding," in *IEEE 45th Vehicular Technology Conference*, pp. 825–829, Chicago, July 1995.
56. M. Zeng, A. Annamalai and V.K. Bhargava, "Recent Advances in Cellular Wireless Communications," *IEEE Communications Magazine*, vol. 37, no. 9, pp. 28–138, 1999.

Signal Design for Ultra-wideband Radio *

Robert A. Scholtz, P. Vijay Kumar, and Carlos J. Corrada-Bravo

Department of Electrical Engineering
University of Southern California, Los Angeles, CA 90089-2565

Summary. Research in the design of ultra-wideband (UWB) radio systems suggests different kinds of signal design criteria. After an introduction to UWB radio, a variety of sequence design problems, motivated by UWB system constraints, will be presented. Design for time-hopping, for spectral flatness, for rapid acquisition, and for multiple access will be discussed, and some sample designs will be presented.

1 Introduction to UWB Radio

When the 3 dB bandwidth of a radio signal becomes 25% or more of the signal's center frequency, most agree that this radio should be called *ultra-wideband* (UWB). The combination of a relatively large bandwidth at a relatively low center frequency provides two kinds of dividends. First, of all radios at the same center frequency, an ultra-wideband radio should provide the finest time resolution in a well-designed receiver, and hence have potential advantages in ranging and multipath mitigation. Second, of all radios with the same bandwidth, ultra-wideband radios operate in the lowest frequency bands and hence have the best chance to propagate through most materials.

One major problem that UWB radios must solve is the satisfactory coexistence of UWB radio signals with the myriad of other narrowband and wideband signals with which they must simultaneously share their frequency bands [1]. This implies that UWB radios should employ spread-spectrum methods to protect them against the interference that they will inevitably encounter from other radio systems. It also implies that UWB radios can only radiate small amounts of power in each of the narrow frequency bands of other radio systems to avoid interfering with them. This latter issue is a matter for the appropriate regulatory entity to oversee.

Regulation of UWB radio currently is being considered in the United States, and may take the form of an upper bound on the radiated power spectral density of the UWB system. Then, the efficiency and performance of the UWB system will depend to a great extent on the flatness and range of its radiated power spectral density.

* This work was supported in part by the Office of Naval Research under grant N00014-96-1-1192 (subcontract from the University of Puerto Rico), and by the National Science Foundation under grant ANI-9730556.

The technology used to implement UWB radio depends on the frequency band in which the radio must operate. Modulations in UWB radios usually are constructed from trains of very short pulses whose width often is in the range of a few nanoseconds to fractions of a nanosecond, giving bandwidths on the order of gigahertz. Since the ability of an antenna to radiate efficiently decreases as frequency approaches zero, the pulse shapes are generally chosen to have little or no energy content as frequency approaches zero. Hence pulses tend to have balanced positive and negative excursions, e.g., one period of a sinusoid, or the derivative of a Gaussian pulse.

Modulation formats vary, but UWB radios generally transmit several pulses per data bit and use coherent detection of the pulse train. The radios usually produce carrierless signals and do not use mixers for the purpose of changing the frequency band of a signal. Some systems resemble baseband direct-sequence spread-spectrum systems and others resemble baseband time-hopped spread-spectrum signals [2]. Digital modulation of these kinds of signals is accomplished by added time shifting or polarity reversal.

We will now describe several signal-design problems motivated by this description of UWB radio.

2 Time-Hopping Signal Models

Time hopping for spectral spreading may provide implementation advantages and may be desirable in ranging systems because it may be easier to find the leading edge of an isolated received pulse signal. One possible form of an unmodulated time-hopping ultra-wideband signal generator is shown in Figure 1. In this example, before data modulation, the ith time-hopped signal is of the form

$$s^{(i)}(t) = \sum_j p(t - jT_\mathrm{f} - c_j^{(i)}T_\mathrm{c}) = \sum_n a_n^{(i)} p(t - nT_\mathrm{c}), \tag{1}$$

where we assume for simplicity that one frame time T_f is composed of N_f such slots, i.e., $N_\mathrm{f}T_\mathrm{c} = T_\mathrm{f}$. The integer time-hopping code $\{c_j^{(i)}\}$, $0 \leq c_j^{(i)} < N_\mathrm{f}$, has period N. Then the quantity $a_n^{(i)}$ is defined as

$$a_n^{(i)} = \begin{cases} 1 & \text{if there is an integer } j \text{ such that } n = jN_\mathrm{f} + c_j^{(i)}\,, \\ 0 & \text{otherwise.} \end{cases} \tag{2}$$

A signal then is described totally by which time slots are occupied and which slots are not.

Lets assume for simplicity that the time width T_w of a pulse $p(t)$ is less than T_c. Then the normalized periodic correlation between the signals of

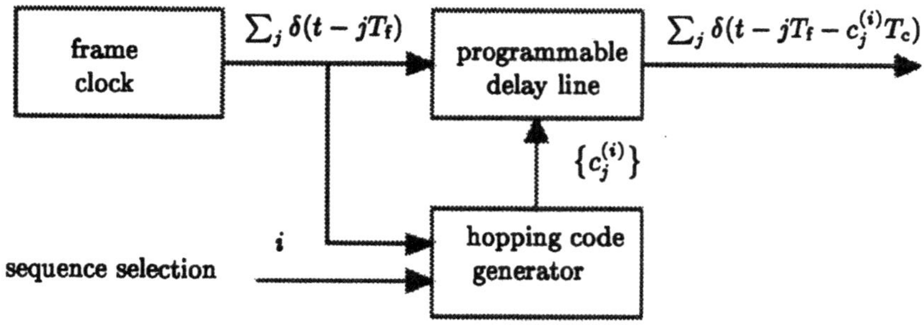

Fig. 1. A framed time-hopping sequence generator.

users i and j is

$$\tilde{R}_{ij}(n_\tau T_c) \triangleq \frac{R_{ij}(n_\tau T_c)}{R_{ii}(0)} = \frac{1}{N} \underbrace{\sum_{n=0}^{NN_f-1} a_n^{(i)} a_{n \ominus n_\tau}^{(j)}}_{\text{coincidence correlation}} . \tag{3}$$

This equation displays the normalized periodic coincidence correlation between the two transmitted signals when one is shifted by an integer number n_τ of slot widths relative to the other. The quantity N is the period of the time-hopping sequence design measured in frame times, and hence the period of any of the binary sequences $\{a_n^{(i)}\}$ is NN_f, i.e., $a_n^{(i)} = a_{n+NN_f}^{(i)}$ for all n and i. The notation $\ominus$ in (3) denotes subtraction modulo NN_f, leading to the periodic nature of the computation. The peak value 1 occurs when $i = j$ and $n_\tau = 0$.

For the moment let's assume that, except for the peak value, the design objective from an auto-correlation viewpoint is to make the periodic coincidence correlation as small as possible. As structured here, the design requires that one and only one slot in each frame be occupied by a pulse. Hence $a_n^{(i)} = 0$ for all but one value of n in each range $jN_f \le n < (j+1)N_f$ for each value of j. Let's display this graphically in an $N_f \times N$ matrix by mapping the sequence $\{a_n^{(i)}\}$ into the matrix $\mathbf{A}^{(i)}$ as follows.

$$a_0^{(i)}, a_1^{(i)}, \ldots, a_{NN_f-1}^{(i)} \Longleftrightarrow \underbrace{\begin{bmatrix} a_0^{(i)} & a_{N_f}^{(i)} & a_{2N_f}^{(i)} & \cdots & a_{(N-1)N_f}^{(i)} \\ a_1^{(i)} & a_{N_f+1}^{(i)} & a_{2N_f+1}^{(i)} & \cdots & a_{(N-1)N_f+1}^{(i)} \\ \vdots & \vdots & \vdots & \ddots & \vdots \\ a_{N_f-1}^{(i)} & a_{2N_f-1}^{(i)} & a_{3N_f-1}^{(i)} & \cdots & a_{NN_f-1}^{(i)} \end{bmatrix}}_{\triangleq \mathbf{A}^{(i)}} \tag{4}$$

That is, one period of the sequence is entered into the matrix $\mathbf{A}^{(i)}$, filling in order the first column, the second column, etc. The constraint of one pulse per frame time translates to exactly one 1 entry and $N_f - 1$ entries which are 0 in each column of the matrix $\mathbf{A}^{(i)}$. So for example, the sequence with period 16 and 4 slots per frame, with slots 0, 7, 9, and 14, occupied by pulses is represented by the matrix

$$N_f = N = 4 \text{ and } a_0^{(i)} = a_7^{(i)} = a_9^{(i)} = a_{14}^{(i)} = 1 \iff \begin{bmatrix} 1&0&0&0 \\ 0&0&1&0 \\ 0&0&0&1 \\ 0&1&0&0 \end{bmatrix} . \tag{5}$$

This is reminiscent of frequency-hopping designs in which the column index represents time and the row index represents frequency, but in this development both the row and column indices represent time shifts.

The sequence notation is clumsy for matrices, so let's doubly index the elements of the matrix by defining $b_{jk}^{(i)} = a_{kN_f+j}$ where $0 \leq j < N_f - 1$ and $0 \leq k < N$. Equivalently,

$$\mathbf{A}^{(i)} \triangleq \begin{bmatrix} b_{00}^{(i)} & b_{01}^{(i)} & b_{02}^{(i)} & \cdots & b_{0,N-1}^{(i)} \\ b_{10}^{(i)} & b_{11}^{(i)} & b_{12}^{(i)} & \cdots & b_{1,N-1}^{(i)} \\ \vdots & \vdots & \vdots & \ddots & \vdots \\ b_{N_f-1,0}^{(i)} & b_{N_f-1,1}^{(i)} & b_{N_f-1,2}^{(i)} & \cdots & b_{N_f-1,N-1}^{(i)} \end{bmatrix} \tag{6}$$

Now let's define the inner product of two arbitrary real $N_f \times N$ matrices $\mathbf{A}$ and $\mathbf{B}$ of the same dimensions to be

$$(\mathbf{A}, \mathbf{B}) \triangleq \sum_m \sum_n a_{mn} b_{mn} \tag{7}$$

where the sums cover the full range of indices of the matrices. If we compute the inner product of two of the binary signal matrices, the result is a recognizable correlation, namely

$$\left(\mathbf{A}^{(i)}, \mathbf{A}^{(j)}\right) = \sum_{m=0}^{N_f-1} \sum_{n=0}^{N-1} b_{mn}^{(i)} b_{mn}^{(j)} = \sum_{n=0}^{NN_f-1} a_n^{(i)} a_n^{(j)} = N \cdot \tilde{R}_{ij}(0). \tag{8}$$

That is, this matrix inner product is the normalized correlation between signals i and j at zero shift (see (3)).

If we could move the elements of one of the matrices around to simulate the process of time shifting the signal, it would be possible to construct values of $\tilde{R}_{ij}(n_\tau T_c)$ for other values of n_τ. Specifically, suppose that signal j is shifted by n_τ slot widths relative to signal i so that $a_{n_\tau}^{(i)}$ is multiplied by $a_0^{(j)}$ in the coincidence correlation of (3). The matrix coordinates of $a_{n_\tau}^{(i)}$ are given by the pair β, α of integers which are determined by Euclidean division from n_τ and N_f to satisfy the equation

$$n_\tau = \alpha N_f + \beta, \qquad \text{where } 0 \leq \beta < N_f. \tag{9}$$

That is, the time shift n_τ corresponds to a shift of α full frame times plus a fraction of a frame time covering β slots. To accomplish the computation of $\tilde{R}_{ij}(n_\tau T_c)$, the elements of $\mathbf{A}^{(j)}$ must be relocated by a "helical" shift to form the matrix

$$\mathbb{T}_{n_\tau}\mathbf{A}^{(j)} = \begin{bmatrix} a^{(j)}_{\ominus n_\tau} & a^{(j)}_{N_f\ominus n_\tau} & a^{(j)}_{2N_f\ominus n_\tau} & \cdots & a^{(j)}_{((N-1)N_f)\ominus n_\tau} \\ a^{(j)}_{1\ominus n_\tau} & a^{(j)}_{(N_f+1)\ominus n_\tau} & a^{(j)}_{(2N_f+1)\ominus n_\tau} & \cdots & a^{(j)}_{((N-1)N_f+1)\ominus n_\tau} \\ \vdots & \vdots & \vdots & \ddots & \vdots \\ a^{(j)}_{(N_f-1)\ominus n_\tau} & a^{(j)}_{(2N_f-1)\ominus n_\tau} & a^{(j)}_{(3N_f-1)\ominus n_\tau} & \cdots & a^{(j)}_{(NN_f-1)\ominus n_\tau} \end{bmatrix}$$

$$= \left[\begin{array}{cccccc} b^{(j)}_{N_f-\beta,N-\alpha-1} & \cdots & b^{(j)}_{N_f-\beta,N-1} & b^{(j)}_{N_f-\beta,0} & \cdots & b^{(j)}_{N_f-\beta,N-\alpha-2} \\ \vdots & \ddots & \vdots & \vdots & \ddots & \vdots \\ b^{(j)}_{N_f-1,N-\alpha-1} & \cdots & b^{(j)}_{N_f-1,N-1} & b^{(j)}_{N_f-1,0} & \cdots & b^{(j)}_{N_f-1,N-\alpha-2} \\ \hline b^{(j)}_{0,N-\alpha} & \cdots & b^{(j)}_{0,0} & b^{(j)}_{0,1} & \cdots & b^{(j)}_{0,N-\alpha-1} \\ \vdots & \ddots & \vdots & \vdots & \ddots & \vdots \\ b^{(j)}_{N_f-\beta-1,N-\alpha} & \cdots & b^{(j)}_{N_f-\beta-1,0} & b^{(j)}_{N_f-\beta-1,1} & \cdots & b^{(j)}_{N_f-\beta-1,N-\alpha-1} \end{array}\right] \quad (10)$$

Here we have used the operator $\mathbb{T}_{n_\tau}$ to represent the effect of moving the entries of the matrix as indicated in (10).

Because of the way in which this matrix has been constructed, it follows immediately that

$$\left(\mathbf{A}^{(i)}, \mathbb{T}_{n_\tau}\mathbf{A}^{(j)}\right) = \sum_{n=0}^{NN_f-1} a^{(i)}_n a^{(j)}_{n\ominus n_\tau} = N \cdot \tilde{R}_{ij}(n_\tau T_c). \quad (11)$$

Notice that all elements in the same row of $\mathbb{T}_{n_\tau}\mathbf{A}^{(j)}$ have the same row index. On the other hand, $b^{(j)}_{mn}$ elements in the same column of $\mathbb{T}_{n_\tau}\mathbf{A}^{(j)}$ have the same column index only if they are on the same side of the line drawn above the row with index 0 in the last matrix in (10). In the special case $\beta = 0$, then all the entries in a column of $\mathbb{T}_{n_\tau}\mathbf{A}^{(j)}$ have the same column index.

2.1 Characterization of a Doubly Periodic Array Design

Typical mathematical array designs that can be found in the literature are constructed to minimize the doubly periodic correlation properties of the arrays. The doubly periodic correlation computation is slightly different than the calculations of the previous section. Let $\mathbf{A}^{(i)}$ and $\mathbf{A}^{(j)}$ be two matrices as defined in (6). The doubly periodic cross-correlation between these two matrices can be defined as

$$P_{\mathbf{A}^{(i)}\mathbf{A}^{(j)}}(n_r, n_c) \triangleq \sum_{m=0}^{N_f-1} \sum_{n=0}^{N-1} b_{m,n}^{(i)} b_{m\ominus n_r, n\ominus n_c}^{(j)} \tag{12}$$

for $0 \le n_r < N_f$ and $0 \le n_c < N$. Here the computation $\ominus$ is modulo N_f in the first subscript and modulo N in the second subscript. To compute this doubly periodic correlation using a matrix inner product, we must transform $\mathbf{A}^{(j)}$ so that its entries are properly positioned to accomplish the computation in (12). This can be done using the $N_f \times N_f$ and $N \times N$ permutation matrices. Let's define an $m \times m$ matrix $\mathbf{P}_m$ to be

$$\mathbf{P}_m \triangleq \begin{bmatrix} 0\,0 \dots 0\,1 \\ 1\,0 \dots 0\,0 \\ 0\,1 \dots 0\,0 \\ \vdots\ \vdots \ddots \vdots\ \vdots \\ 0\,0 \dots 1\,0 \end{bmatrix} \quad \text{with} \quad \mathbf{P}_m^{-1} = \begin{bmatrix} 0\,1\,0 \dots 0 \\ 0\,0\,1 \dots 0 \\ \vdots\ \vdots\ \vdots \ddots \vdots \\ 0\,0\,0 \dots 1 \\ 1\,0\,0 \dots 0 \end{bmatrix}. \tag{13}$$

Multiplication by $\mathbf{P}_m^n$ on the left of a matrix with m rows cyclically permutes the rows of the matrix n positions downward in the matrix. Similarly, multiplication by $\mathbf{P}_m^{-n}$ on the right of a matrix with m columns cyclically permutes the rows of the matrix n positions to the right in the matrix. Using this notation, it follows immediately that

$$\mathbf{P}_{N_f}^{n_r}\, \mathbf{A}^{(j)}\, \mathbf{P}_N^{-n_c} = \tag{14}$$

$$\left[\begin{array}{cccccc} b_{N_f-n_r,N-n_c}^{(j)} & \cdots & b_{N_f-n_r,0}^{(j)} & b_{N_f-n_r,1}^{(j)} & \cdots & b_{N_f-n_r,N-n_c-1}^{(j)} \\ \vdots & \ddots & \vdots & \vdots & \ddots & \vdots \\ b_{N_f-1,N-n_c}^{(j)} & \cdots & b_{N_f-1,0}^{(j)} & b_{N_f-1,1}^{(j)} & \cdots & b_{N_f-1,N-n_c-1}^{(j)} \\ \hline b_{0,N-n_c}^{(j)} & \cdots & b_{0,0}^{(j)} & b_{0,1}^{(j)} & \cdots & b_{0,N-n_c-1}^{(j)} \\ \vdots & \ddots & \vdots & \vdots & \ddots & \vdots \\ b_{N_f-n_r-1,N-n_c}^{(j)} & \cdots & b_{N_f-n_r-1,0}^{(j)} & b_{N_f-n_r-1,1}^{(j)} & \cdots & b_{N_f-n_r-1,N-n_c-1}^{(j)} \end{array}\right]$$

and it follows that

$$\left(\mathbf{A}^{(i)}, \mathbf{P}_{N_f}^{n_r}\mathbf{A}^{(j)}\mathbf{P}_N^{-n_c}\right) = P_{\mathbf{A}^{(i)}\mathbf{A}^{(j)}}(n_r, n_c)\,. \tag{15}$$

2.2 Relation between Time-Hopping and a Doubly Periodic Array Design

A comparison of (10) and (15) indicates that the bottom $N_f - \beta$ rows of $\mathbb{T}_{n_r}\mathbf{A}^{(j)}$ are identical to the bottom $N_f - \beta$ rows $\mathbf{P}_{N_f}^{\beta}\mathbf{A}^{(j)}\mathbf{P}_N^{-\alpha}$. We can write this mathematically as

$$\mathbf{L}_{N_t-\beta\times N_t}\left[\mathbb{T}_{n_\tau}\mathbf{A}^{(j)}\right] = \mathbf{L}_{N_t-\beta\times N_t}\mathbf{P}_{N_t}^{\beta}\mathbf{A}^{(j)}\mathbf{P}_N^{-\alpha} \tag{16}$$

where generally

$$\mathbf{L}_{N_1\times N_2} = [\mathbf{O}_{N_1\times N_2-N_1} \quad \mathbf{I}_{N_1}] \tag{17}$$

operates on the left of a matrix with N_2 rows to reduce the matrix to its last N_1 rows. Of course, generally $\mathbf{I}_N$ denotes an $N\times N$ identity matrix and $\mathbf{O}_{N_1\times N_2}$ denotes an all-zeros matrix of the indicated dimension. A similar comparison of the upper β rows of $\mathbb{T}_{n_\tau}\mathbf{A}^{(j)}$ and $\mathbf{P}_{N_t}^{\beta}\mathbf{A}^{(j)}\mathbf{P}_N^{-(\alpha+1)}$ in (10) and (15) respectively yields

$$\mathbf{U}_{\beta\times N_t}\left[\mathbb{T}_{n_\tau}\mathbf{A}^{(j)}\right] = \mathbf{U}_{\beta\times N_t}\mathbf{P}_{N_t}^{\beta}\mathbf{A}^{(j)}\mathbf{P}_N^{-(\alpha+1)}, \tag{18}$$

where generally

$$\mathbf{U}_{N_1\times N_2} = [\mathbf{I}_{N_1} \quad \mathbf{O}_{N_2-N_1}] \tag{19}$$

selects the upper N_1 rows on which it operates. Together (16) and (18) create a representation for $\mathbb{T}_{n_\tau}\mathbf{A}^{(j)}$ in terms of doubly cyclic shifts of $\mathbf{A}^{(j)}$, namely

$$\mathbb{T}_{n_\tau}\mathbf{A}^{(j)} = \begin{bmatrix}\mathbf{U}_{\beta\times N_t}\mathbf{P}_{N_t}^{\beta}\mathbf{A}^{(j)}\mathbf{P}_N^{-(\alpha+1)}\\ \mathbf{L}_{N_t-\beta\times N_t}\mathbf{P}_{N_t}^{\beta}\mathbf{A}^{(j)}\mathbf{P}_N^{-\alpha}\end{bmatrix} \tag{20}$$

Substituting the above representation into (11) and simplifying with (15) gives

$$\begin{aligned}
\tilde{R}_{ij}(n_\tau T_c) &= \frac{1}{N}\left(\mathbf{A}^{(i)}, \begin{bmatrix}\mathbf{U}_{\beta\times N_t}\mathbf{P}_{N_t}^{\beta}\mathbf{A}^{(j)}\mathbf{P}_N^{-(\alpha+1)}\\ \mathbf{L}_{N_t-\beta\times N_t}\mathbf{P}_{N_t}^{\beta}\mathbf{A}^{(j)}\mathbf{P}_N^{-\alpha}\end{bmatrix}\right)\\
&= \frac{1}{N}\left(\begin{bmatrix}\mathbf{U}_{\beta\times N_t}\mathbf{A}^{(i)}\\ \mathbf{L}_{N_t-\beta\times N_t}\ \mathbf{A}^{(i)}\end{bmatrix}, \begin{bmatrix}\mathbf{U}_{\beta\times N_t}\mathbf{P}_{N_t}^{\beta}\mathbf{A}^{(j)}\mathbf{P}_N^{-(\alpha+1)}\\ \mathbf{L}_{N_t-\beta\times N_t}\mathbf{P}_{N_t}^{\beta}\mathbf{A}^{(j)}\mathbf{P}_N^{-\alpha}\end{bmatrix}\right)\\
&= \frac{1}{N}\left(\mathbf{U}_{\beta\times N_t}\mathbf{A}^{(i)}, \mathbf{U}_{\beta\times N_t}\mathbf{P}_{N_t}^{\beta}\mathbf{A}^{(j)}\mathbf{P}_N^{-(\alpha+1)}\right)\\
&\quad + \frac{1}{N}\left(\mathbf{L}_{N_t-\beta\times N_t}\mathbf{A}^{(i)}, \mathbf{L}_{N_t-\beta\times N_t}\mathbf{P}_{N_t}^{\beta}\mathbf{A}^{(j)}\mathbf{P}_N^{-\alpha}\right)\\
&\le \frac{1}{N}\left(\mathbf{A}^{(i)}, \mathbf{P}_{N_t}^{\beta}\mathbf{A}^{(j)}\mathbf{P}_N^{-(\alpha+1)}\right) + \frac{1}{N}\left(\mathbf{A}^{(i)}, \mathbf{P}_{N_t}^{\beta}\mathbf{A}^{(j)}\mathbf{P}_N^{-\alpha}\right)\\
&= \frac{1}{N}\left[P_{\mathbf{A}^{(i)}\mathbf{A}^{(j)}}(\beta,\alpha+1) + P_{\mathbf{A}^{(i)}\mathbf{A}^{(j)}}(\beta,\alpha)\right].
\end{aligned} \tag{21}$$

where α and β are determined by (9). The inequality comes from the fact that all entries in these matrices are non-negative, and multiplication by the row selectors $\mathbf{U}_{\beta \times N_f}$ and $\mathbf{L}_{N_f-\beta \times N_f}$ simply reduces the range of the row index in the inner product computation.

Notice that (21) is a bound on time hopping correlation values in terms of a sum of two doubly periodic correlation values.

Hence, a set of M matrices $\mathbf{A}^{(i)}$, $i = 1, \ldots, M$, with non-matched doubly periodic correlation at most

$$P_{\max}\left(\{\mathbf{A}^{(i)}\}_{i=1}^{M}\right) \triangleq \max_{\substack{i,\,j,\,n_r,\,n_c:\\ i \neq j \text{ or } n_r \neq 0 \text{ or } n_c \neq 0}} P_{\mathbf{A}^{(i)}\mathbf{A}^{(j)}}(n_r, n_c), \tag{22}$$

can be mapped into a set of M time-hopping sequences (using (4)) with the normalized correlation bound

$$\max_{\substack{i,\,j,\,n_\tau:\\ i \neq j \text{ or } n_\tau \neq 0}} \tilde{R}_{ij}(n_\tau T_c) \leq \frac{2}{N} P_{\max}\left(\{\mathbf{A}^{(i)}\}_{i=1}^{M}\right). \tag{23}$$

When necessary, it is not difficult to adapt this correlation bounding approach to include guard times between pulse frames and other implementation constraints, by inserting mandatory rows of zeros in the $\mathbf{A}^{(i)}$ matrices.

2.3 The Johnson Bound

The Johnson bound (see [3] page 327, and [4]) is a bound on the number of constant-weight words that can be achieved in the design of a cyclic code with a prescribed minimum Hamming distance. Viewing our sequences $\{a_n^{(i)}\}_{i=1}^{M}$ and their cyclic shifts as constant weight cyclic code words over the binary field of two elements, we can transform the Johnson bound to one on the number of time-hopping sequences that can be designed with a prescribed upper bound on auto- and cross-correlation.

Clearly, the design objective is to minimize the periodic correlations $N\tilde{R}_{i,j}(n_\tau T_w)$ except when $i = j$ and $n_\tau = 0$ in which case $N\tilde{R}_{i,j}(n_\tau T_w) = N$. Let there be M time-hopping sequences in all. If we impose the condition

$$N\tilde{R}_{i,j}(n_\tau T_w) \;\leq\; \lambda, \text{ when either } i \neq j \text{ or } n_\tau \neq 0, \tag{24}$$

by specification of the parameter λ, then the Johnson bound states that

$$M \leq \left\lfloor \frac{1}{N} \left\lfloor \frac{NN_f - 1}{N-1} \cdots \left\lfloor \frac{NN_f - (\lambda-1)}{N-(\lambda-1)} \left\lfloor \frac{NN_f - \lambda}{N-\lambda} \right\rfloor \right\rfloor \cdots \right\rfloor \right\rfloor \tag{25}$$

where N is the weight of the sequence, NN_f is the sequence period, and $\lfloor a \rfloor$ denotes the largest integer $\leq a$. When both N and N_f are large and λ is small, the bound can be approximated by

$$M \leq \frac{(NN_f)^\lambda}{N^{\lambda+1}} \;=\; \frac{N_f^\lambda}{N}. \tag{26}$$

Note that unless $N_f >> N$, setting $\lambda = 1$ would result in a small number of time-hopping sequences. Thus the Johnson bound indicates that $\lambda = 2$ is the smallest value for which a multiple-access signal design may be feasible.

2.4 A Time-Hopping Construction with $\lambda = 2$

Let GF(p) denote the finite field with arithmetic modulo a prime p. Let $\mathcal{F}$ denote the set of all polynomials of degree at most 2 with coefficients in GF(p). That is, the i-th polynomial in $\mathcal{F}$ is of the form

$$f^{(i)}(x) = f_2^{(i)}x^2 \oplus f_1^{(i)}x \oplus f_0^{(i)}, \qquad f_2^{(i)}, f_1^{(i)}, f_0^{(i)} \in \mathrm{GF}(p), \tag{27}$$

where $\oplus$ denotes addition modulo p. To each polynomial $f^{(i)}(x)$ in $\mathcal{F}$ we associate a $p \times p$ matrix $\tilde{\mathbf{A}}^{(i)} = [\tilde{b}_{m,n}^{(i)}]$ with elements defined by

$$\tilde{b}_{m,n}^{(i)} = \begin{cases} 1 & \text{if } f^{(i)}(n) = m, \quad 0 \le m \le p-1,\ 0 \le n \le p-1 \\ 0 & \text{otherwise.} \end{cases} \tag{28}$$

That is, the row location m of the 1 in column n in $\tilde{\mathbf{A}}^{(i)}$ is given by $f^{(i)}(n)$. As a result, the matrix $\tilde{\mathbf{A}}^{(i)}$ will have a single 1 in each column.

Let $\tilde{\mathbf{A}}^{(i)}$ and $\tilde{\mathbf{A}}^{(j)}$ denote two matrices corresponding to polynomials $f^{(i)}(x)$ and $f^{(j)}(x)$ belonging to $\mathcal{F}$. Following the development of (15), the doubly periodic correlation $P_{\tilde{\mathbf{A}}^{(i)}\tilde{\mathbf{A}}^{(j)}}(n_r, n_c)$ of the matrices $\tilde{\mathbf{A}}^{(i)}$ and $\tilde{\mathbf{A}}^{(j)}$ is given by

$$\begin{aligned} P_{\tilde{\mathbf{A}}^{(i)}\tilde{\mathbf{A}}^{(j)}}(n_r, n_c) &= \left(\tilde{\mathbf{A}}^{(i)}, \mathbf{P}_{N_a}^{n_r}\tilde{\mathbf{A}}^{(j)}\mathbf{P}_N^{-n_c}\right) \\ &= \left|\{x :\ f^{(i)}(x) = f^{(j)}(x \ominus n_c) \oplus n_r,\ 0 \le x \le p-1\}\right| \end{aligned} \tag{29}$$

The last equality above comes from counting the number of column indices x in which $\tilde{\mathbf{A}}^{(i)}$ and $\mathbf{P}_{N_a}^{n_r}\tilde{\mathbf{A}}^{(j)}\mathbf{P}_N^{-n_c}$ have identical row locations for their 1's. Each such correspondence contributes a 1 to the correlation value. Since $f^{(i)}(x)$ and $f^{(j)}(x)$ are both of degree 2 over a field, the row matching equation on the right in (29) can have at most two solutions except when the polynomials on the two sides of the equation above are identical. We now insure that the latter situation can happen only when $i = j$ and $n_c = n_r = 0$.

The condition for row matching can be reduced as follows:

$$\begin{aligned} & f^{(i)}(x) = f^{(j)}(x \ominus n_c) \oplus n_r \\ \iff\ & f_2^{(i)}x^2 \oplus f_1^{(i)}x \oplus f_0^{(i)} = f_2^{(j)}(x \ominus n_c)^2 \oplus f_1^{(j)}(x \ominus n_c) \oplus f_0^{(j)} \oplus n_r \\ \iff\ & \begin{cases} \left(f_2^{(i)} \ominus f_2^{(j)}\right)x^2 \oplus \left(f_1^{(i)} \oplus 2f_2^{(j)}n_c \ominus f_1^{(j)}\right)x & \\ \qquad \oplus \left(f_0^{(i)} \ominus f_2^{(j)}n_c^2 \oplus f_1^{(j)}n_c \ominus f_0^{(j)} \ominus n_r\right) = 0 & \text{for } i \ne j \\ \left(2f_2^{(j)}n_c\right)x \oplus \left(\ominus f_2^{(j)}n_c^2 \oplus f_1^{(j)}n_c \ominus n_r\right) = 0 & \text{for } i = j\,. \end{cases} \end{aligned} \tag{30}$$

(a) $\#\mathcal{T}_3 = (2^{2m} - 2^m + t2^m + t^2 - t)/3$;
(b) *each sequence in* $\mathcal{T}_3$ *is of period* $2^m + 1 + t$;
(c)

$$\ell_{\min}(\mathcal{T}_3) \geq 2^{m/2-2} - 2 - \frac{1+2t}{2^{m/2+2}};$$

(d)

$$c_{\max}(\mathcal{T}_3) \leq 14 \cdot 2^{m/2} + |t|.$$

Proof. Note the fact that

$$\frac{B_q(2,t)}{q+1-t} = \frac{q-t}{2} = \frac{2^m - t}{2}$$

and $q+t+1$ is an odd number. Taking $d = 2$ in Theorem 6 gives the results of part (i).

Note the fact that

$$\frac{B_q(3,t)}{q+1-t} = \frac{q^2 - q + qt + t^2 - t}{3} = \frac{(2^{2m} - 2^m + t2^m + t^2 - t)}{3}$$

and $(3, q+1+t) = (3, t+1+(-1)^m)$. Taking $d = 3$ in Theorem 6 gives the results of part (ii).

References

1. M. Eichler, *Introduction to the Theory of Algebraic Numbers and Functions,* Academic Press, New York, 1951.
2. T. Helleseth and P. V. Kumar, *Sequences with Low Correlation,* a chapter in: Handbook of Coding Theory, edited by V. Pless and C. Huffman, Elsevier Science Publishers, 1998.
3. D. Kohel, S. Ling, and C.P. Xing, *Explicit sequence expansions,* In: Sequences and Their Applications (C. Ding, T. Helleseth, and H. Niederreiter, eds.), pp. 308–317, Springer, London, 1999.
4. H. Niederreiter, *Continued fractions for formal power series, numbers, and linear complexity of sequences,* In: Contributions to General Algebra 5, (Proc. Salzburg Conf., 1986), pp. 221–233, Teubner, Stuttgart, 1987.
5. H. Niederreiter, *Sequences with almost perfect linear complexity profile,* In: Advances in Cryptology - EUROCRYPT '87(D. Chaum and W.L. Price, eds.), Lecture Notes in Computer Science, Vol. **304**, pp. 37–51, Springer, Berlin, 1988.
6. H. Niederreiter, *The probabilistic theory of linear complexity,* In: Advances in Cryptology – EUROCRYPT '88 (C.G. Günther, ed.), Lecture Notes in Computer Science, Vol. **330**, pp. 191–209, Springer, Berlin, 1988.
7. H. Niederreiter, *Keystream sequences with a good linear complexity profile for every starting point,* In: Advances in Cryptology - EUROCRYPT '89 (J.-J. Quisquater and J. Vandewalle, eds.), Lecture Notes in Computer Science, Vol. **434**, pp. 523–532, Springer, Berlin, 1990.
8. H. Niederreiter, *Finite fields and cryptology,* In: Finite Fields, Coding Theory, and Advances in Communications and Computing (G. L. Mullen and P.J.-S. Shiue, eds.), pp. 359–373, Dekker, New York, 1993.

9. H. Niederreiter, *Some computable complexity measures for binary sequences,* In: Sequences and Their Applications (C. Ding, T. Helleseth, and H. Niederreiter, eds.), pp. 67–78, Springer, London, 1999.
10. H. Niederreiter and M. Vielhaber, *Linear complexity profiles: Hausdorff dimensions for almost perfect profiles and measures for general profiles,* J. Complexity, **13**(1997), 353–383.
11. H. -G. Rück, *A Note on Elliptic Curves over Finite Fields,* Math. of Comp., **49**(1987), 301-304.
12. R.A. Rueppel, *Analysis and Design of Stream Ciphers,* Springer, Berlin, 1986.
13. R.A. Rueppel, Stream ciphers, *Contemporary Cryptology: The Science of Information Integrity* (G.J. Simmons, ed.), pp. 65–134, IEEE Press, New York, 1992.
14. J. H. Silverman, *The Arithmetic of Elliptic Curves,* Springer-Verlag, New York, 1986.
15. H. Stichtenoth, *Algebraic Function Fields and Codes,* Springer, Berlin, 1993.
16. M.A. Tsfasman and S.G. Vlădut, *Algebraic-Geometric Codes,* Kluwer, Dordrecht, 1991.
17. C. P. Xing, *Multi-sequences with almost perfect linear complexity profile and function fields over finite fields,* Journal of Complexity, to appear.
18. C. P. Xing, V. J. Kumar and C. S. Ding, *Low correlation, large linear span sequences from function fields,* Preprint, 2000.
19. C.P. Xing and K.Y. Lam, *Sequences with almost perfect linear complexity profiles and curves over finite fields,* IEEE Trans. Inform. Theory, **45**(1999), 1267–1270.
20. C. P. Xing and N. Niederreiter, *Applications of algebraic curves to constructions of codes and almost perfect sequences,* to appear in Proceedings of the 5th International Conference on Finite Fields and Applications.
21. C.P. Xing, H. Niederreiter, K.Y. Lam, and C.S. Ding, *Constructions of sequences with almost perfect linear complexity profile from curves over finite fields,* Finite Fields Appl. **5**, 301–313 (1999).
22. C. P. Xing, K. Y. Lam and Z. H. Wei, *A class of explicit perfect multi-sequences,* In: Advanced in Cryptology –Asiacrypt'99 (K. Y. Lam, E. Okamoto, C. P. Xing, eds.), Lecture Notes in Computer Science, Vol. 1716, pp. 299-305, Springer-Verlag, Berlin, 1999.

Description of Binary Sequences Based on the Interval Linear Complexity Profile

Vladimir B. Balakirsky

Euler Institute for Discrete Mathematics and its Applications (EIDMA)
c/o Eindhoven University of Technology
P.O.Box 513, 5600 MB Eindhoven, The Netherlands

Summary. We consider the problem of partial approximation of binary sequences by the outputs of linear feedback shift registers. A generalization of the linear complexity profiles of binary sequences leads to a sequence that is regarded as the profile of interval linear complexity. This concept is useful if we want to identify all fragments of the given binary sequence having length at least $2m$ which can be described by $m+n$ bits, where $n < m$. We show that a fixed j–th component belongs to such a fragment with probability $\sim 1/3$ in the ensemble of fair coin–tossing sequences. We also apply the interval linear complexity approach to identify corrupted positions of the sequence generated by a linear feedback shift register with known feedback polynomial, where "corruption" of the sequence means either deletion, insertion or inversion of its components.

1 Introduction

There is a class of problems, relevant to decoding, cryptology, and data compression, where we are interested in a proper partitioning of a given binary sequence into different fragments. The goal of this partitioning is to represent some fragments as the output sequences of different linear feedback shift registers, and we speak about partial linear approximations of the given sequence in this sense. The problems of this class can be effectively attacked using a generalized version of the linear complexity profile of the sequence [1], [2], which was introduced in [3] as the interval linear complexity profile. We will present some basic properties of the interval linear complexity and apply this notion to attack some problems related to effective description of binary sequences.

We will use the following notations. Given a binary sequence $(u_1, \ldots, u_N)$, we denote $\mathbf{u}|_i^j \triangleq (u_i, \ldots, u_j)$ for all i and j such that $1 \leq i \leq j \leq N$.

Definition 1. *Let the integers j and n be fixed in such a way that $j \geq n+1$. A binary sequence $\mathbf{u}|_{n+1}^j$ is generated by the polynomial $f(z) \triangleq 1 + f_1 z + \ldots + f_n z^n$ with $f_1, \ldots, f_n \in \{0, 1\}$ given $\mathbf{u}|_1^n$ if and only if*

$$u_i = f_1 u_{i-1} \oplus \ldots \oplus f_n u_{i-n}, \quad \textit{for all} \;\; i = n+1, \ldots, j.$$

2 Some Properties of the Linear Complexity of Binary Sequences

Definition 2. *The linear complexity of the sequence* $\mathbf{u}|_1^j$, *denoted by* λ_j, *is the smallest integer* n *such that there exists a polynomial*

$$f_j(z) \triangleq 1 + f_{j,1}z + \ldots + f_{j,n-1}z^{n-1} + z^n$$

with $f_{j,1}, \ldots, f_{j,n-1} \in \{0,1\}$ *generating* $\mathbf{u}|_{n+1}^j$ *given* $\mathbf{u}|_1^n$. *The sequence* $(\lambda_1, \ldots, \lambda_N)$ *is referred to as the linear complexity profile (LCP) of* $\mathbf{u}|_1^N$.

The LCP of the sequence $\mathbf{u}|_1^N$ can be constructed by the Berlekamp–Massey (BM) algorithm [2]. We present this algorithm to introduce notations needed for further extensions.

1. Set $j = 1$, $(\lambda_0, f_0(z)) = (\lambda, f(z)) = (\lambda', f'(z)) = (0, 1)$, and $\delta = 0$.
2. Increase δ by 1, set

$$\hat{u}_j = \begin{cases} f_1 u_{j-1} \oplus \ldots \oplus f_\lambda u_{j-\lambda}, & \text{if } \lambda > 0 \\ 0, & \text{if } \lambda = 0 \end{cases}$$

$$\delta_j = \delta$$

$$(\lambda_j', f_j'(z)) = (\lambda', f'(z))$$

$$(\lambda_j, f_j(z)) = \begin{cases} (\lambda, f(z)), & \text{if } u_j = \hat{u}_j \\ (\max\{\lambda, \lambda' + \delta\}, f(z) + z^\delta f'(z)), & \text{if } u_j \neq \hat{u}_j. \end{cases}$$

3. If $\lambda_j > \lambda$, then set $(\lambda', f'(z)) = (\lambda, f(z))$, and $\delta = 0$.
4. Set $(\lambda, f(z)) = (\lambda_j, f_j(z))$.
5. Increase j by 1 and if $j \leq N$, then go to 2.
6. End.

Let $0 = \Lambda_0 < \Lambda_1 < \ldots < \Lambda_T$ be the values of the LCP of the sequence $\mathbf{u}|_1^N$ defined by (see Fig.1)

$$j \in \{0, \ldots, N\},\ \lambda_j = j/2 \iff j \in \{2\Lambda_0, 2\Lambda_1, \ldots, 2\Lambda_T\}. \tag{1}$$

For all $t = 0, \ldots, T$, let

$$F_t(z) \triangleq f_{2\Lambda_t}(z), \quad D_t \triangleq \deg F_t(z). \tag{2}$$

Lemma 1 ([1]). *For any binary sequence* $\mathbf{u}|_1^N$,

$$j \in [\Lambda_{t-1} + \Lambda_t, \Lambda_t + \Lambda_{t+1}) \Longrightarrow \lambda_j = \Lambda_t$$
$$j \in [2\Lambda_t, \Lambda_i + \Lambda_{t+1}) \Longrightarrow f_j(z) = F_t(z).$$

Furthermore, for all $t = 1, \ldots, T$, *the polynomial* $F_t(z)$ *is uniquely determined by the vector* $(u_1, \ldots, u_{2\Lambda_t})$.

Lemma 2. *For any binary sequence* $\mathbf{u}|_1^N$,

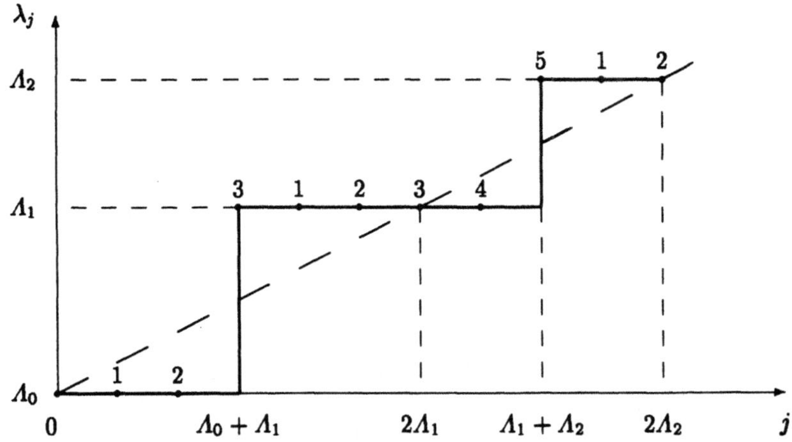

Fig. 1. The LCP of a binary sequence and the values $\delta_j, j = 1, 2, \ldots$

(a) $D_{t-1} = \Lambda_{t-1},\ D_t < \Lambda_t \implies D_{t+1} = \Lambda_{t+1}$;
(b) $D_{t-1} = \Lambda_{t-1},\ D_t = \Lambda_t \implies D_{t+1} = \Lambda_{t+1}$ *or* $D_{t+1} < \Lambda_{t+1}$ *depending on* $u_{2\Lambda_{t+1}}$;
(c) *there are no binary sequences with* $D_{t-1} < \Lambda_{t-1}$ *and* $D_t < \Lambda_t$

for all $t = 1, \ldots, T$, *where* $D_{-1} \triangleq \Lambda_{-1} \triangleq 0$.

Proof : The polynomial $f_j(z)$, constructed by the BM algorithm, can be expressed as

$$f_j(z) = \begin{cases} f_{j-1}(z), & \text{if } u_j = \hat{u}_j \\ f_{j-1}(z) + z^{\delta_j} f_j'(z), & \text{if } u_j \neq \hat{u}_j. \end{cases}$$

Let us fix $t \in \{1, \ldots, T\}$ and note that

$$j \in (\Lambda_{t-1} + \Lambda_t, \Lambda_t + \Lambda_{t+1}] \implies \begin{cases} \delta_j = j - (\Lambda_{t-1} + \Lambda_t) \\ f_j'(z) = F_{t-1}(z). \end{cases} \tag{3}$$

Denote $i = \Lambda_t + \Lambda_{t+1}$. Then $f_i(z) = f_{i-1}(z) + z^{\delta_i} f_i'(z)$, where $\delta_i = \Lambda_{t+1} - \Lambda_{t-1}$ and $f_i'(z) = F_{t-1}(z)$. Therefore

$$\deg[z^{\delta_i} f_i'(z)] = \Lambda_{t+1} - \Lambda_{t-1} + D_{t-1}$$

and since $\deg f_{i-1}(z) \leq \Lambda_t < \Lambda_{t+1}$, we have

$$D_{t-1} = \Lambda_{t-1} \implies \deg f_i(z) = \Lambda_{t+1}.$$

Similarly to (3),

$$j \in (\Lambda_t + \Lambda_{t+1}, 2\Lambda_{t+1}] \implies \begin{cases} \delta_j = j - (\Lambda_t + \Lambda_{t+1}) \\ f_j'(z) = F_t(z) \end{cases}$$

Thus,

$$\deg[z^{\delta_j} f'_j(z)] = j - (\Lambda_t + \Lambda_{t+1}) + D_t = j - \Lambda_{t+1} - (\Lambda_t - D_t)$$

and

$$j \in (\Lambda_t + \Lambda_{t+1}, 2\Lambda_{t+1}) \implies j - \Lambda_{t+1} - (\Lambda_t - D_t) < \Lambda_{t+1}.$$

As a result,

$$j \in (\Lambda_t + \Lambda_{t+1}, 2\Lambda_{t+1}) \implies \deg f_j(z) = \Lambda_{t+1}.$$

If $D_t < \Lambda_t$, then this equality is also valid for $j = 2\Lambda_{t+1}$. Otherwise, if $D_t = \Lambda_t$, then $D_{t+1} = \Lambda_{t+1}$ for $u_j = \hat{u}_j$ and $D_{t+1} < \Lambda_{t+1}$ for $u_j \neq \hat{u}_j$.

These considerations prove first two statements. Suppose the third conclusion is false and let t be the least value so that $D_t < \Lambda_t$ and $D_{t+1} < \Lambda_{t+1}$. Then $t > 0$ since $D_0 = \Lambda_0$, so $D_{t-1} = \Lambda_{t-1}$. But then conclusion (*a*) of the lemma implies that $D_{t+1} = \Lambda_{t+1}$, a contradiction.

Lemma 3. *Let $K_{2n}^{(i)}$ be the number of binary sequences $\mathbf{u}|_1^{2n}$ such that*

$$u_1 = \ldots = u_i = 0, \quad u_{i+1} = 1 \tag{4}$$

and

$$\lambda_{2n} = \deg f_{2n}(z) = n. \tag{5}$$

Then, for all $n > i$,

$$K_{2n}^{(i)} = \frac{1}{6} \cdot 2^{2n-i} + \frac{1}{3} \cdot 2^i. \tag{6}$$

Proof : Let $K_{2n}^{(i)\prime}$ be the number of binary sequences of length $2n$ satisfying (4) such that $\lambda_{2n} = n$ and $\deg f_{2n}(z) < n$. Let us consider binary sequences $\mathbf{u}|_1^{2n}$ satisfying (4) whose LCP $\lambda_1, \lambda_2, \ldots$ crosses the line with the slope 1/2 in the neighboring points $2n - 2l$ and $2n$, i.e., $\lambda_{2n-2l} = n - l$, $\lambda_{2n} = n$, and $\lambda_j \neq j/2$ for all $j = 2n - 2l + 1, \ldots, 2n - 1$. In other words, $\Lambda_t = n - l$ and $\Lambda_{t+1} = n$ for some index t in the sequence of even indices j with the property $\lambda_{j/2} = j$. Then $\lambda_{(2n+2n-2l)/2} = \lambda_{2n-l} = n$ by Lemma 1 and this equation is valid if we fix only first $2n - l$ bits of the sequence $\mathbf{u}|_1^{2n}$ in a certain way : for any $\mathbf{u}|_{2n-l+1}^{2n}$, the LCP of $\mathbf{u}|_1^{2n}$ at position $2n$ is equal to n. However, the degree of the feedback polynomial at position $2n$ (this polynomial is unique by Lemma 1) can be either less than n or equal to n. If $\mathbf{u}|_1^{2n-l}$ is fixed in such a way that $D_t < \Lambda_t$, then these l bits can be chosen in arbitrary way and $D_{t+1} = n$ (statements (a), (c) of Lemma 2). If $\mathbf{u}|_1^{2n-l}$ is fixed in such a way that $D_t = \Lambda_t$, then, for any $l - 1$ bits $u_{2n-l+1}, \ldots, u_{2n-1}$, there exists exactly one assignment for u_{2n} such that $D_{t+1} = n$ and exactly one assignment such that $D_{t+1} < n$ (statement (b) of Lemma 2). Thus, the contribution of sequences $\mathbf{u}|_1^{2n}$ under considerations

Table 1. Constructing of the pair $(\mu_j, g_j(z))$ for $j = 13$ and the sequence $\mathbf{u}|_1^{13} =$ (1000010111001) when $m = 5$.

t	u_{j-t}	$\hat{u}_{j-t}$	λ_t	$f_t(z)$	λ'_t	$f'_t(z)$	δ_t
0	1	0	1	$1+z$	0	1	0
1	0	1	1	1	0	1	1
2	0	0	1	1	0	1	2
3	1	0	3	$1+z^3$	1	1	0
4	1	0	3	$1+z^2+z^3$	1	1	1
5	1	1	3	$1+z^2+z^3$	1	1	2
6	0	0	3	$1+z^2+z^3$	1	1	3
7	1	1	3	$1+z^2+z^3$	1	1	4

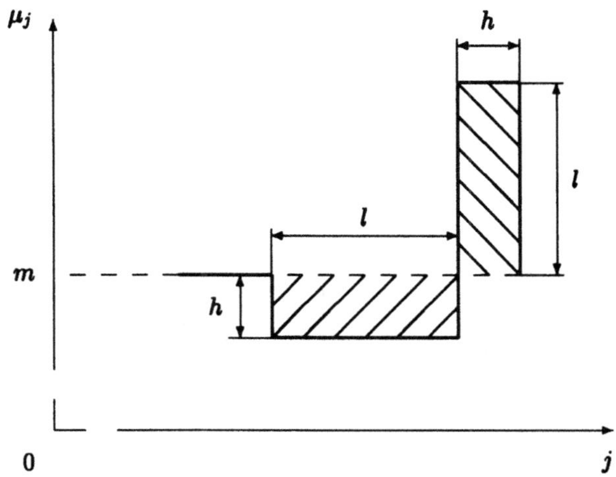

Fig. 2. Fragment of the m–LCP of any binary sequence.

where $h \triangleq m - n$. *Furthermore, if* $j > 2m$ *and* $u_2, \ldots, u_j$ *are i.i.d. random variables chosen from* $\{0,1\}$ *with probability* $1/2$, *then*

$$\Pr\{\, \mu_j = 0 \,\} = \frac{1}{2^m} \tag{10}$$

$$\Pr\{\, \mu_j = n \,\} = \frac{2^n - 2^{-n}}{3 \cdot 2^m}, \quad \textit{for all } n = 1, \ldots, m \tag{11}$$

and

$$\Big(\Pr\{\, \mu_j < m \,\},\ \Pr\{\, \mu_j = m \,\},\ \Pr\{\, \mu_j > m \,\}\Big)$$

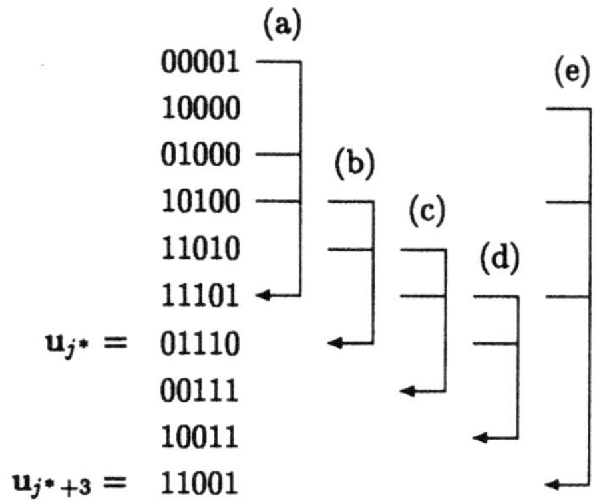

$$\begin{aligned}
\text{(a)}: &\quad g(z) = 1 + z^2 + z^3 + z^5 \\
\text{(b),(c),(d)}: &\quad g(z) = 1 + z^2 + z^3 \\
\text{(e)}: &\quad g(z) = 1 + z^4(1 + z^2 + z^4) \\
&\qquad + a(z)z(1 + z^2 + z^3), \ \deg a(z) \le 2.
\end{aligned}$$

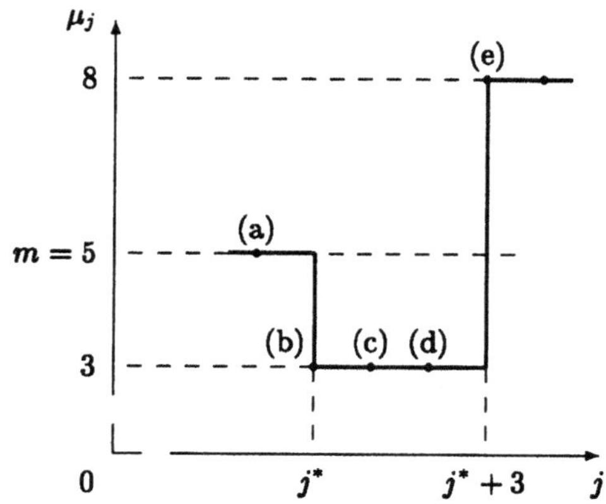

Fig. 3. Fragment of the 5–LCP of the sequence $(u_{j^*-10}, u_{j^*-9}, \ldots) = 10000101110011\ldots$

$$= \left(\frac{1}{3} + \frac{2}{3} \cdot 2^{-2m}, \frac{1}{3} - \frac{1}{3} \cdot 2^{-2m}, \frac{1}{3} - \frac{1}{3} \cdot 2^{-2m}\right). \tag{12}$$

Proof : Suppose that the sequence $\mathbf{u}|_{j-m+1}^{j}$ is generated by a feedback polynomial of degree $n = \mu_j$ given $\mathbf{u}|_{j-m-n+1}^{j-m}$. Then the sequence $u_{j-m+1-n}, \ldots, u_j$ of length $m + n$ can be specified by $2n$ bits : n bits determine coefficients of the feedback polynomial and n bits determine the initial content of the shift register. All other bits are computed or "generated".

(this code is by definition the set of such words of length 2^n obtained from affine functions). The weights of the elements of this coset, i.e. the Hamming distances $d(f,l)$, where $l(x) = a \cdot x \oplus \epsilon$ ($a \in F_2^n$, $\epsilon \in F_2$), are related to the Walsh coefficients of f as follows:

$$d(f,l) = 2^{n-1} - \frac{(-1)^\epsilon}{2}\widehat{\chi_f}(a). \tag{6}$$

In the next theorem, we improve upon Sarkar's and Maitra's result on the divisibility of these weights by powers of 2.

Theorem 2. *Let f be any m-resilient function on F_2^n ($0 \leq m \leq n-2$). Denote by d its algebraic degree. Then for every affine function l on F_2^n, the Hamming distance between f and l is divisible by $2^{m+1+\lfloor\frac{n-m-2}{d}\rfloor}$.*

Proof. f being balanced, its degree d is at least 1. If d equals 1, then the Hamming distance between f and any affine function is 0, 2^{n-1} or 2^n. Thus, it is divisible by 2^{n-1}.
We assume now that $d > 1$. Denote by g the function $f(x) \oplus x_1 \oplus \cdots \oplus x_n$. Recall that, for every $a \in F_2^n$, we have: $\widehat{\chi_f}(a) = \widehat{\chi_g}(\bar{a})$, where $\bar{a} = a + (1,\cdots,1)$. Thanks to relation (6), we deduce that for every affine function $l(x) = a \cdot x \oplus \epsilon$ ($a \in F_2^n$, $\epsilon \in F_2$), we have:

$$d(f,l) = 2^{n-1} - \frac{(-1)^\epsilon}{2}\widehat{\chi_g}(\bar{a}) = 2^{n-1} - 2^{n-1}(-1)^\epsilon\delta_0(\bar{a}) + (-1)^\epsilon\, \widehat{g}(\bar{a}).$$

Thus, for any $j \leq n-1$, the number $d(f,l)$ is divisible by 2^j if and only if $\widehat{g}(\bar{a})$ is divisible by 2^j.
To complete the proof, let us prove that every Fourier coefficient of g is divisible by $2^{m+1+\lfloor\frac{n-m-2}{d}\rfloor}$. According to relation (4) applied to g, it is sufficient to prove that for every $u \in F_2^n$, the number $2^{n-w(u)}\lambda_u$, where λ_u is the coefficient of x^u in the N.N.F. of g, is divisible by this same power of 2.
According to Proposition 2, g has numerical degree at most $n-m-1$. Since $d > 1$, g has same algebraic degree d as f. According to Proposition 1 applied to g, every λ_u is a multiple of $2^{\lfloor\frac{w(u)-1}{d}\rfloor}$. Thus, for every $u \in F_2^n$, the number $2^{n-w(u)}\lambda_u$ is divisible by $2^{n-w(u)+\lfloor\frac{w(u)-1}{d}\rfloor}$. Since $w(u) \leq n-m-1$ for every non-zero coefficient λ_u and since the expression $n - w(u) + \left\lfloor\frac{w(u)-1}{d}\right\rfloor$ does not increase when $w(u)$ does, we deduce that every non-zero term $2^{n-w(u)}\lambda_u$ is divisible by $2^{m+1+\lfloor\frac{n-m-2}{d}\rfloor}$ and the proof is complete.

Remark:
1. Xiao-Massey characterization [21] recalled in section 1 is equivalent to saying that f is m-resilient if and only if the Hamming distance between f and any affine function $l(x) = a \cdot x \oplus \epsilon$ such that $w(a) \leq m$ is equal to 2^{n-1}. It does not state anything about this distance when $w(a) > m$. Theorem 2 gives more insight.

2. The smaller the algebraic degree of the function is, and the smaller its resiliency order is, the larger is the term $\lfloor\frac{n-m-2}{d}\rfloor$ in the exponent of 2 that we have in addition to Sarkar's and Maitra's one.
3. The exponent $m+1$ obtained by Maitra and Sarkar can be smaller than $\lfloor\frac{n-1}{d}\rfloor$ (the exponent given by Theorem 1) when d and/or m is small. Our exponent is always greater than $\lfloor\frac{n-1}{d}\rfloor$, if $d>1$.
4. Theorem 2 and Remark 2 of section 2 imply $d \leq n-m-1-\lfloor\frac{n-m-2}{d}\rfloor$. One could think that this inequality improves upon Siegenthaler's inequality. It does not, since the inequality $d \leq n-m-1$ always implies $d \leq n-m-1-\lfloor\frac{n-m-2}{d}\rfloor$: this is clear if $d=n-m-1$; if $1<d\leq n-m-2$, then we have $d^2-(n-m-1)d+(n-m-2)=d^2-(n-m-1)(d-1)-1 \leq d^2-(d+1)(d-1)-1=0$ which implies $d \leq n-m-1-\frac{n-m-2}{d}$.
5. For $m=1$, the divisibility result given in Theorem 2 has been independently obtained by Anne Canteaut in [2].

Corollary 1. *Let f be any m-resilient function on F_2^n ($m \leq n-2$). Assume there exists an affine function l whose Hamming distance to f is not divisible by 2^{m+2}. Then f has algebraic degree $n-m-1$.*

Proof. The hypothesis implies $\lfloor\frac{n-m-2}{d}\rfloor=0$, i.e. $d>n-m-2$. According to Siegenthaler's inequality, this is possible if $d=n-m-1$ only.

We show now that our divisibility bound is tight:

Theorem 3. *For every $n \geq 2$, every non-negative $m \leq n-2$ and every positive $d \leq n-m-1$, there exists an m-resilient Boolean function on F_2^n of algebraic degree d and an affine function l such that $d(f,l)$ is not divisible by $2^{m+2+\lfloor\frac{n-m-2}{d}\rfloor}$.*

Proof. The assertion is obvious for $d=1$ since $d(f,l)$ is divisible by 2^n if and only if $f \oplus l$ is constant. So we assume that $d>1$. Let h be any Boolean function on F_2^{n-m-1}. The function

$$f(x)=h(x_1,\cdots,x_{n-m-1}) \oplus x_{n-m} \oplus \cdots \oplus x_n \tag{7}$$

is m-resilient (cf. [1]), since for every $a \in F_2^n$ we have:

$$\widehat{\chi_f}(a)=\begin{cases} 2^{m+1}\ \widehat{\chi_h}(a_1,\ldots,a_{n-m-1}) \text{ if } a_{n-m}=\cdots=a_n=1; \\ 0 \text{ otherwise} \end{cases} \tag{8}$$

As recalled in section 2, we know that for every $d \geq 1$, there exists a Boolean function h of degree d on F_2^{n-m-1} whose weight is not divisible by $2^{\lfloor\frac{n-m-2}{d}\rfloor+1}$. The sum $\sum_{x_1,\cdots,x_{n-m-1}\in F_2}(-1)^{h(x_1,\cdots,x_{n-m-1})}$ is then not divisible by $2^{\lfloor\frac{n-m-2}{d}\rfloor+2}$ (according to relation (6) applied to h) and the number $\widehat{\chi_f}(0,\cdots,0,1,\cdots,1)$ is not divisible by $2^{m+3+\lfloor\frac{n-m-2}{d}\rfloor}$. Thus, the Hamming distance between f and the affine function $l(x)=x_{n-m}\oplus\cdots\oplus x_n$ is not divisible by $2^{m+2+\lfloor\frac{n-m-2}{d}\rfloor}$.

Theorem 2 can be improved if the number of terms of highest degree in the A.N.F. of f is small enough:

Theorem 4. *Let f be any m-resilient function of algebraic degree $d > 1$ on F_2^n (with $0 \le m \le n-2$). Denote by r the number of monomials of degree d in the A.N.F. of f. Then for every affine function l on F_2^n, the Hamming distance between f and l is divisible by $2^{m+1+\lfloor \frac{n-m-r-2}{d-1} \rfloor}$.*

Notice that $\frac{n-m-2}{d} < \frac{n-m-r-2}{d-1}$ if and only if $r < \frac{n-m-2}{d}$. Thus Theorem 4 improves more likely upon Theorem 2 when d and m are small.

Proof. Let us denote again by g the function $f(x) \oplus x_1 \oplus \cdots \oplus x_n$. Its algebraic degree is d and the number of monomials of degree d in its A.N.F. is r. According to Proposition 1 applied to g, for every $u \in F_2^n$, the number $2^{n-w(u)}\lambda_u$ is divisible by $2^{n-w(u)+\lfloor \frac{w(u)-r-1}{d-1} \rfloor}$. Since $w(u) \le n-m-1$ for every non-zero coefficient λ_u and since the expression $n - w(u) + \left\lfloor \frac{w(u)-r-1}{d-1} \right\rfloor$ does not increase when $w(u)$ does, we deduce that every non-zero term of the sum $\sum_{u \in F_2{}^n \mid a \preceq u} 2^{n-w(u)}\lambda_u$ is divisible by $2^{m+1+\lfloor \frac{n-m-r-2}{d-1} \rfloor}$ and the proof is complete thanks to relation (4) applied to g.

4 Consequences on the nonlinearity of resilient functions

An upper bound on the nonlinearity of resilient functions can be deduced from the divisibility bound of the previous section (to be precise: to prove the nonlinearity bound, we need our divisibility bound only when $\frac{n}{2} - 1 \ge m + 1 + \lfloor \frac{n-m-2}{d} \rfloor$; see the next remark).

Theorem 5. *Let f be any m-resilient function of algebraic degree $d \ge 1$ on F_2^n ($0 \le m \le n-2$). The nonlinearity of f is divisible by $2^{m+1+\lfloor \frac{n-m-2}{d} \rfloor}$ and is smaller than or equal to $2^{n-1} - 2^{m+1+\lfloor \frac{n-m-2}{d} \rfloor}$ if $\frac{n}{2} - 1 < m+1+\lfloor \frac{n-m-2}{d} \rfloor$, to $2^{n-1} - 2^{\frac{n}{2}-1} - 2^{m+1+\lfloor \frac{n-m-2}{d} \rfloor}$ if n is even and $\frac{n}{2} - 1 \ge m+1+\lfloor \frac{n-m-2}{d} \rfloor$ and to $2^{n-1} - 2^{m+1+\lfloor \frac{n-m-2}{d} \rfloor} \left\lceil 2^{n/2-m-2-\lfloor \frac{n-m-2}{d} \rfloor} \right\rceil$ if n is odd and $\frac{n}{2} - 1 \ge m+1+\lfloor \frac{n-m-2}{d} \rfloor$.*

Moreover, assume that $d > 1$ and denote by r the number of monomials of degree d in the A.N.F. of f. Then the nonlinearity of f is divisible by $2^{m+1+\lfloor \frac{n-m-r-2}{d-1} \rfloor}$ and is smaller than or equal to $2^{n-1} - 2^{m+1+\lfloor \frac{n-m-r-2}{d-1} \rfloor}$ if $\frac{n}{2} - 1 < m+1+\left\lfloor \frac{n-m-r-2}{d-1} \right\rfloor$, to $2^{n-1} - 2^{\frac{n}{2}-1} - 2^{m+1+\lfloor \frac{n-m-r-2}{d-1} \rfloor}$ if n is even and $\frac{n}{2} - 1 \ge m+1+\left\lfloor \frac{n-m-r-2}{d-1} \right\rfloor$ and to $2^{n-1} - 2^{m+1+\lfloor \frac{n-m-r-2}{d-1} \rfloor} \left\lceil 2^{n/2-m-2-\lfloor \frac{n-m-r-2}{d-1} \rfloor} \right\rceil$ if n is odd and $\frac{n}{2} - 1 \ge m+1+\left\lfloor \frac{n-m-r-2}{d-1} \right\rfloor$.

Proof. According to Theorem 2 (resp. Theorem 4), and since $\min(d(f,l), d(f,l\oplus 1)) \leq 2^{n-1}$ for every affine function l, we have $\mathcal{NL}(f) = 2^{n-1} - k\,2^{m+1+\lfloor \frac{n-m-2}{d} \rfloor}$, $k \in \mathbf{Z}$ (resp. $\mathcal{NL}(f) = 2^{n-1} - k\,2^{m+1+\lfloor \frac{n-m-r-2}{d-1} \rfloor}$, $k \in \mathbf{Z}$).
We know that the nonlinearity of any balanced function is smaller than $2^{n-1} - 2^{\frac{n}{2}-1}$, since bent functions are never balanced. Thus $k > 2^{n/2-m-2-\lfloor \frac{n-m-2}{d} \rfloor}$ (resp. $k > 2^{n/2-m-2-\lfloor \frac{n-m-r-2}{d-1} \rfloor}$).

Obviously, when n is odd and $\frac{n}{2} - 1 \geq m + 1 + \lfloor \frac{n-m-2}{d} \rfloor$, this bound can be improved if the best possible nonlinearity of Boolean functions of degree d in n variables is known.

Remark:

1) When $\frac{n}{2} - 1 < m + 1 + \lfloor \frac{n-m-2}{d} \rfloor$, Theorem 5 can be proved directly without using the results of the previous section. Indeed, it is not necessary to know all about the Walsh spectrum of a function to derive such a bound. Let f be any m-resilient (and not $m+1$-resilient) function. Let a be any word of Hamming weight $m+1$ such that $\widehat{\chi_f}(a) \neq 0$. Using the relation $\sum_{u \preceq a} \widehat{\chi_f}(u) = 2^{m+1}\ \widehat{\chi_{f'}}(0)$ where f' denotes the restriction of f to the set of words covered by $a + (1, \ldots, 1)$ and applying Theorem 1 we deduce that $\widehat{\chi_f}(a)$ is divisible by $2^{m+1+\lfloor \frac{n-m-2}{d} \rfloor}$, which is enough to prove the theorem, according to relation (2).

2) Let f be any m-resilient function on F_2^n. Assume its nonlinearity is not divisible by 2^{m+2}. Then f has algebraic degree $n - m - 1$ according to Corollary 1 (it was already observed by Tarannikov that if the nonlinearity of f achieves his bound, then f has algebraic degree $n - m - 1$). Thus, Theorem 5 shows that to have a chance to have nonlinearity $2^{n-1} - 2^{m+1}$, an m-resilient function must have high algebraic degree. This situation seems unusual since, in general, different cryptographic criteria (e.g. degree and resiliency) behave in opposite ways. Here, nonlinearity and algebraic degree go in the same direction. This is in fact not astonishing since these criteria both express a property of confusion of the cryptosystem using function f.

The bound given by the first part of Theorem 5 is tight for $d = n - m - 1$ and for any n, m such that $\frac{2n-7}{3} \leq m \leq n - 2$, since for these values of m, Tarannikov exhibited in [20] m-resilient functions whose nonlinearity is $2^{n-1} - 2^{m+1}$.

From Tarannikov's result, it is a simple matter to deduce that the bound given by the first part of Theorem 5 is tight for $d = n - m - 2$ and for any n, m such that $\frac{2n-9}{3} \leq m \leq n - 3$. Indeed, let f_1 be any m-resilient function in $(n-1)$ variables, of degree $n - m - 2$ and with nonlinearity $2^{n-2} - 2^{m+1}$ (such function exists thanks to Tarannikov's result). Then the function $f(x_1, \cdots, x_n) = f_1(x_1, \cdots, x_{n-1})$ has nonlinearity $2^{n-1} - 2^{m+2}$, is m-resilient and has degree $n - m - 2$. Thus it achieves the bound.

Let us check that this bound is also tight for $d = 1$ and $d = 2$ (for every n, m). The condition $\frac{n}{2} - 1 < m + 1 + \lfloor \frac{n-m-2}{d} \rfloor$ is then always satisfied. The number $2^{n-1} - 2^{m+1+\lfloor \frac{n-m-2}{d} \rfloor}$ equals 0 if $d = 1$ (i.e. if f is affine) and $2^{n-1} - 2^{\lfloor \frac{n+m}{2} \rfloor}$ if $d = 2$ (i.e. if f is quadratic). The nonlinearity of any affine function is obviously null. To show that some quadratic m-resilient function achieving nonlinearity $2^{n-1} - 2^{\lfloor \frac{n+m}{2} \rfloor}$ exists, consider again function (7). If $n - m - 1$ is even, choose h quadratic bent (i.e. with nonlinearity $2^{n-m-2} - 2^{\frac{n-m-3}{2}}$) on F_2^{n-m-1} (such function exists, cf. [13]). If $n - m - 1$ is odd, choose h with nonlinearity $2^{n-m-2} - 2^{\frac{n-m-2}{2}}$ (see [13] as well). Then, according to relations (6) and (8), f achieves nonlinearity $2^{n-1} - 2^{m+1+\frac{n-m-3}{2}} = 2^{n-1} - 2^{\lfloor \frac{n+m}{2} \rfloor}$ if $n - m - 1$ is even and $2^{n-1} - 2^{m+1+\frac{n-m-2}{2}} = 2^{n-1} - 2^{\lfloor \frac{n+m}{2} \rfloor}$ if $n - m - 1$ is odd.

It is difficult to say if the bound is still tight when d equals 3 or more.

5 About correlation-immune functions

In [16], Sarkar and Maitra also obtained divisibility and nonlinearity bounds for m-th order correlation-immune functions. The exponent of 2 they obtained is then m instead of $m + 1$. Similarly, we can modify Theorems 2, 4 and 5 so that they apply to correlation-immune functions. The exponent is then also smaller by 1 than in the case of resilient functions.

In [22], Y. Zheng and X.-M. Zhang showed that, for $m \geq 0.6\, n$, the nonlinearity of f is always smaller than or equal to $2^{n-1} - 2^{m+1}$ (as for resilient functions). We show in the next theorem that a stronger statement can be proved once again using the properties of the N.N.F.

Theorem 6. *Let f be any m-th order correlation-immune function of algebraic degree $d \geq 1$ on F_2^n ($n \geq 2$, $0 \leq m \leq n - 1$).*
For every affine function l on F_2^n, the Hamming distance between f and l is divisible by $2^{n-1} - 2^{m+\lfloor \frac{n-m-1}{d} \rfloor}$.
If $m \geq 0.6\, n$, then the nonlinearity of f is smaller than or equal to $2^{n-1} - 2^{m+1+\lfloor \frac{n-m-2}{d} \rfloor}$.

Proof. The proof of the fact that for every affine function l on F_2^n, the Hamming distance between f and l is divisible by $2^{n-1} - 2^{m+\lfloor \frac{n-m-1}{d} \rfloor}$ is similar to the proof of Theorem 2. We assume that f is not equal to $x_1 \oplus \cdots \oplus x_n$ plus a constant. Denote by g the function $f(x) \oplus x_1 \oplus \cdots \oplus x_n$. The functions f and g have the same algebraic degree. Denote by λ_u the coefficient of x^u in the N.N.F. of g. According to Proposition 2, f is m-th order correlation immune if and only if for every u such that $w(u) \geq n - m$:

$$\lambda_u = 2^{-n}(-2)^{w(u)}\widehat{g}(1, \cdots, 1).$$

According to relation (4) applied to g, we deduce that for every word a we have:

$$(-1)^{w(a)}\,\widehat{g}(a) = \left(\sum_{u\in F_2{}^n \,|\, a\preceq u;\ w(u)\leq n-m-1} 2^{n-w(u)}\lambda_u\right)$$

$$+\left(\sum_{u\in F_2{}^n \,|\, a\preceq u;\ w(u)\geq n-m} (-1)^{w(u)}\right)\widehat{g}(1,\cdots,1).$$

According to Proposition 1 applied to g, the first of these two sums is always divisible by $2^{m+1+\lfloor\frac{n-m-2}{d}\rfloor}$.

According to the (original) definition of correlation-immune functions (cf. introduction), the weight of f is equal to 2^m times the weight of any restriction of f obtained by keeping constant m coordinates of x. Thus, according to Theorem 1 applied to this restriction, $w(f)$ is divisible by $2^{m+\lfloor\frac{n-m-1}{d}\rfloor}$. Thus $\widehat{g}(1,\cdots,1)$ is divisible by this same number.

If $m+1+\lfloor\frac{n-m-2}{d}\rfloor = m+\lfloor\frac{n-m-1}{d}\rfloor$, i.e. if $\lfloor\frac{n-m-2}{d}\rfloor = \lfloor\frac{n-m-1}{d}\rfloor - 1$ then the nonlinearity of f is smaller than or equal to $2^{n-1}-2^{m+1+\lfloor\frac{n-m-2}{d}\rfloor}$. So, we assume now that $\lfloor\frac{n-m-2}{d}\rfloor = \lfloor\frac{n-m-1}{d}\rfloor$. Let us prove that no m-th order correlation-immune function of algebraic degree d on F_2^n can have nonlinearity $2^{n-1}-2^{m+\lfloor\frac{n-m-2}{d}\rfloor}$ (this nonlinearity being divisible by $2^{m+\lfloor\frac{n-m-2}{d}\rfloor}$, this will complete the proof). So let f be an m-th order correlation-immune function of algebraic degree d on F_2^n with nonlinearity $2^{n-1}-2^{m+\lfloor\frac{n-m-2}{d}\rfloor}$. Since $\mathcal{NL}(g) = 2^{n-1}-2^{m+\lfloor\frac{n-m-2}{d}\rfloor}$, one of the Fourier coefficients of g is not divisible by $2^{m+1+\lfloor\frac{n-m-2}{d}\rfloor}$. Thus, according to the expression of $(-1)^{w(a)}\,\widehat{g}(a)$ above, the number $\widehat{g}(1,\cdots,1)$ is not divisible by $2^{m+1+\lfloor\frac{n-m-2}{d}\rfloor}$, and thanks to this same expression, any Fourier coefficient $\widehat{g}(a)$ is divisible by $2^{m+1+\lfloor\frac{n-m-2}{d}\rfloor}$ if and only if the set $\{u\in F_2{}^n \,|\, a\preceq u;\ w(u)\geq n-m\}$ has even size, i.e. if and only if the number $\sum_{i=n-m-w(a)}^{n-w(a)}\binom{n-w(a)}{i}$ is even.

Since for every word $a\in F_2^n$, the Walsh coefficient $\widehat{\chi_f}(a) = \widehat{\chi_g}(\bar{a}) = 2^n\delta_0(\bar{a}) - 2\widehat{g}(\bar{a})$ is divisible by $2^{m+1+\lfloor\frac{n-m-2}{d}\rfloor}$, according to relation (6), and has maximum magnitude $2^{m+1+\lfloor\frac{n-m-2}{d}\rfloor}$, according to relation (2), it must be equal to 0 (when $\sum_{i=0}^{m}\binom{w(a)}{i}$ is even, i.e. when $\binom{w(a)-1}{m}$ is even, according to the well-known property $\binom{j}{i} = \binom{j-1}{i} + \binom{j-1}{i-1}$) or to $\pm 2^{m+1+\lfloor\frac{n-m-2}{d}\rfloor}$ (when $\binom{w(a)-1}{m}$ is odd). According to Lucas' Theorem, $\binom{w(a)-1}{m}$ is odd if and only if the binary expansion of $w(a)-1$ covers that of m.

As in Zheng-Zhang's paper, the contradiction comes now from Parseval's relation: according to the properties of g shown above, relation (1) applied to g or to f states that the number 2^{2n} equals $2^{2m+2+2\lfloor\frac{n-m-2}{d}\rfloor}$ times the size of the set S_m of all words $a \in F_2^n$ such that $\binom{w(a)-1}{m}$ is odd. This is impossible if the size of S_m, i.e. the Hamming weight of the function $\binom{w(a)-1}{m}$ mod 2, is greater than $2^{2n-2m-2}$. Notice that S_m contains all words of weight $m+1$ and thus, its size is greater than or equal to $\binom{n}{m+1}$. As shown by Y. Zheng and X.-M. Zhang, if $m \geq 0.6\, n$, then $\binom{n}{m+1}$ is greater than $2^{2n-2m-2}$.

Notice that if there is no integer between $m+1$ and $n-1$ whose binary expansion covers that of m (e.g. when $m+1$ is a power of 2 and $n \leq 2m$), then S_m contains only the words of weight $m+1$.

Acknowledgement

The author thanks Anne Canteaut and Pascale Charpin for helpful discussions.

References

1. P. Camion, C. Carlet, P. Charpin, N. Sendrier, *On correlation-immune functions*, Advances in Cryptology: Crypto '91, Proceedings, Lecture Notes in Computer Science, V. 576 (1991), pp. 86–100.
2. A. Canteaut, *On the Weight Distributions of Optimal Cosets of the First-Order Reed-Muller Codes*, IEEE Transactions on Info Theory, vol. 47, No. 1 (2001), pp. 407-413.
3. A. Canteaut, C. Carlet, P. Charpin, C. Fontaine, *Propagation characteristics and correlation-immunity of highly nonlinear Boolean functions*, Advances in Cryptology - EUROCRYPT'2000, Lecture Notes in Computer Science 1807 (2000), pp. 507-522.
4. A. Canteaut, M. Trabbia, *Improved fast correlation attacks using parity-check equations of weight 4 and 5*, Advanced in Cryptology-EUROCRYPT 2000. Lecture notes in computer science 1807 (2000), pp. 573-588.
5. C. Carlet, *Two new classes of bent functions*, EUROCRYPT' 93, Advances in Cryptology, Lecture Notes in Computer Science 765, Springer Verlag (1994), pp.77-101.
6. C. Carlet, P. Guillot, *A new representation of Boolean functions*, Proceedings of AAECC'13, LNCS 1719 (1999), pp. 94-103.
7. C. Carlet, P. Guillot, *Bent, resilient functions and the Numerical Normal Form*, DIMACS Series in Discrete Mathematics and Theoretical Computer Science, Vol 56, pp. 87-96 (2001)
8. E. Filiol, C. Fontaine, *Highly Nonlinear Balanced Boolean Functions with a Good Correlation Immunity*, Advanced in Cryptology, Eurocrypt '98, Helsinki, Finland, Lecture Notes in Computer Sciences, Vol. 1403 (1998), pp. 475–488.
9. I. Honkala, A. Tietäväinen, *Codes and number theory*, In "Handbook of Coding Theory", Part 2, chapter 16, V. S. Pless, W. C. Huffman, Eds, R. A. Brualdi, assistant editor, Amsterdam, the Netherlands: Elsevier, 1998.

10. T. Johansson, F. Jönsson, *Improved fast correlation attack on stream ciphers via convolutional codes*, Advances in Cryptology - EUROCRYPT'99, number 1592 in Lecture Notes in Computer Science (1999), pp. 347–362.
11. T. Johansson, F. Jönsson *Fast correlation attacks based on turbo code techniques*, Advances in Cryptology - CRYPTO'99, number 1666 in Lecture Notes in Computer Science (1999), pp. 181–197.
12. R. J. McEliece, *Weight congruences for p-ary cyclic codes*, Discrete Math., 3 (1972), pp. 177-192.
13. F. J. MacWilliams, N. J. A. Sloane, *The theory of error-correcting codes*, North–Holland, Amsterdam, 1977.
14. S. Maitra, P. Sarkar, *Highly nonlinear resilient functions optimizing Siegenthaler's Inequality*, Crypto'99, Lecture Notes in Computer Science, Vol. 1666 (1999), pp. 198–215.
15. W. Meier, O. Staffelbach, *Fast correlation attack on certain stream ciphers*, Jounal of Cryptology (1989), pp. 159-176.
16. P. Sarkar, S. Maitra, *Nonlinearity Bounds and Constructions of Resilient Boolean Functions*, CRYPTO 2000, LNCS Vol. 1880, ed. Mihir Bellare, pp. 515-532 (2000).
17. C.E. Shannon, *Communication theory of secrecy systems*, Bell system technical journal, 28 (1949), pp. 656-715.
18. T. Siegenthaler, *Correlation-immunity of nonlinear combining functions for cryptographic applications*, IEEE Transactions on Information theory, V. IT-30, No 5 (1984), pp. 776-780.
19. T. Siegenthaler, *Decrypting a Class of Stream Ciphers Using Ciphertext Only*, IEEE Transactions on Computer, V. C-34, No 1 (1985), pp. 81-85.
20. Y. Tarannikov, *On resilient Boolean functions with maximal possible nonlinearity*, Cryptology ePrint archive (http://eprint.iacr.org/), Report 2000/005, 18 pp. (2000); and Proceedings of Indocrypt 2000, Springer Verlag, LNCS 1977 (2000) pp. 19-30.
21. Xiao Guo-Zhen, J. L. Massey, *A Spectral Characterization of Correlation-Immune Combining Functions*, IEEE Trans. Inf. Theory, Vol IT 34, n° 3 (1988), pp. 569-571.
22. Y. Zheng, X.-M. Zhang, *Improving Upper Bound on Nonlinearity of High Order Correlation Immune Functions*, Proceedings of The Seventh Annual Workship on Selected Areas in Cryptography (SAC 2000), (2000), pp. 258-269.

On the Linear Complexity of Generalised Legendre Sequence

Zongduo Dai[1], Junhui Yang[2], Guang Gong[3], and Ping Wang[1]

[1] State Key Laboratory of Information Security, Graduate School at Beijing, University of Science and Technology of China, 100039-08, Beijing, China
[2] Institute of Software, Academic Sinica, 100080, Beijing, China
[3] Department of Combinatorics and Optimization, University of Waterloo, Waterloo, Ontario N2L 3G1, Canada

Summary. The linear complexity of a class of generalised Legendre sequences is characterized and determined. The result in this paper improves some related results in [6].

1 Introduction

The design for the key sequences with perfect performance is always the hotspot in the digital communication and stream cipher research. A good key sequence often has large linear complexity and strong pseudorandomness. Legendre sequences look random with respect to elementary statistical tests and also quite good from the linear complexity viewpoint ([2], [4]). About various statistical properties and linear complexity of the Legendre sequence, we refer to [1]-[5] for detail. Recently, the so-called generalised Legendre sequence $S_{\underline{b}}(p, R)$ is defined and its linear complexity is discussed in [6], where $R = r^t$ is a prime power and p is a prime such that $R|p-1$, and $\underline{b}$ is an arrangement of all elements in the finite field F_R of order R. In [6] the linear complexity of the sequence $S_{\underline{b}}(p, R)$ is determined for the case when $R = 3$ and 4, and some partial results for the case when $R = 8$ or an odd prime r. Basing on [6], the linear complexity of the sequence $S_{\underline{b}}(p, R)$ is determined for the case when $R = r^t, r \nmid t\ (> 2)$ and $\underline{b}$ is some natural arrangement. In particular, when $R = r$ or $R = 2^3$, our results strengthen or imply some related results in [6].

2 Definition of Generalised Legendre Sequence

In this section, we restate the definition of the generalised Legendre sequence([6]), and then discuss some invariable properties of its linear complexity.

[1] Supported by 973-Foundation (No. 01999035804) and National Natural Science Foundation of China (No. 69773015).

Let $R = r^t$ be a prime power and p a prime, where $R|p-1$. Set $\underline{b}$ be an arrangement of all elements of the field F_R written as:

$$\underline{b} = (b_0 b_1 \cdots b_j \cdots b_{R-1}). \tag{1}$$

Definition. Let g be a generator of the multiplicative subgroup $F_p^* (= F_p \backslash \{0\})$ of the field F_p. The generalised Legendre sequence, or the R–th residual sequence of period p, denoted by $S_{\underline{b},g}(p, R)$, is a sequence $(s_0 s_1 \cdots s_i \cdots) \in F_R^\infty$, where

$$s_i = \begin{cases} 0 \text{ if } i \equiv 0 (\text{mod } p), \\ b_j \text{ if } i \equiv g^{j+Rk} (\text{mod } p), 0 \leq j < R. \end{cases}$$

In this paper, we fix p and R, and simply denote $S_{\underline{b},g}(p, R)$ by $S_{\underline{b},g}$.

Remark 1. Our definition about the generalised Legendre sequence is of different form from that in [6]. For the fixed g, our condition '$i = g^{j+Rk} (\text{mod } p)$' is denoted by '$\log_p i = j$' in [6]. But, it is easy to verify that they are equivalent to each other.

A problem is how $S_{\underline{b},g}$ changes when the generator g and the arrangement $\underline{b}$ are changed. Explicitly, how large is the linear complexity of $S_{\underline{b},g}$ for the different g and $\underline{b}$?

Before continuing for our results, we introduce some notations. Let μ be an integer such that $(\mu, R) = 1$. We define the μ–decimation of $\underline{b}$ as

$$\underline{b}^{(\mu)} = (b_0 b_\mu \cdots b_{\mu j} \cdots b_{\mu(R-1)}), \text{ where if } j' \equiv j (\text{mod } R), \text{ then } b_{j'} = b_j.$$

The arrangement $L^\lambda \underline{b} = (b_\lambda b_{\lambda+1} \cdots b_{p-1} b_0 \cdots b_{\lambda-1})$ is called the left shift of $\underline{b}$. With the map $x \to x^r$ acting on each component of $\underline{b}$, we get another arrangement $\underline{b}^r = (b_0^r b_1^r \cdots b_{R-1}^r)$. By multipling each component of $\underline{b}$ with $\alpha \in F_R$, $\alpha\underline{b} = (\alpha b_0\ \alpha b_1 \cdots \alpha b_{R-1})$. Clearly, $\underline{b}^{(\mu)}$ $(r \not| \mu)$, $L^\lambda \underline{b}$, $\underline{b}^r$ and $\alpha\underline{b} (\alpha \neq 0)$ are still the arrangements over F_R. Similarly, we can define the corresponding generalised Legendre sequences $S_{\underline{b}^{(\mu)},g}$, $S_{\underline{b}^r,g}$ and so on.

About the preceding sequences, we have some lemmas as follows.

Lemma 1. Let λ and μ be two integers such that $(\lambda, p-1) = 1$ and $\lambda\mu \equiv 1 (\text{mod } R)$. Then, $S_{\underline{b},g^\lambda} = S_{\underline{b}^{(\mu)},g}$.

Proof. Let $\underline{b}$ be as that in (1). We can assume that $S_{\underline{b},g^\lambda} = (a_0 a_1 \cdots a_i \cdots)$, $\underline{b}^{(\mu)} = (b'_0 b'_1 \cdots b'_{R-1})$ and $S_{\underline{b}^{(\mu)},g} = (c_0 c_1 \cdots c_i \cdots)$. When $i \equiv 0 (\text{mod } p)$, by the definitions of these sequences in the preceding paragraph, $c_i = a_i = 0$. When $i \not\equiv 0 (\text{mod } p)$, we can assume $i \equiv g^{\lambda(j+Rk)} (\text{mod } p)$. By their definitions again, $c_i = b'_{\lambda j} = b_{\mu\lambda j} = b_j = a_i$. ■

By Lemma 1, to obtain all the generalised Legendre sequences, it is sufficient to pick up any given generator g of F_p^*. So, here and hereafter we fix g, and simply denote $S_{\underline{b},g}$ by $S_{\underline{b}}$.

Lemma 2. $S_{\underline{b}}^{(g^\lambda)} = S_{L^\lambda \underline{b}}$, $S_{\alpha\underline{b}} = \alpha S_{\underline{b}}$ if $\alpha \neq 0$, $S_{\underline{b}^r} = S_{\underline{b}}^r$.

Proof. Let $\underline{b}$ be as that in (1), and $S_{\underline{b}} = (a_0 a_1 \cdots a_i \cdots)$. Denoting $S_{\underline{b}}^{(g^\lambda)} = (c_0 c_1 \cdots c_i \cdots)$, we have $c_i = a_{ig^\lambda}$. Let $L^\lambda \underline{b} = (b'_0 \cdots b'_j \cdots b'_{R-1})$ and $S_{L^\lambda \underline{b}} = (d_0 d_1 \cdots d_i \cdots)$. If $i \equiv 0 (\text{mod } p)$, then $c_i = a_{ig^\lambda} = 0 = d_i$. If $i \not\equiv 0 (\text{mod } p)$, we can assume $i \equiv g^{j+Rk} (\text{mod } p)$. Then $ig^\lambda \equiv g^{j+\lambda+Rk} (\text{mod } p)$, and $c_i = a_{ig^\lambda} = b_{j+\lambda} = b'_j = d_i$. So, the first equation holds. It is obvious that other two equations hold. ∎

About some invariable properties of linear complexity $L(S_{\underline{b}})$ of $S_{\underline{b}}$, by Lemma 2, we get the following results.

Corollary 1.

1. $L(S_{L^\lambda \underline{b}}) = L(S_{\alpha \underline{b}}) = L(S_{\underline{b}^r}) = L(S_{\underline{b}})$.
2. For any arrangement $\underline{b}$ over F_R, there always exists an arrangement $\underline{b}' = (01 b'_2 \cdots b'_{R-1})$ such that $L(S_{\underline{b}}) = L(S_{\underline{b}'})$. ∎

3 Main Results

In this paper, we mainly characterize the linear complexity $L(S_{\underline{b}})$ of the sequence $S_{\underline{b}}$ for the case when $R = r^t, r \not| t$ and $\underline{b}$ is some arrangement called the partial nature arrangement.

When $R = r$, the arrangement $(01 \cdots r-1)$ was called the nature arrangement([6]). We generalise the definition of the nature arrangement. When $R = r^t$, the field F_R has a basis $\{\theta_1, \cdots, \theta_t\}$ over F_r and the corresponding basis vector $\theta = (\theta_1, \cdots, \theta_t)$. For an arrangement $\underline{b} = (b_0 b_1 \cdots b_j \cdots b_{R-1})$ over F_R, we call $\underline{b}$ as (θ, a)−nature arrangement where $1 \le a \le t$, if the following condition holds for each $0 \le j < R$:

$$b_j = \sum_{0 \le i < t} b_{ij} \theta_i, 0 \le b_{ij} < r \Longrightarrow j \equiv \sum_{0 \le i < a} b_{ij} r^i (\text{mod } r^a).$$

For the given basis θ, it is clear that the number of (θ, a)-nature arrangements over F_R is $(r^{t-a}!)^{r^a}$. In particular, when $a = t$, there exists only one (θ, a)-nature arrangement. We also call (θ, a)−nature arrangement as a−nature arrangement if θ is fixed. When $t = 1$, 1-nature arrangement is just the nature arrangement $(01 \cdots r-1)$ in [6]. An a−nature arrangement is called the partial nature arrangement, if $a < t$; the whole nature arrangement, if $a = t$. Discussing the invertible transformations acting on the basis vector θ, we know that the number of the whole nature arrangements over F_R is $|GL_t(F_r)| = \prod_{0 \le i < t} (r^t - r^i)$, where $GL_t(F_r)$ is the group consisting of all $t \times t$ invertible matrices over F_r.

Let $F_p^{*R} = \{i^R | i \in F_p^*\} =< g^R >$. The number r can be regarded as an element of F_p^* / F_p^{*R} under the canonical map. We denote the order of i in the group F_p^* / F_p^{*R} by $\text{Ord}_{F_p^*/F_p^{*R}}(i)$ (or simply $\text{Ord}(i)$). Then, the following conclusion is obvious.

Lemma 3. Assume

$$r \equiv g^{r^c\mu+Rk} (\text{mod } p), 1 \leq r^c\mu \leq R, r \nmid \mu, 0 \leq c \leq t. \qquad (2)$$

Then $\text{Ord}(r) = r^{t-c}$.

When $R = 2$, the generalised Legendre sequence is just the Legendre sequence and its linear complexity had been characterized in [4]. So, we only discuss the case when $R > 2$.

We restate the results in [6] as Theorem A- D, then we state our main result as Main Theorem.

Theorem A [6]. When $R = 3$, for any arrangement $\underline{b}$,

$$L(S_{\underline{b}}) = \begin{cases} \frac{2(p-1)}{3}, & \text{if } \text{Ord}(3) = 1, \\ p-1, & \text{if } \text{Ord}(3) = 3. \end{cases}$$

Theorem B [6]. When $R = 4$, for any arrangement $\underline{b}$,

$$L(S_{\underline{b}}) = \begin{cases} \frac{3(p-1)}{4}, & \text{if } p \equiv 1 (\text{mod } 8), \\ p-1, & \text{if } p \equiv 5 (\text{mod } 8). \end{cases}$$

Theorem C [6]. Let $R = 2^3$.

1. If $\text{Ord}(2) = 2$, $L(S_{\underline{b}}) = p - 1$.

2. Assume that $\text{Ord}(2) = 1$. Let β be an element of order p of the algebraic closure $\overline{F}_r$ of F_r, and

$$f_j(\beta) = \sum_{\substack{0<i<p \\ i \equiv g^{j+Rk} (\text{mod } p), 0 \leq k < \frac{p-1}{R}}} \beta^i.$$

2.1 If there exist one or seven 1's in the set $\{f_j(\beta) | 0 \leq j < 8\}$, then $L(S_{\underline{b}}) = \frac{7}{8}(p-1)$.

2.2 If there are three or five 1's in $\{f_j(\beta) | 0 \leq j < 8\}$, and $\underline{b} = (01y \cdots y^6)$ where y is a generator of F_8^*, then $L(S_{\underline{b}}) \geq \frac{p-1}{2}$.

Theorem D [6]. Let $R = r$ be an odd prime and the arrangement $\underline{b} = (012 \cdots r-1)$. Then $L(S_{\underline{b}}) \geq \frac{(r-1)(p-1)}{r}$. If $Ord(r) = 1$, then $L(S_{\underline{b}}) = \frac{(r-1)(p-1)}{r}$.

The following theorem is the main result in this paper, which will be proved in the next section.

Main Theorem. Let $R = r^t > 2$ be a prime power and p an odd prime such that $r \nmid t$ and $R | p-1$. Set $\text{Ord}_{F_p^*/F_p^{*R}}(r) = r^d$. Then we have

1. If $d = t$, then for any arrangement $\underline{b}$ over F_R, $L(S_{\underline{b}}) = p - 1$.

2. If $1 \leq d < t$, then for any $(t-d+1)$–nature arrangement $\underline{b}$ over F_R, $L(S_{\underline{b}}) = p - 1$.

3. If $d = 0$, then for any $t-$nature arrangement $\underline{b}$, $L(S_{\underline{b}}) = p - 1 - \frac{p-1}{R}$.

The main theorem implies Theorem A, supplements Theorem C and strengthens Theorem D.

Remark 2. When $R = r^t, r \not| t$, but $\underline{b}$ is not any partial nature arrangement, or when $R = r^t$ and $r|t$, it seems so hard to decide $L(S_{\underline{b}})$ as to enumerate all cases (see [6]). However, it is an interesting problem worth the further research.

4 Proof of Main Theorem

Let $S_{\underline{b}} = (s_0 s_1 \cdots s_i \cdots)$ and $S_{\underline{b}}(x) = \sum_{0 \le i < p} s_i x^i$. About the linear complexity $L(S_{\underline{b}})$ of $S_{\underline{b}}$, there is the following conclusion:

Lemma 4 ([4],[7]). Let

$$\mathcal{K}_{\underline{b}} = \{i | S_{\underline{b}}(\beta^i) = 0, 0 < i < p\} \subseteq F_p^*, \tag{3}$$

where β is an element of order p of the algebraic closure $\overline{F}_r$ of F_r. Then,

$$L(S_{\underline{b}}) = p - 1 - |\mathcal{K}_{\underline{b}}|.$$

■

The values $f_j(\beta^i)$, $0 \le j < R$, $0 \le i < p$, are considered in [6], where $f_j(x)$ is defined as the following polynomial

$$f_j(x) = \sum_{\substack{0<i<p \\ i \equiv g^{j+Rk} (\mathrm{mod}\ p) \\ 0 \le k < \frac{p-1}{R}}} x^i. \tag{4}$$

We may rewrite the polynomial (4) to

$$f_j(x) = \sum_{h \in <g^R>} x^{g^j h} (\mathrm{mod}\ x^p - 1), \tag{5}$$

where we view the power exponent $g^j h$ of x as the element in the multiplicative group F_p^*. It is easy to see that $f_j(\beta^i)$ is well defined and the two definitions are equivalent to each other if one considers only the values $f_j(\beta^i)$.

In the rest of this paper, we fix β as an element of order p in $\overline{F}_r$. Then, by (5), $f_j(\beta)$ is well defined. The following two lemmas show some properties about $f_j(x)$ and $S_{\underline{b}}(x)$ in [6].

Lemma 5. ([6]). If $\mathrm{Ord}(r) = 1$, then $f_j(\beta^i) \in F_r$, for any i and j.

Lemma 6. For any i, $f_j(\beta^i) \in F_{r^R}$.

Proof. Since $f_j(\beta^i)^{r^R} = \sum_{h\in<g^R>} \beta^{g^j ihr^R} = \sum_{h\in<g^R>} \beta^{g^j ih} = f_j(\beta^i)$, $f_j(\beta^i) \in F_{r^R}$. ■

Lemma 7 ([6]). Let $\underline{b} = (b_0 b_1 \cdots b_{R-1})$. Then,

1. $S_{\underline{b}}(x) = \sum_{0\le j<R} b_j f_j(x) (\mathrm{mod}\ x^p - 1)$.
2. $f_j(\beta^{ih}) = f_j(\beta^i)$, for any $h \in< g^R >, i \in F_p^*$. Moreover, $\sum_{0\le j<R} f_j(\beta) = -1$.
3. $S_{\underline{b}}(\beta^{ih}) = S_{\underline{b}}(\beta^i)$, for any $h \in< g^R >$; $S_{\underline{b}}(\beta^{g^\lambda}) = \sum_{0\le j<R} b_{j-\lambda} f_j(\beta)$, $0 \le \lambda < R$; $S_{\underline{b}}(1) = 0$. ■

The following lemma is an elementary fact in algebra.

Lemma 8. Let c be that defined in (2) and $gcd(t, R) = 1$. Then $< R, g^R >=< g^{r^c} >$. ■

In the sequel, for any $1 \le a \le t$, we define

$$M_{\underline{b},a} = \{\, g^\lambda \,|S_{\underline{b}}(\beta^{g^\lambda}) = 0,\ 0 \le \lambda < r^a \,\},\ 1 \le a \le t. \tag{6}$$

Lemma 9. Assume $r \not| t$. Then

$$\mathcal{K}_{\underline{b}} = \dot{\bigcup}_{g^\lambda \in M_{\underline{b},c}} g^\lambda < g^{r^c} > .$$

In particular,

$$|\mathcal{K}_{\underline{b}}| = \frac{p-1}{r^c} |M_{\underline{b},c}|;\ |M_{\underline{b},c+i}| = r^i |M_{\underline{b},c}|, \forall 0 \le i \le t-c.$$

Proof. First we prove that for any $i \in \mathcal{K}_{\underline{b}}$, both ig^R and iR belong to $\mathcal{K}_{\underline{b}}$. By Lemma 7, $S_{\underline{b}}(\beta^{ig^R}) = S_{\underline{b}}(\beta^i) = 0$. So, $ig^R \in \mathcal{K}_{\underline{b}}$. Since

$$0 = S_{\underline{b}}(\beta^i)^R = \left(\sum_{0\le j<R} b_j f_j(\beta^i)\right)^R = \sum_{0\le j<R} b_j^R f_j(\beta^{iR})$$
$$= \sum_{0\le j<R} b_j f_j(\beta^{iR}) = S_{\underline{b}}(\beta^{iR}),$$

$iR \in \mathcal{K}_{\underline{b}}$. Hence, $\mathcal{K}_{\underline{b}}$ must be the union of some cosets of $< R, g^R >=< g^{r^c} >$ in F_p^*. Since g is a generator of F_p^*, all these cosets are of the form $g^\lambda < g^{r^c} >$ where $g^\lambda \in M_{\underline{b},c}$. So, $\mathcal{K}_{\underline{b}} = \dot{\bigcup}_{g^\lambda\in M_{\underline{b},c}} g^\lambda < g^{r^c} >$ and $|\mathcal{K}_{\underline{b}}| = |< g^{r^c} > ||M_{\underline{b},c}| = \frac{p-1}{r^c}|M_{\underline{b},c}|$.

Note that

$$\mathcal{K}_{\underline{b}} = \dot{\bigcup}_{g^\lambda \in M_{\underline{b},c}} g^\lambda < g^{r^c} >= \dot{\bigcup}_{0\le j<\frac{p-1}{r^c}} g^{jr^c} \cdot M_{\underline{b},c}.$$

Then,

$$M_{\underline{b},c+i} = M_{\underline{b},c+i} \cap \mathcal{K}_{\underline{b}} = \dot{\bigcup}_{0\le j<r^i-1} g^{jr^c} \cdot M_{\underline{b},c}.$$

Hence, $|M_{\underline{b},c+i}| = r^i |M_{\underline{b},c}|$, for any $0 \le i \le t-c$. ∎

Lemma 10. Let $\underline{b}$ be an $a-$nature arrangement over F_R. Let c be the number defined in (2). Then,

1. If λ and λ' are two integers such that $\lambda \not\equiv \lambda' (\text{mod } r^a)$, then $S_{\underline{b}}(\beta^{g^{\lambda}}) \ne S_{\underline{b}}(\beta^{g^{\lambda'}})$. As a consequence, $|M_{\underline{b},a}| = 0$ or 1, and $|M_{\underline{b},a+i}| = 0$ or r^i.
2. $|M_{\underline{b},t}| \le r^{t-a}$. In particular, when $a = t$ and $\text{Ord}(r) = 1$, $|M_{\underline{b},c}| = 1$.

Proof. 1. Let $\Delta \equiv \lambda - \lambda' (\text{mod } R), 0 \le \Delta < R$. By the hypothesis, $\Delta \not\equiv 0 (\text{mod } r^a)$. We can assume that

$$\Delta = \sum_{0 \le i < t} \Delta_i r^i, 0 \le \Delta_i < r, \tag{7}$$

where

$$\Delta_0 = \Delta_1 = \cdots = \Delta_{k-1} = 0, \Delta_k \ne 0, 0 \le k < a.$$

By Lemma 7,

$$S_{\underline{b}}(\beta^{g^{\lambda'}}) = \sum_j b_{j-\lambda'} f_j(\beta) = \sum_j b_{j-\lambda+\Delta} f_j(\beta),$$

$$S_{\underline{b}}(\beta^{g^{\lambda'}}) - S_{\underline{b}}(\beta^{g^{\lambda}}) = \sum_j (b_{j-\lambda+\Delta} - b_{j-\lambda}) f_j(\beta). \tag{8}$$

Since $\underline{b}$ is an $a-$nature arrangement, there exists a basis $\{\theta_i | 0 \le i < t\}$ such that if

$$b_{j-\lambda} = \sum_{0 \le i < t} b_{ij}\theta_i, \qquad 0 \le b_{ij} < r \tag{9}$$

$$b_{j-\lambda+\Delta} = \sum_{0 \le i < t} b'_{ij}\theta_i, \qquad 0 \le b'_{ij} < r \tag{10}$$

then

$$j - \lambda \equiv \sum_{0 \le i < a} b_{ij} r^i (\text{mod } r^a) \tag{11}$$

$$j - \lambda + \Delta \equiv \sum_{0 \le i < a} b'_{ij} r^i (\text{mod } r^a). \tag{12}$$

By (7) and (11),

$$j - \lambda + \Delta = \sum_{0 \le i \le k} b_{ij} r^i + \Delta_k r^k (\text{mod } r^{k+1}), k < a, \tag{13}$$

By (12) and (13) again,

$$\sum_{0 \le i \le k} b'_{ij} r^i \equiv \sum_{0 \le i \le k} b_{ij} r^i + \Delta_k r^k (\text{mod } r^{k+1}).$$

So,

$$b'_{ij} = b_{ij}, 0 \le i < k; \; b'_{kj} \equiv b_{kj} + \triangle_k (\text{mod } r). \tag{14}$$

Then, by (9), (10) and (14),

$$b_{j-\lambda+\triangle} - b_{j-\lambda} = \triangle_k \theta_k + \sum_{k<i<t} (b'_{ij} - b_{ij})\theta_i.$$

By (8),

$$\begin{aligned} S_{\underline{b}}(\beta^{g^{\lambda'}}) - S_{\underline{b}}(\beta^{g^{\lambda}}) &= \sum_j (b_{j-\lambda+\triangle} - b_{j-\lambda}) f_j(\beta) \\ &= \triangle_k \theta_k \sum_j f_j(\beta) + \sum_j \left[\sum_{k<i<t} (b'_{ij} - b_{ij})\theta_i \right] f_j(\beta) \\ &= -\triangle_k \theta_k + \sum_{k<i<t} \theta_i \left[\sum_j (b'_{ij} - b_{ij}) f_j(\beta) \right] \\ &\qquad \text{(by Lemma 7)} \\ &= -\triangle_k \theta_k + \sum_{k<i<t} \theta_i \delta_i, \end{aligned} \tag{15}$$

where

$$\delta_i = \sum_j (b'_{ij} - b_{ij}) f_j(\beta) \in F_{r^R}. \qquad \text{(by Lemma 6)}$$

Since $(t, R) = 1$, the basis $\{\theta_i | 0 \le i < t\}$ of F_{r^t} over F_r is also a basis of $F_{r^{tR}}$ over F_{r^R}. Hence,

$$S_{\underline{b}}(\beta^{g^{\lambda'}}) - S_{\underline{b}}(\beta^{g^{\lambda}}) = -\triangle_k \theta_k + \sum_{k<i<t} \delta_i \theta_i \ne 0.$$

2. If $S_{\underline{b}}(\beta^{g^{\lambda}}) = 0$ for some $\lambda (0 \le \lambda < r^a)$, by the above conclusion 1,

$$M_{\underline{b},t} \subseteq \{ j | j \equiv \lambda (\text{mod } r^a), 0 \le j < R \} = \{ \lambda + r^a k | 0 \le k < r^{t-a} \}.$$

So, $|M_{\underline{b},t}| \le r^{t-a}$. In particular, when $\text{Ord}(r) = 1$, by Lemma 5, $f_j(\beta^i) \in F_r$. Then, $S_{\underline{b}}(\beta^i) = \sum_j b_j f_j(\beta^i) \in F_R$. Now $\varphi : \lambda \to S_{\underline{b}}(\beta^{g^{\lambda}})$ is a map from the set $\{\lambda | 0 \le \lambda < R\}$ to F_R. When $t = a$, by the above conclusion 1, φ is injective. In particular, there exits a unique λ such that $S_{\underline{b}}(\beta^{g^{\lambda}}) = 0$, i.e., $|M_{\underline{b},t}| = 1$. Hence $|M_{\underline{b},c}| = 1$, since $c = t$ when $Ord(r) = 1$. ■

Proof of Main Theorem. We know $\text{Ord}(r) = r^{t-c}$ by Lemma 3, so $c = t - d$. By Lemma 4 and Lemma 9,

$$L(S_{\underline{b}}) = p - 1 - \frac{p-1}{r^c} \cdot |M_{\underline{b},c}|, \; c = t - d. \tag{16}$$

So, it is sufficient to prove

$$|M_{\underline{b},c}| = \begin{cases} 0 & c < t, \\ 1 & c = t. \end{cases}$$

When $c = 0$, if $|M_{\underline{b},c}| > 0$, then $L(S_{\underline{b}}) = p - 1 - (p-1)|M_{\underline{b},c}| \leq 0$ by (16), a contradiction.

When $0 < c < t$, if $|M_{\underline{b},c}| > 0$, then by Lemma 9 and Lemma 10 (with $a = t - d + 1 = c + 1$), we have

$$r^{t-c} \leq r^{t-c}|M_{\underline{b},c}| = |M_{\underline{b},t}| \leq r^{t-a} = r^{t-c-1},$$

a contradiction.

When $c = t$, we have $d = 0$, then $Ord(r) = 1$. Now $|M_{\underline{b},c}| = 1$ by Lemma 10. Hence $L(S_{\underline{b}}) = p - 1 - \frac{p-1}{r^c}|M_{\underline{b},c}| = p - 1 - \frac{p-1}{r^t}$. ∎

References

1. A. J. Bromfield, F. C. Piper, Linear recursion properties of unrelated binary sequences, Discr. Appl. Math., 1990, **27**: 187-193.
2. I. Damgaard, On the randomness of Legendre and Jacobi sequences, in *Advance in Cryptology: CRYPTO'88,* Spring-Verlag, 1990, 163-172.
3. C. Sh. Ding, The differential cryptanalysis and design of the natrual stream ciphers, in *Fast Software Encryption,* Preneel, B., Ed., Berlin, Germany: Springer-Verlag, LNCS 809, 1994, 101-115.
4. C. Sh. Ding, T. Helleseth, W. Shan, On the complexity of Legendre sequence, IEEE Transactions on IT, May 1998, **44(3)**: 1276-1278.
5. T. Helleseth, Legendre sums and codes related to QR codes, Discr. Appl. Math., 1992, **35**: 107-113.
6. Y. P. Hu, Sh. M. Wei, G. Zh. Xiao, On the linear complexity of generalised Legendre/Jacobi sequences, DianZi XueBao (in Chinese), 2000, **28(2)**: 113-117.
7. R. Lidl, H. Niederreiter, "Finite Fields", in *Encyclopedia of Mathematics and Its Applications,* Vol. 20, Reading, MA: Addison-Wesley, 1983.
8. S. A. Vanstone, R. J. Zucchrato, Elliptic curve cryptosystems using curves of smooth order over ring Z_n, IEEE Transactions on IT, July 1997, **43(4)**: 1231-1237.

Hyper-Cyclotomic Algebra

Solomon W. Golomb[1] and Guang Gong[2]

[1] Department of Electrical Engineering-Systems,
University of Southern California, Los Angeles, CA 90089-2565 USA
[2] Department of Electrical and Computer Engineering,
University of Waterloo, Waterloo, Ontario N2L 3G1 CANADA

Summary. Two new operators imposed on the set of cyclotomic cosets modulo $p^n - 1$ and their algebraic structure are discussed. These two new operators can be performed on the cyclotomic algebra in a field. The resulting algebraic structure is called *hyper-cyclotomic algebra*. It is shown that there exist distributive laws which connect these new operators to the ordinary operations of a field. An application of this new algebraic system to cyclic invariant sequences is provided.

1 Introduction

The cyclotomic algebra was studied by Golomb in [2] [3] for use with computation of correlation of binary sequences having period $2^n - 1$. In this paper, we investigate two new operators imposed on the set of cyclotomic cosets modulo $p^n - 1$ and their algebraic structure. These two new operators can be performed on the cyclotomic algebra in a field. The resulting algebraic structure is called a *hyper-cyclotomic algebra*. We show that there are distributive laws which connect these new operators to the ordinary operations of a field. An application of this new algebraic system is to facilitate computation of the linear span and the correlation of sequences having period $p^n - 1$ via the computation with *Gauss periods*, the elements in the cyclotomic algebra, using these new operators.

The following notation will be used throughout this paper.

- $\mathcal{F}$, a field.
- $\mathbb{Z}$ represents the integer ring.
- p is a prime number, n a positive integer, and $q = p^n$.
- $\mathbb{F}_{p^n} = GF(p^n)$.
- $\mathbb{Z}_m$ is the ring of integers modulo m.

2 Cyclotomic Set Monoid

The following definition is taken from [1]

Definition 1. A *monoid* is a triple $(M, o, 1)$ in which M is a non-vacuous set, o is an associative binary composition (or product) in M, and 1 is an element of M such that

$$o(1, a) = a = o(a, 1) \text{ for all } a \in M. \tag{1}$$

An element 1 of M such that (1) holds is called a *unit* in (M, o). A unit in M is unique.

A cyclotomic coset C_s modulo $q-1$ (with respect to p) is defined by

$$C_s = \{s, sp, \cdots, sp^{n_s - 1}\},$$

where n_s is the smallest positive integer such that $s \equiv sp^{n_s} \pmod{q-1}$. The subscript s is chosen as the smallest integer in C_s, and s is called the *coset leader* of C_s. Let

$$\Omega_n = \{\text{ all cosets } C_s \text{ in } Z_{q-1}\ \}, \text{ and}$$
$$\Omega_n^- = \{C_s \in \Omega_n | s \neq 0\}.$$

We define two operations in Ω_n as follows:

$$C_r \oplus C_s = C_{r+s}$$
$$C_r \circ C_s = C_{rs}$$

where $s+r$ and sr are in the cosets containing $s+r$ and sr respectively. We still call these operations *addition* and *multiplication* in Ω_n.

Note that the product $C_r \circ C_s = C_{rs}$, when $r \in C_r$, $s \in C_s$, and $rs \in C_{rs}$, is independent of which representative elements are selected from the cosets C_r, C_s and C_{rs}. To see this, suppose that we have picked r' from C_r and s' from C_s. Then $r' = m_1 r$ and $s' = m_2 s$, whre m_1 and m_2 are multiples which do not effect coset membership. (A cyclotomic coset is, by construction, a minimal set which is closed under multiplication by multipliers.) Then $r's' = (m_1 m_2) rs$, which is in the same coset as rs because $m_1 m_2$ is again a multiplier. (The multipliers form a subgroup of the multiplicative group of the integers modulo $q-1$.) On the other hand, the sum $C_r \oplus C_s = C_{r+s}$ depends on the representatives picked from C_r and C_s. We achieve uniqueness by requiring r and s to be the leaders of the cosets C_r and C_s, respectively, and then interpreting C_{r+s} to be the coset containing the element $r+s$. Thus we have proved the following lemma.

Lemma 1. *The operators $\oplus$ and $\circ$ are well defined in Ω_n.*

Theorem 1. *With the above notation, both $(\Omega_n, \oplus, C_0)$ and $(\Omega_n^-, \cdot, C_1)$ are monoids. Moreover, these two operations satisfy the following distributive law*

$$C_r \circ (C_s \oplus C_t) = (C_r \circ C_s) \oplus (C_r \circ C_t).$$

Proof. From Lemma 1, the operators $\oplus$ and $\circ$ are well defined in Ω_n. It is easy to check C_0 and C_1 are the additive unit and multiplicative unit of Ω_n respectively. According to Definition 1, both $(\Omega_n, \oplus, C_0)$ and $(\Omega_n^-, \cdot, C_1)$ are monoids. For the distributive law, since

$$C_r \circ (C_s \oplus C_t) = C_r \circ C_{s+t} = C_{r(s+t)}$$
$$(C_r \circ C_s) \oplus (C_r \circ C_t) = C_{rs} \oplus C_{rt} = C_{rs+rt},$$

and $r(s+t)$ and $rs + rt$ belong to the same coset, the assertion is established. □

Remark 1. The multiplicative group in Ω_n is $(\Omega_n^*, \cdot, C_1)$, where Ω_n^*, is the set of nonzero elements in Ω_n with indices coprime to $p^n - 1$.

We will call Ω_n a *cyclotomic coset monoid,* which is different from the cyclotomic algebra in [3] which we will discuss in the next section.

In the following, we will introduce an isomorphic image of Ω_n. Let $I_n = \{r \in \mathbb{Z}_{q-1} | r \text{ is a coset leader }\}$ and let

$$\sigma : \Omega_n \to I_n \text{ by } \sigma : C_r \mapsto r. \tag{2}$$

The following proposition is immediate from the definition of σ.

Proposition 1. *$\sigma : \Omega_n \to I_n$ is a one-to-one correspondence between Ω_n and I_n, and I_n is the isomorphic image of Ω_n under σ, i.e., $\sigma(\Omega_n) = I_n$.*

The fact that $\sigma(\Omega_n) = I_n$ induces that $(I_n, \oplus, 0)$ and $(I_n^-, \circ, 1)$ are also monoids where the two operations $\oplus$ and $\circ$ are induced from the operations in Ω_n. I.e., we have

$$r \oplus s = t \tag{3}$$

where t is the coset leader of the coset C_{r+s} and the multiplication is the ordinary one since the product rs is also a coset leader modulo $q-1$ if both r and s are coset leaders modulo $q-1$, and I_n^* is the multiiplication group in I_n.

In the following, we will consider that I_n and Ω_n are interchangeable.

Example 1. Let $p = 2$ and $n = 4$. Then $q = 2^4 - 1 = 15$. The elements in Ω_4 are:

$$\begin{aligned} C_0 &= \{0\} \\ C_1 &= \{1, 2, 4, 8\} \\ C_3 &= \{3, 6, 12, 9\} \\ C_5 &= \{5, 10\} \\ C_7 &= \{7, 14, 13, 11\} \end{aligned}$$

and $I_4 = \{0, 1, 3, 5, 7\}$. We have the following addition table and multiplication table for I_4.

Table 1

$\oplus$	0	1	3	5	7
0	0	1	3	5	7
1	1	1	1	3	1
3	3	1	3	1	5
5	5	3	1	5	3
7	7	1	5	3	7

$\circ$	1	3	5	7
1	1	3	5	7
3	3	3	0	3
5	5	0	5	5
7	7	3	5	1

3 Cyclotomic Algebra Ψ_n and New Operators on Ψ_n

Let ξ be a primitive $(q-1)$st root of unity in $\mathcal{F}$. For any $r \in I_n$, let

$$\eta_r = \sum_{i=0}^{n_r - 1} \xi^{rp^i} \tag{4}$$

where $n_r = |C_r|$. This notation was introduced by Gauss for r coprime to $q-1$, and is called a *Gauss period* in [6]. Here we will consider all coset leaders modulo $q-1$. Let

$$\Psi_n = \{\eta_r | r \in I_n\}. \tag{5}$$

Thus the elements in Ψ_n have the following ordinary addition and multiplication in $\mathcal{F}$.

$$\eta_r + \eta_s = \sum_{i=0}^{n_r-1} \alpha^{rp^i} + \sum_{i=0}^{n_s-1} \alpha^{sp^i} \tag{6}$$

$$\eta_r \cdot \eta_s = \sum_{i=0}^{n_r-1} \alpha^{rp^i} \cdot \sum_{i=0}^{n_s-1} \alpha^{sp^i} \tag{7}$$

$$= \sum_{k=0}^{M} c_k \eta_k \tag{8}$$

where c_k are non-negative integers. Thus Ψ_n forms an algebra under $+$ and $\cdot$. Sometimes we also write the product of η_r and η_s as $\eta_r\eta_s$ for short. Golomb studied this product $\eta_r\eta_s$ in [2] [3]. Here we will give an explicit formula to compute c_k's in the above representation.

Theorem 2. *Let $n_j = |C_j|, j \in I_n$. Then the indices k and the coefficients of η_k in the representation of $\eta_r \cdot \eta_s$ can be determined as follows.*

Case 1. *One of n_r and n_s is equal to n.*

1. $n_r = n_s = n$

$$\eta_r \cdot \eta_s = \sum_{k=0}^{n-1} c_k \eta_{r+p^k s}, \text{ where } c_k = n/n_{t+p^k s}.$$

2. $n_r | n_s = n$ *and* $n_r < n$.

$$\eta_r \cdot \eta_s = \sum_{k=0}^{n_r-1} c_k \eta_{r+p^k s}, \text{ where } c_k = n/n_{t+p^k s}.$$

Case 2. *Both n_r and n_s are less than n.*

1. $(n_r, n_s) = 1$. *Then* $\eta_r \cdot \eta_s = \frac{n_r n_s}{n_{r+s}} \eta_{r+s}$.

2. $(n_r, n_s) = d > 1$. *If* $n_r = n_s = d$, *then*

$$\eta_r \cdot \eta_s = \sum_{k=0}^{d-1} c_k \eta_{r+p^k s}, \text{ where } c_k = d/n_{r+p^k s}.$$

If $n_r \neq n_s$, *then*

$$\eta_r \cdot \eta_s = \sum_{k=0}^{d-1} c_k \eta_{r+p^k s}, \text{ where } c_k = lcm(n_r, n_s)/n_{r+p^k s}.$$

Proof. Let $t_k = t + p^k s$.

Case 1. One of n_r and n_s is equal to n.

1. $n_r = n_s = n$. In this case,

$$C_{t_k} \subset \{p^j t_k | 0 \leq j < n\}.$$

Hence all $C_{t_k}, k = 0, \cdots, n-1$ are different elements in Ω_n and the coset size of C_{t_k} is a factor of n. Thus the result follows.

2. If $n_r | n_s = n$ and $n_r < n_s$, then

$$C_{t_k} \subset \{p^i t_{k+jn_r} | 0 \leq j \leq n/n_r, 0 \leq i < n_r\}$$

where $0 \leq k < n_r$. Applying the same argument as Case 1-1, the result follows.

Case 2. Both n_r and n_s are less than n.

1. $(n_r, n_s) = 1$. We have $n_r n_s | n$ and

$$C_{r+s} \subset \{p^j(r + sp^i) | 0 \leq j \leq n_r; 0 \leq i \leq n_s\}.$$

Thus $n_{r+s} | n_r n_s$ which establishes the assertion.

2. If $(n_r, n_s) = d > 1$ and $n_r = n_s$, this is similar as in Case 1-1. If $n_r \neq n_s$, then

$$C_{t_k} \subset \{p^{id} t_{k+jd} | 0 \leq j < n_s/d; 0 \leq i < n_r\}$$

where $0 \leq k < d$. Thus the result follows. □

From the proof of Theorem 2, the computation of $\eta_r \eta_s$ is essentially carried out on the coset monoid Ω_n. We have the following two new operators which are induced from the monoid Ω_n.

$$\eta_r \oplus \eta_s = \eta_{r+s} \tag{9}$$

$$\eta_r \circ \eta_s = \eta_{rs} = \sum_{j=0}^{n_{rs}-1} \xi^{rsp^j} \tag{10}$$

For the sum of any $m > 0$ elements in Ψ_n, we define

$$\eta_r \oplus \left(\sum_{i=1}^{m} \eta_i \right) = \sum_{i=1}^{m} (\eta_r \oplus \eta_i) \tag{11}$$

$$\eta_r \circ \left(\sum_{i=1}^{m} \eta_i \right) = \sum_{i=1}^{m} (\eta_r \circ \eta_i) \tag{12}$$

Now we have four operations in Ψ_n. How does $\oplus$ interact with $+$ and $\cdot$? How does $\circ$ interact with $\oplus$, $+$ and $\cdot$? What distributive laws (if any) connect $\oplus$ and $\circ$ with $+$ and $\cdot$? From the identities (11) and (12), both the new operators $\oplus$ and $\circ$ satisfy the distributive law for the ordinary addition operator. Next we will prove that

Theorem 3. *The operator $\circ$ in Ψ_n has the following distributive law for the ordinary multiplication operation and $\oplus$:*

$$\eta_r \circ (\eta_s \cdot \eta_t) = (\eta_r \circ \eta_s) \cdot (\eta_r \circ \eta_t), \text{ for } r \in \Omega_n^*, s, t \in \Omega_n \tag{13}$$

$$\eta_r \circ (\eta_s \oplus \eta_t) = (\eta_r \circ \eta_s) \oplus (\eta_r \circ \eta_t), r, s, t \in \Omega_n. \tag{14}$$

In order to prove this result, we need the following lemma. We write $\eta_s = \eta_s(\xi)$ for its dependence on ξ.

Lemma 2. *If* $(r, q-1) = 1$, *then*

$$\eta_s(\xi^r) = \eta_{rs}(\xi) = \eta_{rs}.$$

Proof. Note that

$$\eta_s(\xi^r) = \sum_{i=0}^{n_s-1} \xi^{rsp^i}.$$

Since $(r, q-1) = 1$, the cosets containing s and rs have the same size. I.e., $n_s = n_{rs}$. Substituting this into the above identity, we get

$$\eta_s(\xi^r) = \sum_{i=0}^{n_{rs}-1} \xi^{rsp^i} = \eta_{rs}(\xi) = \eta_{rs}.$$

Proof of Theorem 3. From Theorem 2, we can write

$$\eta_s \cdot \eta_t = \sum_k c_k \eta_k$$

where the c_k's are positive integers. Thus

$$\begin{aligned}\eta_r \circ (\eta_s \cdot \eta_t) &= \eta_r \circ \left(\sum_k c_k \eta_k\right) \\ &= \sum_k c_k (\eta_r \circ \eta_k) \\ &= \sum_k c_k \eta_{rk}\end{aligned}$$

where the second to last identity is from the definition (11). On the other hand, for $r \in \Omega_n^*$, according to Lemma 2,

$$\begin{aligned}(\eta_r \circ \eta_s) \cdot (\eta_r \circ \eta_t) &= \eta_{rs} \cdot \eta_{rt} \\ &= \eta_s(\xi^r)\eta_t(\xi^r) \\ &= \sum_k c_k \eta_k(\xi^r) \\ &= \sum_k c_k \eta_{rk}\end{aligned}$$

which establishes the assertion (13). The second distributive law is a direct result from Theorem 1. □

Example 2. We will use Example 1. Here ξ is a primitive 15th root of unity, and

$$\begin{aligned}\eta_0 &= 1 \\ \eta_1 &= \xi + \xi^2 + \xi^4 + \xi^8 \\ \eta_3 &= \xi^3 + \xi^6 + \xi^{12} + \xi^9 \\ \eta_5 &= \xi^5 + \xi^{10} \\ \eta_7 &= \xi^7 + \xi^{14} + \xi^{13} + \xi^{11}\end{aligned}$$

The multiplication tables for $\cdot$ and $\circ$ of Ψ_4 are shown as Tables 2 and 3.

Table 2

$\cdot$	η_0	η_1	η_3	η_5	η_7
η_0	η_0	η_1	η_3	η_5	η_7
η_1	η_1	$\eta_1 + 2\eta_3 + 2\eta_5$	$\eta_1 + 2\eta_5 + 2\eta_7$	$\eta_3 + \eta_7$	$4\eta_0 + \eta_1 + \eta_3 + \eta_7$
η_3	η_3	$2\eta_7 + \eta_1 + 2\eta_5$	$4\eta_0 + 3\eta_3$	$\eta_1 + \eta_7$	$2\eta_1 + 2\eta_5 + \eta_7$
η_5	η_5	$\eta_3 + \eta_7$	$\eta_1 + \eta_7$	$2\eta_0 + \eta_5$	$\eta_1 + \eta_3$
η_7	η_7	$4\eta_0 + \eta_1 + \eta_3 + \eta_7$	$2\eta_1 + 2\eta_5 + \eta_7$	$\eta_1 + \eta_3$	$2\eta_3 + 2\eta_5 + \eta_7$

Table 3

$\circ$	η_1	η_3	η_5	η_7
η_1	η_1	η_3	η_5	η_7
η_3	η_3	η_3	η_0	η_3
η_5	η_5	η_0	η_5	η_5
η_7	η_7	η_3	η_5	η_1

From Tables 2 and 3, we have

$$\eta_7 \circ (\eta_3 \cdot \eta_7) = \eta_7 \circ (2\eta_1 + 2\eta_5 + \eta_7) = 2\eta_7 + 2\eta_5 + \eta_1$$

and

$$(\eta_7 \circ \eta_3) \cdot (\eta_7 \circ \eta_7) = \eta_3 \cdot \eta_1 = 2\eta_7 + 2\eta_5 + \eta_1$$

Thus

$$\eta_7 \circ (\eta_3 \cdot \eta_7) = (\eta_7 \circ \eta_3) \cdot (\eta_7 \circ \eta_7)$$

as stated in Theorem 3 since $7 \in \Omega_4^*$.

Remark 2. Theorem 3 shows that the new operator $\circ$ in the cyclotomic algebra Ψ_n is analogous to the exponential operation in the field $\mathcal{F}$. I.e., it is analogous to $(a \cdot b)^c = a^c \cdot b^c$ for $a, b \in \mathcal{F}$ and c an integer.

4 Applications to Cyclic Invariant Sequences

Before we introduce this connection, we restate here that the number of elements in Ψ_n is equal to the number of elements in Ω_n [2]. Let $I_p(d)$ be the number of irreducible polynomials over $\mathbb{F}_p$ of degree d. Then

$$|\Omega_n| = \sum_{d|n} I_p(d)$$

and

$$|\Omega_n^*| = \phi(p^n - 1)/n.$$

where $\phi(x)$ is the Euler function, representing the number of integers coprime to $p^n - 1$ in the range of 1 to $p^n - 1$.

From the theory of linear feedback shift register (LFSR) sequences, $|\Omega_n|$ is the number of shift distinct linear feedback shift register (LFSR) sequences over $\mathbb{F}_p$ with irreducible feedback polynomials of degree dividing n. Moreover $|\Omega_n^*|$ is the number of all shift distinct m-sequences over $\mathbb{F}_p$ of degree n. Any sequence over $\mathbb{F}_p$ can be decomposed into a sum of phase shifts of these LFSR sequences.

Let $A = \{a_i\}$ be a sequence over $\mathbb{F}_p$ of period dividing $p^n - 1$. Then the sequence A has the following trace representation:

$$a_i = f(\xi^i) \text{ where } f(x) = \sum_{r \in \Omega_n} Tr_p^{n_r}(\beta_r x^{ri})$$

where ξ is a primitive element in $\mathbb{F}_{p^n}$, $\beta_r \in \mathbb{F}_{p^{n_r}}$, and $Tr_p^{n_r}(x)$ is the trace function from $\mathbb{F}_{p^{n_r}}$ to $\mathbb{F}_p$. Under this assumption, we can take $\mathcal{F} = \mathbb{F}_{p^n}$. From Section 3, we have

$$\eta_r = \eta_r(\xi) = Tr_p^{n_r}(\xi)$$

If the elements of A satisfy the cyclic invariant condition, i.e., $a_i = a_{ip}, i = 0, 1, \cdots$, then we say that A is *cyclic invariant*. In this case, the trace represenation of A becomes

$$f(x) = \sum_{r \in \Omega_n} d_r Tr_p^{n_r}(x^{ri}), \ d_r \in \mathbb{F}_p.$$

(This can be verified directly using the trace representation and the cyclic invariant condition, so we omit it here.)

Decimation Operator and Shift Operator. Let s be an integer. A sequence $B = \{b_i\}$ whose elements are given by

$$b_i = a_{si}, i = 0, 1, \cdots,$$

is called an s-decimation of A, denoted by $B = A^{(s)}$. For the case $(s, q-1) = 1$, applying Theorem 3 to the s-decimation of A, we have

$$\begin{aligned} b_i &= a_{si} = f(\xi^{si}) \\ &= \sum_{r \in \Omega_n} d_r Tr_p^{n_r}(\xi^{sri}) \\ &= \sum_{r \in \Omega_n} c_{r,i} Tr_p^{n_{ri}}(\xi^{sri}), \ \text{where } c_{r,i} = d_r \frac{n_r}{n_{ri}} \\ &= \sum_{r \in \Omega_n} c_{r,i} \eta_{sri} \\ &= \sum_{r \in \Omega_n} c_{r,i} (\eta_s \circ \eta_{ri}) \\ &= \eta_s \circ \left(\sum_{r \in \Omega_n} c_{r,i} \eta_{ri} \right). \end{aligned}$$

Thus

$$a_{si} = \eta_s \circ \left(\sum_{r \in \Omega_n} c_{r,i} \eta_{ti} \right), i = 0, 1, \cdots . \tag{15}$$

Hence the new operator $\circ$ on the cyclotomic algebra Ψ_n corresponds to the decimation operator on cyclic invariant sequences. The formula (15) provides a new insight for computing correlation between the sequence and its decimation in terms of the cyclotomic formulas provided in [3].

Next we will show that the new operator $\oplus$ on Ψ_n corresponds to the (left) shift operator on the sequence A. Let L be the shift operator, i.e., $L^t(A) = a_t, a_{t+1}, \cdots$. Here we only present the special case of n is prime. The same approach used here can apply to general n, but it needs a more complicated formula for justification of the coefficients. Applying (11), we now have

$$\begin{aligned} a_{t+i} &= f(\xi^{i+t}) \\ &= \sum_{r \in \Omega_n} d_r Tr_p^{n_r}(\xi^{r(i+t)}) \\ &= \sum_{r \in \Omega_n} d_r \eta_{ri+rt} \\ &= \sum_{r \in \Omega_n} d_r (\eta_{rt} \oplus \eta_{ri}) \\ &= \sum_{r \in \Omega_n} d_r ((\eta_t \circ \eta_r) \oplus \eta_{ri}) \end{aligned}$$

i.e., we have

$$a_{t+i} = \sum_{r \in \Omega_n} d_r ((\eta_t \circ \eta_r) \oplus \eta_{ri}), i = 0, 1, \cdots . \tag{16}$$

Remark 3. Any 2-level autocorrelation sequence over $\mathbb{F}_p$ is a cyclic invariant sequence for $p = 2$ [4], [5], and for $p > 2$, we have found that all known 2-level autocorrelation sequences are cyclic invariant. However, it has not been proved whether the 2-level autocorrelation condition implies the cyclic invariant condition. The formulae (15) and (16) provide another representation of the decimated sequences and the shifted sequences in terms of cyclotomic algebra with the new operators $\oplus$ and $\circ$. These representations will facilitate computation of the linear span and the correlation of cyclic invariant sequences and may enable one to characterize families of 2-level autocorrelation sequences and low crosscorrelation sequences.

References

1. N. Jacobson, *Basic Algebra I*, W.H. Freeman and Company, San Francisco, 1974.
2. S.W. Golomb, *Shift Register Sequences,* Holden-Day, Inc., 1967. Revised edition, Aegean Park Press, 1982.
3. S.W. Golomb, Theory of transformation groups of polynomials over $GF(2)$ with applications to linear shift register sequences, *Information Sciences*, vol.1, no. 1, December 1968, pp.87-109.
4. S. W. Golomb, "On the classification of balanced binary sequences of period $2^n - 1$," *IEEE Trans. Inform. Theory*, vol. IT-26, pp. 730-732, Nov. 1980.
5. G. Gong and S.W. Golomb, Binary sequences with two-level autocorrelation, *IEEE Trans. on Inform. Theory*, vol. 45, No.2, March 1999, pp.692-693.
6. B.L. van der Waerden, *Modern Algebra*, Volume I, Frederick Ungar Publishing Co, 1950.

Cyclic Projective Planes, Perfect Circular Rulers, and Good Spanning Rulers

Solomon W. Golomb[1] and Herbert Taylor[2]

[1] Department of Electrical Engineering-Systems, University of Southern California, Los Angeles, CA 90089-2565 USA
[2] Box 234008, Leucadia, CA 92023 USA

Summary. A *spanning ruler with n marks* is a set of n integers $0 = d_1 < d_2 < \cdots < d_n$ such that all $\binom{n}{2}$ differences $d_j - d_i$, with $i < j$, are distinct. The spanning ruler is called *perfect* if these distinct differences are the consecutive integers from 1 to $\binom{n}{2}$, occurring in any order. Two proofs are given that there are no perfect spanning rulers when $n > 4$. The length of the shortest spanning ruler with n marks, denoted $L(n)$, has now been determined by exhaustive computer search for all $n \leq 23$.

Two constructions for very good, though not necessarily best, spanning rulers are described which are obtained from *cyclic projective planes*, i.e. cyclic $(v, k, 1)$ difference sets. (One of these constructions has long been used to construct spanning rulers, often under the name "difference triangle sets" [4].) Methods of obtaining upper and lower bounds on $L(n)$ are discussed.

Applications are mentioned to such areas as pulse radar (for the pulse patterns), radio astronomy (for siting the receiving antennas), and xray diffraction crystallography (for reconstructing the bonding angles of the crystal from the scattering angles of the xrays).

Introduction

The literature on *spanning rulers* (under various names) as a topic in recreational mathematics goes back to the nineteenth century, and a summary of results on this and similar ruler problems was compiled by J. Leech (see [1]). There are several early references (pre-1980) to their application to patterns for pulse radar (see Bloom and Golomb [2]). In one of his *Scientific American* columns, Martin Gardner (see [3]) called the shortest spanning ruler with n marks a *Golomb ruler*, a term which was widely adopted. A mathematically equivalent problem involves the construction of *difference triangle sets* [4].

In this paper, two proofs are given of *Theorem 1:* "For $n > 4$, no perfect ruler exists." The first proof can already be found in Golomb [14]. (See also [2] and [5].) The second proof is new, to the best of our knowledge. Shortest spanning rulers up to $n = 19$ marks were published in several papers (see, e.g. [6] and [7]); and a world-wide distributed effort on the internet (see [8]) has produced the results for $n \leq 23$ shown in **Table 1** below. (This effort is still ongoing.)

The method described in Section 2 of this paper involving "perfect circular rulers" can be traced back implicitly to J. Singer [9], and semi-explicitly (to the extent of producing an asymptotic upper bound) to P. Erdös [10]. Good upper and lower bounds on difference triangle sets (and therefore on shortest spanning rulers) were given in [4]. The second proof of Theorem 1 leads to a new approach to lower bounds on $L(n)$, the length of the shortest spanning ruler on n marks. The explicit values actually obtained are not as good as those in [4], but we believe that the full power of this method has not yet been reached. In Section 7, we give a new and apparently highly efficient construction for "spanning ruler sets".

In addition to the application to pulse radar patterns, several other important applications of spanning rulers are mentioned in Section 8.

1 Pulse Patterns and "Optimal Rulers"

A pulse radar is able to send one or more pulses of radio-frequency energy toward a target. It is a convenient and rather realistic assumption to require all the pulses to be the same in both duration and amplitude. The signal design problem then reduces to devising ***patterns*** of these identical pulses, so that the autocorrelation function of the pattern is as impulse-like as possible.

The problem is usually restated as follows: For each positive integer n, what is the shortest length $L = L(n)$ for which there is a sequence $\{a_1, a_2, \ldots, a_n\}$ with $0 = a_1 < a_2 < \ldots < a_n = L$, such that the set of $\binom{n}{2}$ differences $\{a_j - a_i\}$, with $1 \le i < j \le n$, are all distinct?

The model underlying this restatement is the following. At each integer a_i in the sequence, there is a pulse of brief duration and unit amplitude. Thus, the sequence corresponds to a pulse pattern with n pulses, spread out over a total duration of L (or, more precisely L plus one pulse duration) with the unnormalized autocorrelation function $K(\tau)$ satisfying $K(0) = n$, $0 \le K(\tau) \le 1$ for $1 \le \tau \le L$, and $K(\tau) = 0$ for $\tau > L$. (The distinctness of the differences $\{a_j - a_i\}$ guarantees $K(\tau) \le 1$ for all $\tau, |\tau| \ge 1$.) Finding the ***shortest*** length $L = L(n)$ for the sequence achieves the desired signal parameters with the shortest duration for the pulse pattern.

The sequence model is also described in terms of a certain class of ***rulers*** (the measuring devices, not the monarchs or autocrats).

A ruler of length L has only n marks on it, at integer positions $a_1, a_2, \ldots, a_n$, where $a_1 = 0$ and $a_n = L$ are the two endpoints of the ruler. If every integer distance $d, 1 \le d \le L$, can be measured in one and only one way as a distance between two of the n marks, then the ruler is called a ***perfect ruler***. For a perfect ruler, $L = \binom{n}{2}$, since there are exactly $\binom{n}{2}$ distances between the n marks, and these must be some permutation of $\{1, 2, 3, \ldots, L\}$. In Figure 1, we see perfect rulers for $n = 2, 3$, and 4, with the corresponding radar pulse patterns and their autocorrelation functions $K(\tau)$.

Unfortunately, for $n > 4$, there are no perfect rulers.

Theorem 1. *For $n > 4$, no perfect rulers exist.*

Proof. A ruler with n marks has $n-1$ *intervals* between marks. For a *perfect* ruler, these $n-1$ intervals must all have distinct positive integer lengths, and the sum of these lengths must be $L = \binom{n}{2}$. Hence these intervals must be (in some order) $1, 2, 3, \dots, n-1$, since any *other* set of $n-1$ distinct positive integers will have a larger sum than $\binom{n}{2}$. Since *all* distances between marks on the ruler must be distinct, the interval of length 1 cannot be next to an interval of length $\leq n-2$ (since the sum of two consecutive intervals is a measured distance of the ruler, and every length from 1 to $n-1$ is *already* measured, as a single interval). This can only be achieved if the interval of length 1 is at one end of the ruler, and is immediately followed by the interval of length $n-1$. Similarly, the interval of length 2 cannot be next to any interval of length $\leq n-3$ (to avoid two consecutive intervals with a total length $\leq n-1$, equaling the length of a single interval), nor can it be next to the interval of length $n-2$ (since $2 + (n-2) = n = 1 + (n-1)$, and the distance n would be measured in more than one way). This requires that the interval of length 2 must also be at an end of the ruler, and must also be next to the interval of length $n-1$. But then the entire ruler consists of only three intervals: $1, n-1, 2$, which means there are only four marks altogether, and $n = 4$. (Note that with $n = 4$, there really is a perfect ruler with the consecutive intervals of lengths 1,3,2.) □

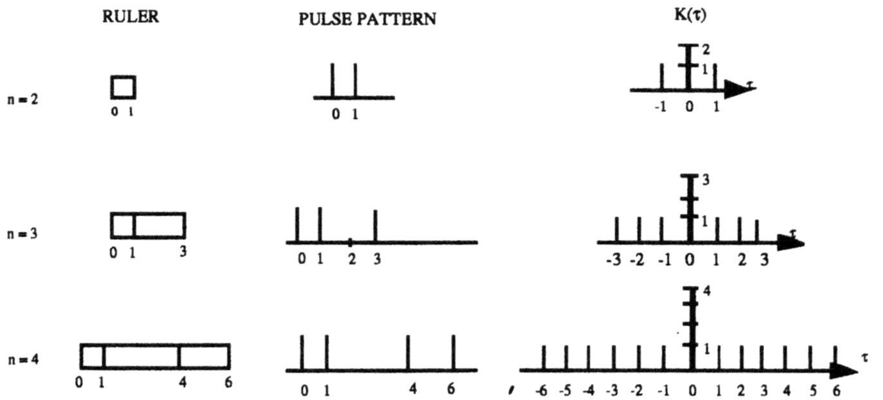

Fig. 1. Perfect rulers for $n = 2, n = 3$, and $n = 4$, with the corresponding pulse radar patterns and their unnormalized autocorrelation functions.

There are two obvious ways to relax the requirement on a perfect ruler to get objects which exist for all n. A ***covering ruler*** with n marks and length L measures every distance from 1 to L, as a distance between two marks on the ruler, in ***at least one way***; while a ***spanning ruler*** with n marks and length L measures every distance from 1 to L, as a distance between two marks on the ruler, in ***at most one way***. The interesting combinatorial problems are to determine the ***longest covering ruler*** with n marks, and the ***shortest spanning ruler*** with n marks, for each positive integer n. Both of these problems have long histories in the combinatorial literature. However, the application to pulse radar involves only finding the ***shortest spanning ruler*** for each n. (Martin Gardner [3] termed these objects "Golomb rulers", a name which seems subsequently to have been widely adopted.)

The behavior of $L(n)$ as a function of n, for the shortest spanning ruler, is quite erratic in detail, although it has been shown [10] that asymptotically $L(n) \sim n^2$ as $n \to \infty$. The value of $L(n)$ has been determined by exhaustive computer search for all $n \leq 23$. Lengths 15 and 16 were done by [6], lengths 17 and 18 by W.O. Sibert in 1993, and length 19 by [7]. In addition to left-right reversal of the ruler, these rulers are not unique for several of the smaller values of n. One example of a spanning ruler of length $L(n)$, for each $n \leq 23$, is shown in Table 1. For $n \geq 20$, a distributed worldwide computer effort has been underway for several years, so that $L(20), L(21), L(22)$, and $L(23)$ are now known (see [8]).

Fig. 2. Radar pulse pattern, and its autocorrelation, for $n = 5$ pulses.

TABLE 1. Table of the Shortest Spanning Rulers

n	$L(n)$	m	Sequence of Marks
2	1	1	0,1
3	3	1	0,1,3
4	6	1	0,1,4,6
5	11	2	0,1,4,9,11
6	17	4	0,1,4,10,12,17
7	25	5	0,1,4,10,18,23,25
8	34	1	0,1,4,9,15,22,32,34
9	44	1	0,1,5,12,25,27,35,41,44
10	55	1	0,1,6,10,23,26,34,41,53,55
11	72	2	0,1,4,13,28,33,47,54,64,70,72
12	85	1	0,2,6,24,29,40,43,55,68,75,76,85
13	106	1	0,2,5,25,37,43,59,70,85,89,98,99,106
14	127	1	0,5,28,38,41,49,50,68,75,92,107,121,123,127
15	151	1	0,6,7,15,28,40,51,75,89,92,94,121,131,147,151
16	177	1	0,1,4,11,26,32,56,68,76,115,117,134,150,163,168,177
17	199	1	0,8,31,34,40,61,77,99,118,119,132,143,147,182,192,194,199
18	216	1	0,2,10,22,53,56,82,83,89,98,130,148,153,167,188,192,205,216
19	246	1	0,1,6,25,32,72,100,108,120,130,153,169,187,190,204,231,233,242,246
20	283	1	0, 24, 30, 43, 55, 71, 75, 89, 104, 125, 127, 162, 167, 189, 206, 215, 272, 275, 282, 283
21	333	1	0, 4, 23, 37, 40, 48, 68, 78, 138, 147, 154, 189, 204, 238, 250, 251, 256, 277, 309, 331, 333
22	356	1	0, 1, 9, 14, 43, 70, 106, 122, 124, 128, 159, 179, 204, 223, 253, 263, 270, 291, 330, 341, 353, 356
23	372	1	0, 6, 22, 24, 43, 56, 95, 126, 137, 146, 172, 173, 201, 213, 258, 273, 281, 306, 311, 355, 365, 369, 372

In Table 1, the quantity "m" is the number of inequivalent rulers of length $L(n)$ which are shortest spanning rulers with n marks. Only one of each set of m rulers is listed explicitly.

2 Perfect Circular Rulers from Cyclic Projective Planes

A systematic construction for *very good* spanning rulers (though not necessarily *best* spanning rulers) with k marks, and length not exceeding $(k-1)^2$, can be obtained from cyclic (v, k, λ)-designs with $v = n^2 + n + 1, k = n + 1, \lambda = 1$ (called *finite projective planes*), which are known to exist whenever n is a prime or a power of a prime. Think of the cyclic $(v, k, 1)$ design D as a circle of circumference v, with k marks at the k positions d_i, for all $d_i \in D$.

On this circle, every integer arc length from 1 to $v - 1$ can be found in one and only one way as a *distance* between two marks. Remove the longest arc between two consecutive marks, and straighten out what is left. This will be a spanning ruler with k marks, and length at most $v - k = n^2 = (k-1)^2$. For example, $D = \{3, 6, 7, 12, 14\}$ is a $(21, 5, 1)$ cyclic difference set, and the longest arc (modulo 21) is from 14 to 3, which we remove. We then look at $D - 3 = \{0, 3, 4, 9, 11\}$; and a ruler with five marks, at 0, 3, 4, 9, and 11, is a shortest spanning ruler, measuring every length from 1 to 11 inclusive except for length 10. (This is the other shortest spanning ruler with five marks, inequivalent to the one shown in Figure 2.)

Figure 3 shows the perfect circular ruler of circumference 13, corresponding to the $(v, k, \lambda) = (13, 4, 1)$ cyclic difference set. Here, $n = k - \lambda = 3$ is a "multiplier"; and the difference set $D = \{0, 2, 5, 6\}$ (mod 13) is a union of the ***cyclotomic cosets*** $\{0\}$ and $\{2, 6, 5\}$

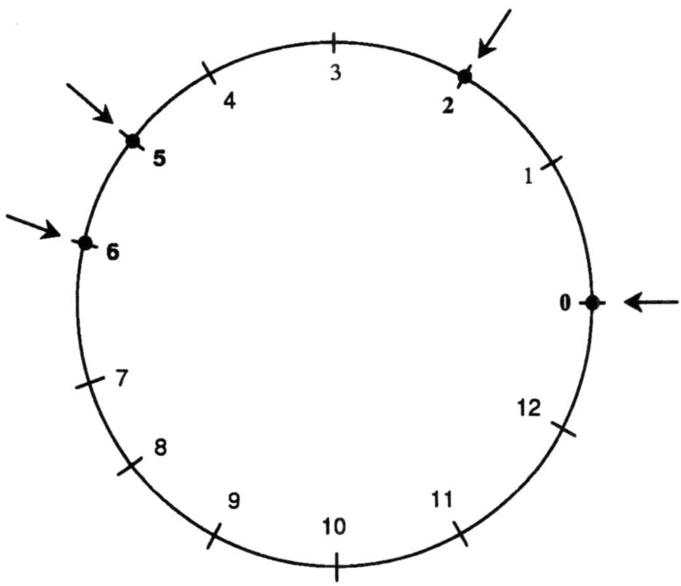

Fig. 3. "Perfect Circular Ruler" modulo 13, from the $(v, k, \lambda) = (13, 4, 1)$ cyclic difference set $D = \{0, 2, 5, 6\}$ (mod 13).

Note that our geometric model uses a circle of circumference v, with "marks" on this circumference at the "points" $d_1, d_2, \ldots, d_k$, where these are measured arc lengths from an arbitrary starting point. By the cyclic difference set property for $\lambda = 1$, the differences $d_j - d_i$, with $i \neq j$, take on each value from 1 to $v - 1$, modulo v, exactly once. Therefore, arcs (or equivalently central angles) of the circle are found in each integer length from 1 to $v - 1$, each in exactly one way, as the circumferential distances between two marks on the circle. (The central angles of the circle, as determined by pairs of marks on the circumference, take on every value $2\pi s/v$ for $s = 1, 2, 3, \ldots, v - 1$.)

It is in this sense that cyclic projective planes correspond not only to cyclic $(v, k, 1)$ difference sets, but also to "perfect circular rulers".

The known constructions for cyclic $(v, k, 1)$ difference sets are for $k-1 = n$ being of the form $q = p^m$ with p prime and m a positive integer. By the Bruck-Ryser-Chowla Theorem, these finite projective planes are known *not* to exist for infinitely many other values of n. However, there are also infinitely many values of n for which the existence of a "plane of order n" remains an open question. [11]

There are two ways to go from a perfect circular ruler to a good (though not necessarily best) spanning ruler. As already noted, one way is to remove the longest arc between consecutive marks from a perfect circular ruler on $k = n + 1$ marks, and straighten out what is left to get a spanning ruler on k marks. (Among all $(v, k, 1)$ cyclic difference sets for given k, select the one whose longest arc between consecutive marks is greatest, to get as short a spanning ruler as possible by this method.)

The other way to go from a perfect circular ruler to a good spanning ruler is to cut the circular ruler at one of the marks, and then straighten out the entire circle to get a spanning ruler. A perfect circular ruler with k marks and circumference $v = k^2 - k + 1$ then yields a spanning ruler with $k + 1$ marks and length $L = k^2 - k + 1$. This is independent of which $(v, k, 1)$ cyclic difference set is chosen, and of which of the k marks is used to cut the circle.

Since these constructions are only known to exist when $k - 1 = q = p^m$, the spacing of such integers q influences the utility of these constructions for obtaining bounds on $L(k)$ for all values of k.

More generally, a "circular spanning ruler" with k marks and circumference v has k marks at the integer positions $0 \leq g_1 < g_2 < \ldots < g_k$ such that the differences $g_j - g_i$ are all distinct modulo v for all $i \neq j$, without the requirement that these differences include all the consecutive integers from 1 to $v - 1$. It is still the case that a "straight" spanning ruler is obtained from a circular spanning ruler in each of the two ways described above for the case of perfect circular rulers; and removing the longest arc results in the shortest spanning ruler from the given circular spanning ruler. For circular spanning rulers, the circumference v is not limited to values of the form $n^2 + n + 1$. Moreover, the correspondence is now bidirectional, because, given any "straight" spanning ruler of length L, a circular spanning ruler of circumference $\leq 2L + 1$ can be obtained by adjoining a "closing arc" of length $\leq L + 1$. (If the "closing arc" has length $L + 1$, it is easily seen that all measured arcs on the circle with lengths less than the circumference of $2L + 1$ will be distinct.) An interesting open question is to determine the smallest circular spanning ruler with k marks which corresponds to the shortest straight spanning ruler with k marks.

3 Difference Triangle Sets

When the successive marks on an n-mark spanning ruler are $d_1 < d_2 < d_3 < \ldots < d_n$, then the sequence of first differences, $d_2 - d_1, d_3 - d_2, \ldots, d_n - d_{n-1}$, forms the top row of a "difference triangle set", DTS, of length $n-1$. There is an extensive literature on DTS's (see, e.g. [4]), a problem isomorphic to the (straight) spanning ruler problem.

The proof of Theorem 1, that there are no perfect rulers with $n > 4$ marks, was based on reasoning involving the entries in the top row of the corresponding difference triangle sets (DTS). We called these entries the "intervals" of the ruler. Another proof of Theorem 1 can be based on the fact that the intervals of a perfect ruler with n marks must sum to $L(n) = \binom{n}{2}$, and then *all* entries in the DTS must sum to $\binom{\binom{n}{2}+1}{2}$. This proof is given in the next section.

Most of the conjectured values of $L(n)$ for $n > 23$ are based on the cyclic projective plane construction where the longest arc is removed from a perfect circular ruler.

A related problem seeks *sets* of rulers which collectively measure distances all of which are distinct. This problem is also discussed in [4]. A new construction for such sets is given in section 7 below.

4 Another Proof of Theorem 1

Here is another way to prove Theorem 1.

Proof #2. As in the first proof, a perfect ruler with n marks has $n-1$ "intervals" (distances between adjacent marks), whose consecutive lengths $a_1, a_2, \ldots, a_{n-1}$ must be a permutation of $\{1, 2, \ldots, n-1\}$ for a total length of $\binom{n}{2}$. These intervals form the top row of the associated "perfect difference triangle". (Here, $a_j = d_{j+1} - d_j$.)

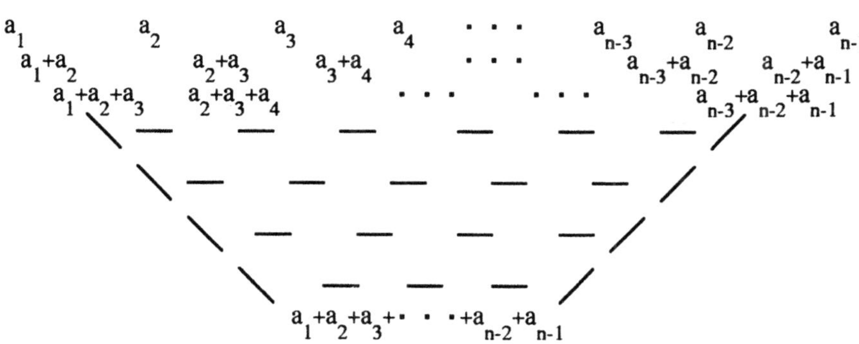

This triangle consists of all the lengths measured by the perfect ruler; so these $\binom{n}{2}$ numbers must therefore be a permutation of $\{1,2,3,\ldots,\binom{n}{2}\}$, and these numbers will sum to $S_1 = \binom{\binom{n}{2}+1}{2} = (n^4 - 2n^3 + 3n^2 - 2n)/8$. On the other hand, by noting how often each of the $n-1$ intervals appears in the difference triangle, we have: $S = (n-1)a_1 + 2(n-2)a_2 + 3(n-3)a_3 + 4(n-4)a_4 + \cdots + (n-1)a_{n-1} = \sum_{j=1}^{n-1} j(n-j)a_j$. The coefficients are left-right symmetric, and increase monotonically from the end values to the middle values. Therefore, a permutation of $\{1,2,\ldots,n-1\}$, as the values of a_j which maximizes this sum, is $S_{\max} \Leftrightarrow \{1,3,5,\ldots,\frac{n-1}{2},\ldots,4,2\}$ if n is odd, and $S_{\max} \Leftrightarrow \{1,3,5,\ldots,\frac{n}{2},\ldots,4,2\}$ if n is even. At $n=4$, the permutation we get of $\{1,2,3,\}$ is $\{1,3,2\}$, and the sum formula gives $1\cdot3\cdot1+2\cdot2\cdot3+3\cdot1\cdot2 = 3+12+6 = 21$ as the maximum possible value of S, which agrees with $\binom{\binom{4}{2}+1}{2} = \binom{7}{2} = 21$, consistent with the existence of a perfect ruler on $n=4$ marks. However, already at $n=5$, we have $S_1 = \binom{\binom{5}{2}+1}{2} = 55$, but $S_{\max} = 4\cdot1\cdot1+3\cdot2\cdot3+2\cdot3\cdot4+1\cdot4\cdot2 = 4+18+24+8 = 54$, which is too small. This discrepancy increases with increasing n. At $n=6$, we have $S_1 = \binom{\binom{6}{2}+1}{2} = 120$, but $S_{\max} = 5\cdot1\cdot1+4\cdot2\cdot3+3\cdot3\cdot5+2\cdot4\cdot4+1\cdot5\cdot2 = 5+24+45+32+10 = 116$. By the time we reach $n=12$, we have $S_1 = \binom{\binom{12}{2}+1}{2} = \binom{67}{2} = 2211$, compared to $S_{\max}(12) = 2011$, a deficit of 200. (One way to express the contrast between the two sequences, $S_1(n) = \binom{\binom{n}{2}+1}{2} = \{0,1,6,21,55,120,231,406,666,1035,1540,2211,\ldots\}$ and $S_{\max}(n) = \{0,1,6,21,54,116,220,382,620,995,1410,2011,\ldots\}$, is that the row of fourth differences of $S_1(n)$ takes the constant value 3, while the row of fourth differences of $S_{\max}(n)$ is the alternating sequence $\{2,3,2,3,2,3,\ldots\}$. In closed form, $S_{\max}(n) = \frac{1}{48}(5n^4 - 4n^3 - 2n^2 + 4n - 3\delta)$, where $\delta = \left\{\begin{matrix} 0, n \text{ even} \\ 1, n \text{ odd}\end{matrix}\right\}$. We then see that $\Delta(n) = S_1(n) - S_{\max}(n) = \frac{1}{48}(n^4 - 8n^3 + 20n^2 - 16n + 3\delta)$, which has $\Delta(0) = \Delta(1) = \Delta(2) = \Delta(3) = \Delta(4) = 0, \Delta(5) = 1$, and $\Delta(n)$ is monotonically increasing for $n \geq 5$, growing asymptotically like $\frac{n^4}{48}$. □

5 A Lower Bound on $L(n)$

From the explicit expression in the previous section, namely

$$\Delta(n) = S_1(n) - S_{\max}(n) = \frac{1}{48}(n^4 - 8n^3 + 20n^2 - 16n + 3\delta),$$

we can obtain a non-trivial lower bound on $L(n)$, as follows.

Suppose $L(n) - \binom{n}{2} = r = r_n$. Then, generalizing $S_{\max}(n)$ to $S_{\max}^{(r)}(n)$ and $S_1(n)$ to $S_1^{(r)}(n)$ in the obvious way, namely assuming that the $n-1$

"intervals" along the ruler have lengths summing to $L(n) = \binom{n}{2} + r_n$, we easily find $S_{\max}^{(r)}(n) \leq S_{\max}(n) + r\left(\frac{n}{2}\right)^2$, because the largest contribution would occur if all of the extra length r were added to the central interval which is multiplied by, at most, $\left(\frac{n}{2}\right)^2$ in $S_{\max}(n)$; and $S_1^{(r)}(n) \geq S_1(n) + r$, because in the new difference triangle, the lowest entry is the length of the ruler, which has been increased by r. From these inequalities, we have $S_1^{(r)}(n) - S_{\max}^{(r)}(n) \geq \Delta(n) - r\left(\frac{n^2-4}{4}\right)$, from which $r_n \geq \frac{4}{n^2-4}\left(\frac{n^4-8n^3+20n^2-16n+3\delta}{4\delta}\right)$, or $r_n \geq \frac{1}{12}\left(n^2 - 8n + 24 - \frac{48}{n} + \frac{96-3\delta}{n^2} + 0\left(\frac{1}{n^3}\right)\right)$. Explicitly, $r_{12} \geq 6, r_{48} \geq 162, r_{120} \geq 1122$; and asymptotically, $\lim_{n\to\infty} \frac{L(n)}{n^2} \geq \frac{7}{12}$.

In Table 2, we see a comparison of the actual difference $L(n) - \binom{n}{2}$ with the lower bound $\lceil \frac{4}{n^2-4}\Delta(n) \rceil$; and the actual ratio $L(n)/n^2$. With extra effort, the method described here can give a somewhat better lower bound on $L(n)$, but may still fall short of the lower bound obtained in [4].

Obtaining an upper bound on $L(n)$ is the subject of the next section.

n	$L(n)$	$L(n) - \binom{n}{2}$	$\lceil \frac{4}{n^2-4}\Delta(n) \rceil$	$L(n)/n^2$
2	1	0	0	0.2500
3	3	0	0	0.3333
4	6	0	0	0.3750
5	11	1	1	0.4400
6	17	2	1	0.4722
7	25	4	1	0.5102
8	34	6	2	0.5313
9	44	8	3	0.5432
10	55	10	4	0.5500
11	72	17	5	0.5950
12	85	19	6	0.5903
13	106	28	8	0.6272
14	127	36	9	0.6480
15	151	46	11	0.6711
16	177	57	13	0.6914
17	199	63	15	0.6808
18	216	63	17	0.6667
19	246	75	20	0.6814
20	283	93	22	0.7075
21	333	123	25	0.7551
22	356	125	28	0.7355
23	372	119	31	0.7032

TABLE 2. On the growth rate of $L(n)$.

6 An Upper Bound on $L(k)$

(In this section, we use $L(k)$ instead of $L(n)$ to conform with the standard notation for (v, k, λ) difference sets.)

Theorem 2. *For those values of $k \geq 2$ for which a cyclic $(v, k, 1)$ difference set exists, $L(k) \leq k^2 - \frac{5}{2}k + 2$.*

Proof. For these values of k, the "perfect circular ruler" has circumference $v = k^2 - k + 1$, divided into k unequal intervals. The average interval length is therefore $\frac{v}{k} = k - 1 + \frac{1}{k}$. If a sum of k unequal integers is $k^2 - k + 1$, the largest of these integers must be at least $(3k-1)/2$ if k is odd, and $(3k-2)/2$ if k is even. (These values are achieved by picking k consecutive integers centered around $\frac{1}{2}(k-1)$, and making the minor adjustments necessary to make the sum equal $k^2 - k + 1$.) Since there is an interval of at least this length, removing an arc of this length leads to a spanning ruler no longer than $(k^2 - k + 1) - \left(\frac{3k}{2} - 1\right) = k^2 - \frac{5}{2}k + 2$. □

Notes.

1. This can be improved somewhat by observing that a "perfect circular ruler" must have some very short intervals, including intervals of lengths 1 and 2, and at least one of the lengths 3 or 4.
2. The spacing of the values of k for which cyclic $(v, k, 1)$ difference sets exist becomes important for obtaining a general upper bound on $L(k)$. With the pessimistic assumption that these difference sets may only exist when $k-1$ is a prime or a power of a prime, it is known that such values of k will always exist between x and $x + 0(x^t)$, for some $t < 1$. This is sufficient to prove that $L(k) \sim k^2$, but a stronger argument would be required to prove $L(k) \leq k^2$ for all $k \geq 1$. We believe this old conjecture to be likely, but do not claim to have a proof.

7 A New Construction For Spanning Ruler Sets

Def. A *spanning ruler set* is a collection of spanning rulers for which no distance is measured by more than one ruler.

Here is a way that a spanning ruler with k marks, $0 = d_1 < d_2 < \cdots < d_k$, can be combined with an n-mark spanning circular ruler, provided that $n > d_k$, to obtain a set of n spanning rulers, each with k marks, which form a spanning ruler set.

Let the spanning circular ruler (which need not be a "perfect circular ruler") have its n marks at $g_1, g_2, \ldots, g_n$, all distinct modulo v, where v is the circumference of the circle. For each $i, 1 \leq i \leq n$, we form a k-mark ruler as indicated:

$$\{g_{i+d_1}, g_{i+d_2}, \ldots, g_{i+d_k}\},$$

where the subscripts are taken modulo n (on the range of 1 to n), but the g_j's themselves are taken modulo v, with the requirement that when regarded as ordinary integers, we must have $g_{i+d_1} < g_{i+d_2} < \cdots < g_{i+d_k}$.

We illustrate this construction with the straight spanning ruler having $k = 3$, namely $\{d_1, d_2, d_3\} = \{0, 1, 3\}$, and the circular spanning ruler having $n = 4$ (noting that $n > d_3$), namely $\{g_1, g_2, g_3, g_4\} = \{0, 2, 5, 6\}$ (modulo 13). The construction then yields the following four spanning rulers having three marks each:

$$\{g_{1+0}, g_{1+1}, g_{1+3}\} = \{g_1, g_2, g_4\} = \{0, 2, 6\}.$$
$$\{g_{2+0}, g_{2+1}, g_{2+3}\} = \{g_2, g_3, g_1\} = \{2, 5, 13\}.$$
$$\{g_{3+0}, g_{3+1}, g_{3+3}\} = \{g_3, g_4, g_2\} = \{5, 6, 15\}.$$
$$\{g_{4+0}, g_{4+1}, g_{4+3}\} = \{g_4, g_1, g_3\} = \{6, 13, 18\}.$$

(Note that the subscripts are treated modulo 4, but the g_j's themselves are taken modulo 13, with the requirement that each of the n resulting rulers must consist of an increasing set of three integers.)

The difference triangles for these four rulers are:

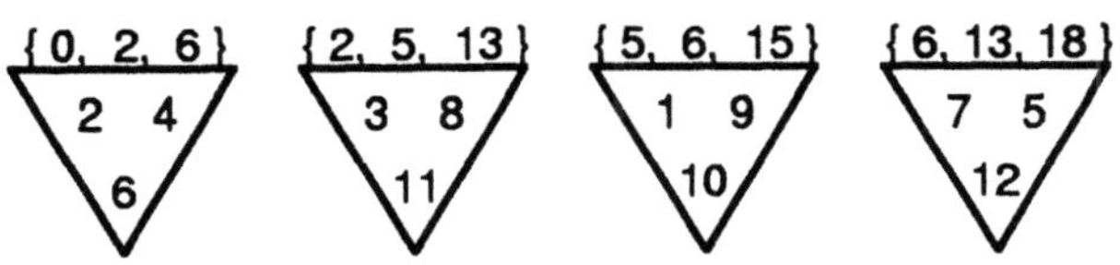

In this example, the difference triangles collectively contain every integer from 1 to 12, each exactly *once*. It helped that we began with a perfect straight ruler and a perfect circular ruler; but all that is guaranteed in general is that the collection of all entries in all the difference triangles will be free from duplications.

We can "normalize" the example just given by subtracting the left-most member of each ruler from the other numbers in the same ruler, to give:

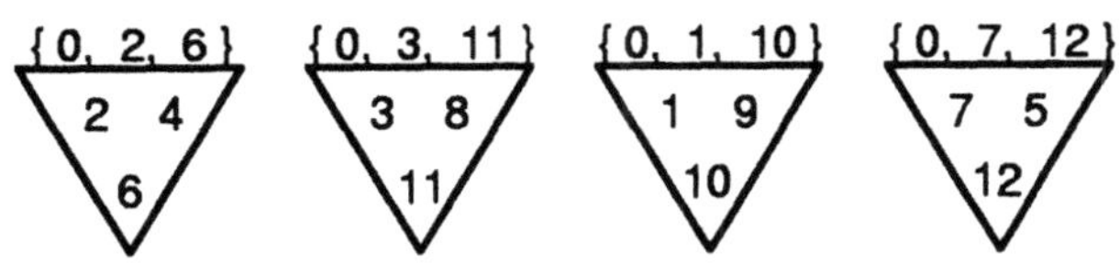

The difference triangles are unchanged, but this succeeds in disguising the construction from which the example arose.

Simple variations of the straight ruler and the circular ruler lead to other examples of sets of four three-mark rulers that measure each length from 1 to 12, each in only one way, as follows:

$$\{d_1, d_2, d_3\} = \{0, 2, 3\},\ \{g_1, g_2, g_3, g_4\} = \{0, 2, 5, 6\} \pmod{13},$$

gives

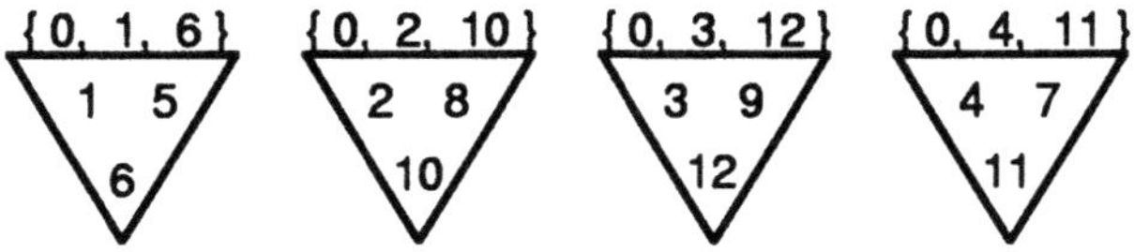

$$\{d_1, d_2, d_3\} = \{0, 1, 3\}, \{g_1, g_2, g_3, g_4\} = \{0, 1, 3, 9\} \pmod{13},$$

gives

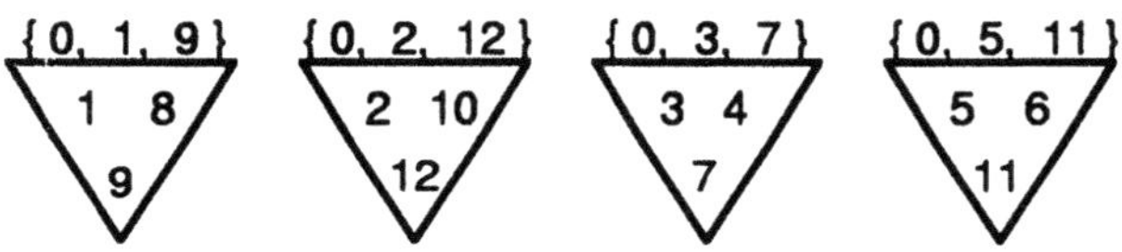

$$\{d_1, d_2, d_3\} = \{0, 2, 3\}, \{g_1, g_2, g_3, g_4\} = \{0, 1, 3, 9\} \pmod{13},$$

gives

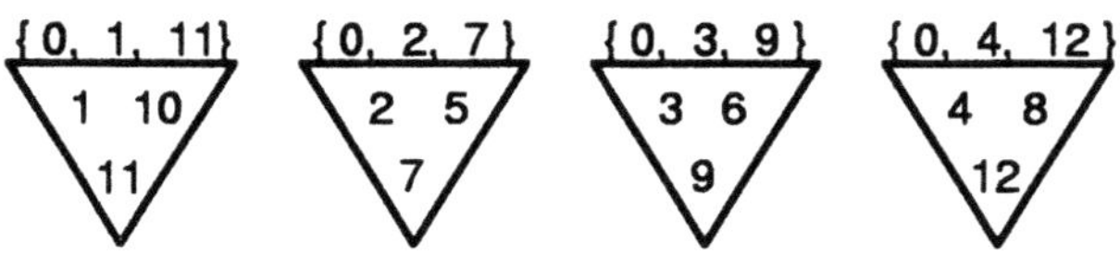

There is a fifth example of such a set of four 3-mark rulers,

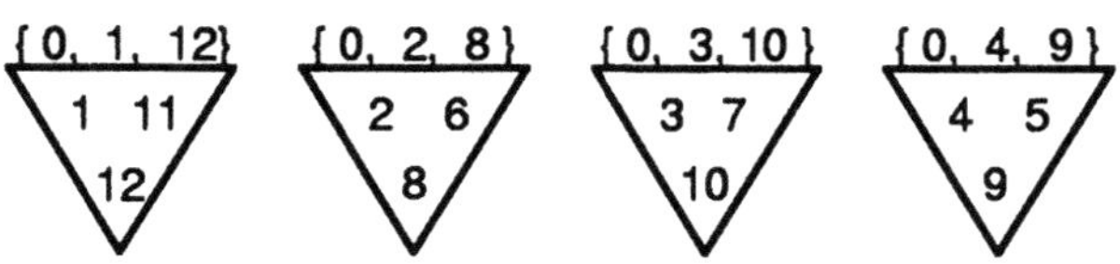

for which a straight ruler/circular ruler derivation has not yet been found. Note that these five examples represent the length "12" in all five possible ways: $1 + 11, 2 + 10, 3 + 9, 4 + 8$, and $5 + 7$.

The lengths from 1 to 12 can also be measured by either of the following two spanning ruler sets:

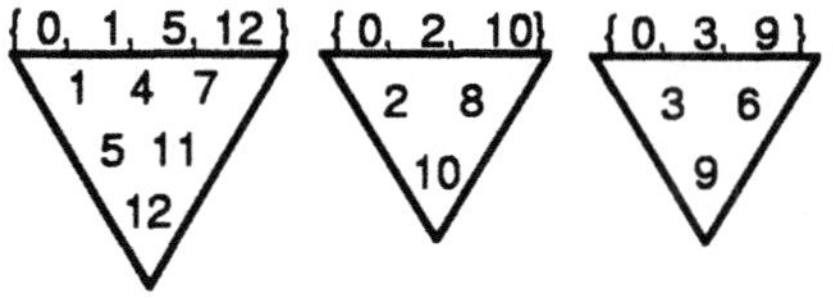

OR

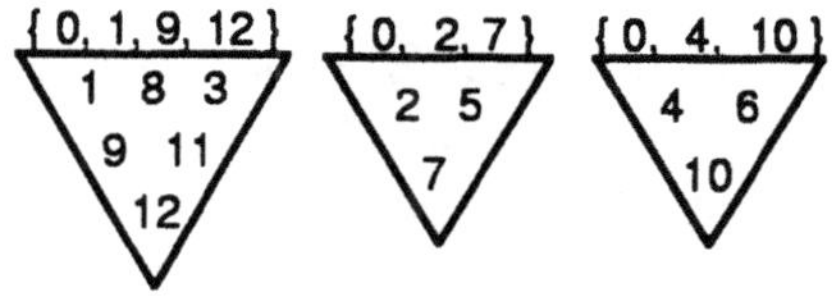

Direct derivations for these examples have also not yet been found.

As a second kind of example, from the 4-mark straight spanning ruler $\{d_1, d_2, d_3, d_4\} = \{0, 1, 4, 6\}$ and the 8-mark circular spanning ruler $\{g_1, g_2, g_3, g_4, g_5, g_6, g_7, g_8\} = \{0, 1, 5, 7, 17, 35, 38, 49\}$ (mod 57), the construction gives the following eight four-mark rulers, shown in normalized form with their difference triangles:

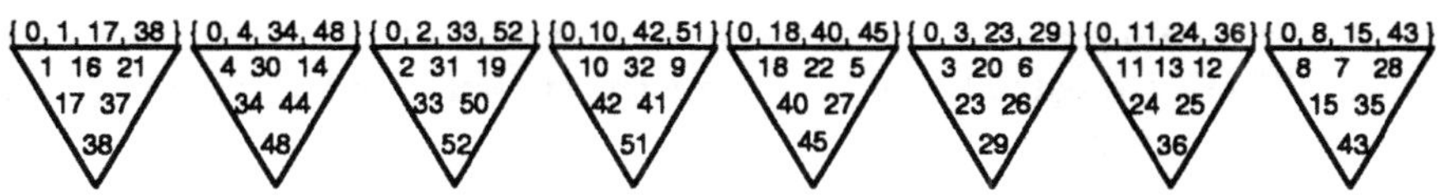

The total number of entries in the difference triangles by this construction is, in general, $\binom{k}{2}n$, which in this case is $6 \cdot 8 = 48$. Here, as promised, they are all distinct, and form a subset of the first 52 positive integers. (The four skipped values, the "unmeasured distances", are 39, 46, 47, and 49.)

The proof that the construction described in this section always succeeds in generating a spanning ruler set is not difficult. Because the g_i's are the marks on an n-mark spanning circular ruler, the distances from a fixed g_i to a "subsequent" g_j (proceeding counterclockwise around the circle) are all distinct from one another. Moreover, with $0 = d_1 < d_2 < \cdots < d_k$ as the marks on a (straight) spanning ruler, with $d_k < n$, the subsets of the n marks on the circular ruler corresponding to $\{g_{i+d_1}, g_{i+d_2}, \cdots, g_{i+d_k}\}$ for each i, $1 \le i \le n$, considered as straight rulers, not merely are spanning rulers individually, but collectively measure a subset of the *distinct* distances $g_j - g_{j'}$, $(j \neq j')$ on the circular ruler.

8 Other Applications

Minimum spanning rulers have another, very different application in radio astronomy. In radio astronomy, only receiving antennas are used, and the spacing between two antennas generates a difference in the time a signal is received, which is used in making interferometry measurements on the signal. If several antennas are constructed along a straight line, they should be spaced along that line in the pattern of a spanning ruler, to get maximum interferometry information by having all the arrival time differences distinct. (See [12].)

The properties of these rulers also play an important role in x-ray diffraction crystallography. In this technique, one attempts to determine the bonding angles of a crystal by shining x-rays at it, and observing the diffraction patterns which emerge. What in fact is measurable by this method is the *differences* of the bonding angles, from which one wishes to reconstruct the bonding angles themselves. This leads to an "inverse problem": given the autocorrelation function $K(\tau)$, what is the set of possible signals which might have produced it? For several decades, crystallographers relied on a "theorem" of S. Piccard which asserted (in our terminology): "If two spanning rulers have the same autocorrelation function $K(\tau)$, the two rulers are either identical or mirror images of each other." A counter-example was found in [2] for 6-mark rulers: $\{0,1,4,10,12,17\}$ and $\{0,1,8,11,13,17\}$, which are in fact *shortest* 6-mark spanning rulers. This generalizes to a two-parameter family of counter-examples, all involving 6-mark spanning rulers. No counter-examples with fewer than 6 marks are possible, and none with more than 6 marks are known. (There are partial results which suggest that counter-examples may occur only in the case $n = 6$; see [13].)

References

1. J. Leech, "On the representation of $1, 2, \ldots, n$ by differences", *J. London Math. Soc.* 31 (1956) 160-169.
2. G.S. Bloom and S.W. Golomb, "Applications of Numbered, Undirected Graphs", *Proceedings of the IEEE*, vol.65, no.4, April, 1977, 562-571.
3. M. Gardner, *Wheels, Life and Other Mathematical Amusements,* W.H. Freeman and Co., New York, 1983; Chapter 15, 152-165.
4. T. Kløve, "Bounds and construction for difference triangle sets", *IEEE Transactions on Information Theory,* vol. 35, no. 4, July, 1989, 879-886.
5. S.W. Golomb, "Construction of signals with favourable correlation properties", in *Surveys in Combinatorics, 1991*, D. Keedwell, Ed., Cambridge University Press, 1991.
6. J.B. Shearer, "Some new optimum Golomb rulers", *IEEE Trans. Inform. Theory* **36**, 183-184, 1990.
7. A. Dollas, W.T. Rankin, and D. McCracken, "A new algorithm for Golomb ruler derivation and proof of the 19 mark ruler", *IEEE Trans. Inform. Theory* **44**, 379-382, 1998.
8. B. Hayes, "Computing Science: Collective Wisdom", *American Scientist,* vol. 89, no. 2, March-April 1998, 118-122.
9. J. Singer, "A theorem in finite projective geometry and some applications to number theory", *Trans. Amer. Math. Soc.* 43 (1938), 377-385.
10. P. Erdös, "On a problem of Sidon in additive number theory and on some related problems – Addendum", *J. London Math. Soc.* 19 (1944), 208.

11. J.H. van Lint and R.M. Wilson, *A Course in Combinatorics*, Cambridge University Press, 1992, especially Chapters 19 and 26-28.
12. A.K. Dewdney, "COMPUTER RECREATIONS: The search for an invisible ruler that will help radio astronomers to measure the earth", *Scientific American* 253, no. 6 (Dec. 1985) 16-26.
13. G. Yovanof and S.W. Golomb, "The polynomial model in the study of counterexamples to S. Piccard's Theorem", *Ars Combinatorica* **48**, 43-63, 1998.
14. S.W. Golomb, "How to Number a Graph", in *Graph Theory and Computing*, R.C. Read, editor, Academic Press, 1972, 23-37 (Specifically, Theorem 4 on pp. 27-28). (Paper presented at the conference on "Graph Theory and Computing", Kingston, Jamaica, January, 1969.)

Linear Recursive Sequences over Elliptic Curves

Guang Gong[1] and Charles C.Y. Lam[2]

[1] Department of Electrical and Computer Engineering
University of Waterloo
Waterloo, Ontario N2L 3G1, CANADA
ggong@cacr.uwaterloo.ca

[2] Department of Combinatorics & Optimization
University of Waterloo
Waterloo, Ontario N2L 3G1, CANADA
cy3lam@cacr.uwaterloo.ca

Summary. In this paper, we introduce linear feedback shift register sequences (LFSR) over the group of the elliptic curve points, and a construction of binary sequences obtained from these LFSR sequences. The former is called *LFSR-EC sequences*. Properties on representation, period, and linear span of these two types of sequences are discussed. Also, the even case for the elliptic curve sequence proposed in [5] is analysed.

1 Introduction

Since Koblitz and Miller [11,15] proposed using elliptic curves over $\mathbb{F}_q$ to construct public-key cryptosystems, elliptic curve public-key cryptosystems over $\mathbb{F}_{2^l}$ or $\mathbb{F}_q$ have become widely used in applications such as smart cards which provide limited space for implementation of modular computations. Currently, the operations of points on elliptic curves over $\mathbb{F}_{2^l}$ or $\mathbb{F}_q$ have a well-developed technology in both hardware and software implementations.

Linear recursive sequences (or equivalently, linear feedback shift register sequences) over $\mathbb{F}_p$ have a significant influence in the development of communications and cryptography. In this work, we combine elliptic curves and linear feedback shift registers to generate pseudorandom sequences which provide strong potential applications in stream ciphers and pseudorandom number generators. The motivation for such a construction comes from constrained systems or devices, for example, wireless devices which employ an elliptic curve public-key cryptosystem, then such a construction automatically provide a pseudorandom sequence (or number) generator. If these sequences have good randomness, then it can be used for generating random numbers in the elliptic curve digital signature algorithm and session key in their encryption phases.

Elliptic curve sequences with strong cryptographic properties have been studied in the literature. Kaliski [10] used the randomness criteria based on the computational complexity of the discrete log problem over the elliptic curves to construct a class of elliptic curve pseudorandom sequences. Gong, Berson and Stinson [5] discussed the linear span, period and $0-1$ distribution of the one-stage LFSR over supersingular elliptic curves over finite fields of characteristic 2. Mahassni and Shparlinski [9] obtained a general bound for the $0-1$ distribution of the one-stage LFSR on elliptic curves over finite fields. Hallgren [8] discussed some heuristics with regard to the linear congruential generator over elliptic curves. Doumen and Beelen [1] has recently proposed a method on generating binary sequences using linear recurrence relations on elliptic curves.

The paper is organised as follows. In Section 2, we introduce a construction of linear recursive sequences over the group of points on an elliptic curve over finite fields, denoted by *EC sequences* for short, and discuss their representation, decomposition and period. In Section 3, we give a construction of binary sequences obtained from EC sequences and derive their randomness properties. In Section 4, we analyse a case for the supersingular elliptic curve sequences proposed in [5].

We conclude this section with the definition of elliptic curves over finite fields and linear recursive sequences.

A. Elliptic Curves over Finite Fields

Consider $\mathbb{F}_q$ where q is a power of a prime. If the characteristic of the field is 2, an elliptic curve E over $\mathbb{F}_q$ can be written in one of the following standard forms: [14]

$$E: \; y^2 + c_3 y = x^3 + c_4 x + c_6, c_i \in \mathbb{F}_q, \text{ if } E \text{ is supersingular, or}$$
$$E: \; y^2 + xy = x^3 + c_2 x^2 + c_6, c_i \in \mathbb{F}_q, \text{ if } E \text{ is non-supersingular.}$$

If the characteristic is greater than 3, E can be written in the following form

$$E: y^2 = x^3 + ax + b, \; a, b \in \mathbb{F}_q.$$

The points $P = (x, y)$, $x, y \in \mathbb{F}_q$ that satisfy the equation, together with a "point at infinity" denoted by $\mathcal{O}$, form an abelian group $(E, +, \mathcal{O})$ with identity element $\mathcal{O}$ (for the case where the characteristic is 3, the reader is referred to [14].)

B. Linear Recursive Sequences over rings

Let r be a positive integer and let $\mathbb{Z}_r$ be the integer residue ring modulo r. Let $f(x) = x^n - \sum_{i=0}^{n-1} c_i x^i$, $c_i \in \mathbb{Z}_r$ be a monic polynomial over $\mathbb{Z}_r$, and $\underline{b} = \{b_i\}$, $b_i \in \mathbb{Z}_r$. $\underline{b}$ is called an n-th order linear recursive sequence or linear feedback shift register over $\mathbb{Z}_r$ (generated by $f(x)$) if for a given n-tuple $(b_0, b_1, \ldots, b_{n-1}) \in \mathbb{Z}_r^n$,

$$b_{k+n} = \sum_{i=0}^{n-1} c_i b_{k+i}, \; k = 0, 1, \ldots$$

$f(x)$ is called a *characteristic polynomial* of $\underline{b}$ over $\mathbb{Z}_r$ and $(b_0, b_1, \ldots, b_{n-1})$ is called an *initial state* of $\underline{b}$. The characteristic polynomial of $\underline{b}$ having the least degree is called the *minimal polynomial* of $\underline{b}$ and the degree of the minimal polynomial of $\underline{b}$ is called the *linear span* of $\underline{b}$.

The period of $f(x)$ is defined as the smallest integer t such that $f(x) \mid x^t - 1$. Let $G_{\mathbb{Z}_r}(f)$ represent the set consisting of sequences with characteristic polynomial $f(x)$. It is easy to see that for any $\underline{b} \in G_{\mathbb{Z}_r}(f)$, the period of $\underline{b}$ is a factor of the period of $f(x)$.

Let $V(\mathbb{Z}_r) = \{(h_0, h_1, \ldots) \mid h_i \in \mathbb{Z}_r\}$. Then $V(\mathbb{Z}_r)$ is a linear space over $\mathbb{Z}_r$ and $G_{\mathbb{Z}_r}(f)$ is a subspace of $V(\mathbb{Z}_r)$.

Fact 1 *$G_{\mathbb{Z}_r}(f)$ is a linear space over $\mathbb{Z}_r$.*

C. Trace Representation

Let $\mathrm{Tr}_p^{p^k}(x) = x + x^p + \cdots + x^{p^{k-1}}$ be a trace function from $\mathbb{F}_{p^k}$ to $\mathbb{F}_p$.

Fact 2 *Let $f(x)$ be an irreducible polynomial over $\mathbb{F}_p$ of degree n and γ be a root of $f(x)$ in an extension field of $\mathbb{F}_p$. Then for any sequence $\underline{b} = \{b_i\} \in G_{\mathbb{F}_p}(f)$, there exists some $\beta \in \mathbb{F}_{p^n}$ such that*

$$b_i = \mathrm{Tr}_p^{p^n}(\beta\gamma^i), \ i = 0, 1, \ldots$$

For the theory of linear recursive sequences over rings, the reader is referred to [4] and [20]. For the treatment of linear recursive sequences over finite fields the reader is referred to [6] and [12].

2 Construction and Basic Properties of LFSR Sequences over E

In this section, we will introduce a concept of linear (inhomogeneous) recurring sequences of points in the group of points on an elliptic curve over $\mathbb{F}_q$ and discuss some basic properties on the representation and period.

Definition 1 *Let E be an elliptic curve over $\mathbb{F}_q$ of order r, $f(x) = x^n - c_{n-1}x^{n-1} - \cdots - c_1 x - c_0 \in \mathbb{Z}_r[x]$ be a monic polynomial over $\mathbb{Z}_r$ and $\underline{P} = \{P_k\}$ where $P_k = (x_k, y_k) \in E$. For fixed $c_{-1} \in \mathbb{Z}_r$ and given $(n+1)$-tuple $(Q, P_0, \ldots, P_{n-1}) \in E^{n+1}$, if the sequence satisfies the following linear recursive relation:*

$$P_{n+k} = \sum_{i=0}^{n-1} c_i P_{i+k} + c_{-1} Q, \ k = 0, 1, \ldots,$$

then $\underline{P}$ is called an nth-order linear recursive sequence over E (generated by $f(x)$)*, or traditionally, an* nth-order linear feedback shift register [6] (LFSR) sequence over the elliptic curve E (generated by $f(x)$)*, or simply* EC sequence *for short; $f(x)$ is called a* characteristic polynomial *of $\underline{P}$ over $\mathbb{Z}_r$ and $(Q, P_0, P_1, \ldots, P_{n-1})$ is called an initial state of $\underline{P}$.*

We denote $G_E(f)$ the set consisting of all EC sequences over E with characteristic polynomial $f(x)$.

Theorem 1 *For any* $\underline{\mathbf{P}} = \{P_k\} \in G_E(f)$ *with initial state* $(\mathcal{O}, P_0, P_1, \ldots, P_{n-1})$, *the elements of* $\underline{\mathbf{P}}$ *can be represented by*

$$P_k = \sum_{i=0}^{n-1} b_{k,i} P_i, \;\; k = 0, 1, \ldots$$

where for each j *with* $0 \le j \le n-1$, $\underline{\mathbf{b}}_j = \{b_{k,j}\} \in G_{\mathbf{Z}_r}(f)$ *with initial states given by* $b_{k,j} = \delta_{k,j}$ *for* $0 \le k \le n-1$.

Proof: We have

$$P_{n+k} = \sum_{i=0}^{n-1} c_i P_{i+k}, \;\; c_i \in \mathbb{Z}_r, \;\; k = 0, 1, \cdots$$

Suppose

$$P_k = \sum_{i=0}^{n-1} b_{k,i} P_i, \;\; b_{k,i} \in \mathbb{Z}_r, \;\; k = 0, 1, \ldots$$

Obviously, initial conditions are determined by the terms $b_{k,j}$ with $0 \le k, j \le n-1$, where

$$b_{k,j} = \delta_{k,j} \quad \text{for } 0 \le k, j \le n-1.$$

Then,

$$P_k = \sum_{i=0}^{n-1} c_i P_{k-n+i} = \sum_{i=0}^{n-1} c_i \sum_{j=0}^{n-1} b_{k-n+i,j} P_j = \sum_{j=0}^{n-1} \left(\sum_{i=0}^{n-1} c_i b_{k-n+i,j} \right) P_j.$$

Therefore, by comparing coefficients, we can take

$$b_{k,j} = \sum_{i=0}^{n-1} c_i b_{k-n+i,j}.$$

Thus $\underline{\mathbf{b}}_j \in G_{\mathbf{Z}_r}(f)$. □

From Theorem 1, if the initial state of $\underline{\mathbf{P}} = \{P_k\} \in G_E(f)$ is given as $(\mathcal{O}, s_0 P, s_1 P, \ldots, s_{n-1} P)$, $s_i \in \mathbb{Z}_r$, then

$$P_k = \sum_{j=0}^{n-1} b_{k,j}(s_j P) = \left(\sum_{j=0}^{n-1} b_{k,j} s_j \right) P = a_k P, \;\; k = 0, 1, \ldots \quad (1)$$

where

$$a_k = \sum_{j=0}^{n-1} b_{k,j} s_j, \;\; k = 0, 1, \ldots$$

Define $\underline{\mathbf{a}} = \{a_k\}$. This yields $\underline{\mathbf{a}} = \sum_{j=0}^{n-1} s_j \underline{\mathbf{b}}_j$. Since $\underline{\mathbf{b}}_j \in G_{\mathbb{Z}_r}(f)$ and from Fact 1, $G_{\mathbb{Z}_r}(f)$ is a linear space, therefore $\underline{\mathbf{a}} = \sum s_j \underline{\mathbf{b}}_j \in G_{\mathbb{Z}_r}(f)$. From (1), P_k is computed by multiplying P by a_k (mod t), where t is the order of P. For this special initial state, we define

$$\underline{\mathbf{a}}P = \{a_i P\}_{i \geq 0} \tag{2}$$

and $G_P(f) = \{\underline{\mathbf{a}}P \mid \underline{\mathbf{a}} \in G_{\mathbb{Z}_r}(f)\}$. We call the EC sequence $\underline{\mathbf{a}}P$ as *a single point EC sequence.*

Let $V_E = \{(a_0, a_1, \ldots) | a_i \in E\}$, then $V(E)$ is a linear space over $\mathbb{Z}_r$. Let $\{\mathcal{O}, P_0, P_1, \ldots, P_{n-1}\}$ be the initial state of the sequence $\underline{\mathbf{P}}$. From Theorem 1 and equation (2), we have

$$\underline{\mathbf{P}} = \sum_{j=0}^{n-1} \underline{\mathbf{b}}_j P_j \text{ or } \underline{\mathbf{P}} = \sum_{j=0}^{n-1} \underline{\mathbf{P}}_j,$$

where $\underline{\mathbf{P}}_j = \{b_{k,j} P_j\}_{i \geq 0}$. That is, an n-th order linear recursive sequence over the points of E with characteristic polynomial $f(x)$ and initial state $(\mathcal{O}, P_0, P_1, \ldots, P_{n-1})$ can be decomposed into the sum of n initial state points, each multiplied by a scalar factor which comes from an n-th order LFSR sequence over $\mathbb{Z}_r$ with characteristic polynomial $f(x)$.

Corollary 1 *The period of $\{P_k\}$ is a factor of the period of $f(x)$.*

Proof: Let per(f) be the smallest integer t such that $f(x) | x^t - 1$. It is easy to see that per$(\underline{\mathbf{b}}_j)$ | per(f). Let $t = \text{per}(f)$, then $b_{k+t,j} = b_{k,j}$, hence

$$P_{k+t} = \sum_{j=0}^{n-1} b_{k+t,j} P_j = \sum_{j=0}^{n-1} b_{k,j} P_j = P_k.$$

Therefore t is a period of $\{P_i\}$. □

In the following, we restrict ourselves into special cases where either the elliptic curves has a prime order, that is, $|E| = r$ is a prime p, or that a given point $P \in E$ has a prime order p, with $f(x)$ being an irreducible polynomial over $\mathbb{F}_p$.

Lemma 1 *Let $f(x)$ be an irreducible polynomial of degree 2 over $\mathbb{F}_p$ and γ be a root of $f(x)$ in an extension field of $\mathbb{F}_p$.*

1. *The order of E over $\mathbb{F}_q$ is p. For any $\underline{\mathbf{P}} \in G_E(f)$, the elements of $\underline{\mathbf{P}}$ can be represented by*

$$P_k = \text{Tr}_p^{p^2}(\alpha\gamma^k) P_1 + \text{Tr}_p^{p^2}(\beta\gamma^k) P_0$$

where

$$\alpha = \frac{-1}{\gamma^p - \gamma} \text{ and } \beta = \frac{-\gamma}{\gamma^p - \gamma},$$

both as elements in the quadratic extension of $\mathbb{Z}_p$.

2. Let $P \in E$ with order p and $P_1 = tP$ and $P_0 = sP$, $t, s \in \mathbb{Z}_p$, then

$$P_k = \mathrm{Tr}_p^{p^2}((t\alpha + s\beta)\gamma^k)P, \ k = 0, 1, \ldots$$

Proof: Note that the two roots of $f(x)$ are γ and γ^p. From Theorem 1,

$$P_k = b_{k,0}P_0 + b_{k,1}P_1,$$

where $\underline{\mathbf{b}}_j = \{b_{k,j}\} \in G_{\mathbb{F}_p}(f)$, $j = 0, 1$. Since $f(x)$ is irreducible, according to Fact 2, $\underline{\mathbf{b}}_j$ has the following trace representation:

$$\begin{aligned} b_{k,0} &= \mathrm{Tr}_p^{p^2}(\rho_0\gamma^k) = \rho_0\gamma^k + \rho_0^p\gamma^{pk} \\ b_{k,1} &= \mathrm{Tr}_p^{p^2}(\rho_1\gamma^k) = \rho_1\gamma^k + \rho_1^p\gamma^{pk}, \ \rho_i \in \mathbb{F}_{p^2}, \ i = 0, 1 \end{aligned}$$

with initial conditions

$$\begin{pmatrix} b_{0,0} \\ b_{1,0} \end{pmatrix} = \begin{pmatrix} 1 \\ 0 \end{pmatrix} \text{ and } \begin{pmatrix} b_{0,1} \\ b_{1,1} \end{pmatrix} = \begin{pmatrix} 0 \\ 1 \end{pmatrix}.$$

Solving, we have

$$\rho_0 = \frac{\gamma^p}{\gamma^p - \gamma} \text{ and } \rho_1 = \frac{-1}{\gamma^p - \gamma}.$$

Therefore

$$\begin{aligned} P_k &= b_{k,1}P_1 + b_{k,0}P_0 \\ &= \left(\frac{-\gamma^k}{\gamma^p - \gamma} + \frac{\gamma^{pk}}{\gamma^p - \gamma}\right)P_1 + \left(\frac{\gamma^p}{\gamma^p - \gamma}\gamma^k + \frac{-\gamma}{\gamma^p - \gamma}\gamma^{pk}\right)P_0 \\ &= \mathrm{Tr}_p^{p^2}(\alpha\gamma^k)P_1 + \mathrm{Tr}_p^{p^2}(\beta\gamma^k)P_0. \end{aligned}$$

□

In a more general case where $f(x)$ is an irreducible polynomial of degree n over $\mathbb{F}_p$, we have the following results.

Theorem 2 *Let $f(x)$ be an irreducible polynomial of degree n over $\mathbb{F}_p$ and γ be a root of $f(x)$ in an extension field of $\mathbb{F}_p$.*

1. The order of E over $\mathbb{F}_q$ is p. For any $\underline{\mathbf{P}} \in G_E(f)$, with initial state $(\mathcal{O}, P_0, P_1, \ldots, P_{n-1})$, the elements of $\underline{\mathbf{P}}$ can be represented as

$$P_k = \sum_{i=0}^{n-1} \mathrm{Tr}_p^{p^n}(\alpha_i\gamma^k)P_i,$$

where $\alpha_i \in \mathbb{F}_{p^n}$, $i = 0, 1, \ldots n-1$, with $(\alpha_0, \alpha_1, \ldots, \alpha_{n-1})$ given by the first row of B^{-1} where

$$B = \begin{bmatrix} 1 & 1 & \cdots & 1 \\ \gamma & \gamma^p & \cdots & \gamma^{p^{n-1}} \\ \vdots & \vdots & \ddots & \vdots \\ \gamma^{n-1} & \gamma^{(n-1)p} & \cdots & \gamma^{(n-1)p^{n-1}} \end{bmatrix}. \quad (3)$$

2. *Let $P \in E$ with order p and $P_i = s_i P$ for $i = 0, 1, \ldots n-1$ and $s_i \in \mathbb{Z}_p$, then*

$$P_k = \mathrm{Tr}_p^{p^n}(\beta\gamma^k)P, \tag{4}$$

where

$$\beta = \sum_{i=0}^{n-1} s_i \mathrm{Tr}_p^{p^n}(\alpha_i).$$

Proof: Using the similar approach as in the proof of Lemma 1, we have for $0 \le i \le n-1$,

$$b_{k,i} = \mathrm{Tr}_p^{p^n}(\alpha_i \gamma^k), \; k = 0, 1, \ldots \tag{5}$$

with $0 \le i \le n-1$. From Theorem 1,

$$b_{k,i} = \delta_{k,i}, \; 0 \le i, k \le n-1. \tag{6}$$

Let $M = (b_{ki})_{n \times n}$. From (5), we have $M = AB$ where

$$A = \begin{bmatrix} \alpha_0 & \alpha_1 & \cdots & \alpha_{n-1} \\ \alpha_0^p & \alpha_1^p & \cdots & \alpha_{n-1}^p \\ \vdots & \vdots & \ddots & \vdots \\ \alpha_0^{p^{n-1}} & \alpha_1^{p^{n-1}} & \cdots & \alpha_{n-1}^{p^{n-1}} \end{bmatrix},$$

and B is given by (3). From (6), we have $M = I$ where I is the $n \times n$ identity matrix. Thus $AB = I$. The assertion is established. □

Note : For a single point EC sequence associated with P, the representation in (4) requires the subgroup generated by the point P has prime order p, instead of requiring the order of E to be prime.

3 Binary Sequences Obtained from EC Sequences

In this section, we give the construction of a class of periodic binary sequences obtained from EC sequences, and derive the linear span for some special characteristic polynomials. We first introdyce two operators on E which map the points of E to $\mathbb{F}_2^2$ and $\mathbb{F}_2$. Let

$$\tau : E \to \mathbb{F}_2, \text{ and } \delta : E \to \mathbb{F}_2^2$$

defined as follows: for $P \in E$, if $P = \mathcal{O}$, then

$$\tau(P) = 0 \text{ and } \delta(P) = (0, 0).$$

Otherwise, let $P = (x, y)$, if E is an elliptic curve over $\mathbb{F}_{2^l}$, then

$$\tau(P) = \mathrm{Tr}_2^{2^l}(y) \text{ and } \delta(P) = (\mathrm{Tr}_2^{2^l}(x), \mathrm{Tr}_2^{2^l}(y));$$

and if E is an elliptic curve over $\mathbb{F}_q$ where q is a prime > 3, then

$$\tau(P) = h(y) \text{ and } \delta(P) = (h(x), h(y)),$$

where $h(x)$ is a function which takes the *most significant bit* of an integer a modulo q:

$$h(a) = \begin{cases} 0, & 0 \leq a \leq \frac{q-1}{2} \\ 1, & \text{otherwise.} \end{cases}$$

Let $|E| = r$, $f(x)$ be a monic polynomial over $\mathbb{Z}_r$ and $\underline{\mathbf{P}} \in G_E(f)$. Recall that $\underline{\mathbf{P}} = \{P_i\}_{i\geq 0}$ with $P_i = (x_i, y_i)$. Define *binary sequences* $\underline{\mathbf{u}} = \{u_i\}$ and $\underline{\mathbf{s}} = \{s_i\}$ as follows:

$$u_i = \tau(P_i), \; i = 0, 1, \ldots$$
$$(s_{2i}, s_{2i+1}) = \delta(P_i), \; i = 0, 1, \ldots$$

We call $\underline{\mathbf{u}}$ an EC bit sequence of y-coordinate (associated with $f(x)$) and $\underline{\mathbf{s}}$ an EC bit sequence (associated with $f(x)$) respectively.

Remark 1 *While $\underline{\mathbf{s}}$ seems to be more efficient over $\underline{\mathbf{u}}$ due to a higher output rate, $\underline{\mathbf{s}}$ is however not desirable for non-supersingular curves E over $\mathbb{F}_{2^l}$. For a non-supersingular curve E over $\mathbb{F}_{2^l}$,*

$$E : y^2 + xy = x^3 + c_2x^2 + c_6, \; c_2, c_6 \in \mathbb{F}_{2^l},$$

from [19, Proposition 91], we know that a point (x, y) on the curve is a double of another point, that is, $(x, y) = 2(z, w)$ for some $(z, w) \in E$ if an only if

$$\mathrm{Tr}_2^{2^l}(x) = \mathrm{Tr}_2^{2^l}(c_2).$$

Hence the subsequence $\{s_{2i}\}_{i\geq 0}$ would be $(0000\cdots)$, $(1111\cdots)$, $(1010\cdots)$, *or* $(0101\cdots)$.

Remark 2 *$\underline{\mathbf{s}}$ is the EC sequence studied in [5] with E being a supersingular curve over $\mathbb{F}_{2^l}$ where l is odd and $f(x) = x - 1$. We will study the $0-1$ distribution of $\underline{\mathbf{s}}$ when l is even in Section 4.*

Example 1 (curve defined on $\mathbb{F}_q$). Consider the curve $E_1 : y^2 = x^3 + x + 4$ over $\mathbb{F}_{23}$. This curve has order 29 and is cyclic. Here $q = 23$ and take $p = 29$. Let $P = (4, 7)$ be a point on E_1. The points, together with the map δ, is listed below.

Choose $f(x) = x^2 - 28x - 26 \in \mathbb{Z}_{29}[x]$. Taking an initial state $Q = \mathcal{O}$, $P_0 = P$ and $P_1 = 2P$, then

$$P_k = 28P_{k-1} + 26P_{k-2} = a_kP, \; k = 2, 3, \ldots \tag{7}$$

where

$$a_k = 28a_{k-1} + 26a_{k-2}, \; k = 2, 3, \ldots$$

i	iP	$\delta(iP)$	i	iP	$\delta(iP)$	i	iP	$\delta(iP)$	i	iP	$\delta(iP)$
1	(4, 7)	(0, 0)	9	(9, 12)	(0, 1)	16	(0, 21)	(0, 1)	23	(1, 12)	(0, 1)
2	(10, 18)	(0, 1)	10	(11, 9)	(0, 0)	17	(14, 5)	(1, 0)	24	(8, 15)	(0, 1)
3	(13, 11)	(1, 0)	11	(17, 9)	(1, 0)	18	(17, 14)	(1, 1)	25	(15, 17)	(1, 1)
4	(15, 6)	(1, 0)	12	(14, 18)	(1, 1)	19	(11, 14)	(0, 1)	26	(13, 12)	(1, 1)
5	(8, 8)	(0, 0)	13	(0, 2)	(0, 0)	20	(9, 11)	(0, 0)	27	(10, 5)	(0, 0)
6	(1, 11)	(0, 0)	14	(22, 5)	(1, 0)	21	(18, 14)	(1, 1)	28	(4, 16)	(0, 1)
7	(7, 20)	(0, 1)	15	(22, 18)	(1, 1)	22	(7, 3)	(0, 0)	29	$\mathcal{O}$	(0, 0)
8	(18, 9)	(1, 0)									

Table 1.

which is an LFSR sequence over $\mathbb{Z}_{29}$ with initial condition $(a_0, a_1) = (1, 2)$. Relation (7) gives the EC sequence

$$\underline{\mathbf{P}} = (P, 2P, 24P, 28P, 16P, 16P, 23P, 16P, 2P, 8P, 15P, 19P, 23P, 7P, \\ 11P, 26P, 28P, 10P, 22P, 6P, 15P, 25P, 17P, 24P, 12P, \ldots)$$

and the binary bit sequence $\underline{\mathbf{u}}$ obtained from $\underline{\mathbf{P}}$ by applying the map τ is

$$\underline{\mathbf{u}} = (0, 1, 1, 1, 1, 1, 1, 1, 1, 0, 1, 1, 1, 1, 0, 1, 1, 0, 0, 0, 1, 1, 0, 1, 1, \ldots).$$

One can obtain the elements of $\underline{\mathbf{u}}$ by looking up Table 1.

Example 2 (curve defined on $\mathbb{F}_{2^l}$). Let t be a root of the polynomial $g(x) = x^5 + x^2 + 1$ and $\mathbb{F}_{2^5}$ be the field generated by t. Consider the non-supersingular curve $E_2 : y^2 + xy = x^3 + t^9x^2 + t^{15}$ over $\mathbb{F}_{2^5}$. This curve has order 26. Let $P = (t, t^7)$ be a point on E_2. The points, together with the map δ, is listed below.

i	iP	$\delta(iP)$	i	iP	$\delta(iP)$	i	iP	$\delta(iP)$
1	(t, t^7)	(0, 0)	10	(t^{24}, t^{17})	(1, 1)	19	(t^{14}, t^{10})	(0, 1)
2	(t^{21}, t)	(1, 0)	11	(t^2, t^{18})	(0, 1)	20	(t^9, t^{29})	(1, 0)
3	(t^{30}, t^{29})	(0, 0)	12	(t^{22}, t^5)	(1, 1)	21	(t^{23}, t^{25})	(0, 0)
4	$(t^{26}, 1)$	(1, 1)	13	$(0, t^{23})$	(0, 0)	22	(t^{26}, t^{28})	(1, 0)
5	(t^{23}, t^{28})	(0, 0)	14	(t^{22}, t^4)	(1, 0)	23	$(t^{30}, t^{16}$	(0, 0)
6	(t^9, t^{17})	(1, 1)	15	(t^2, t^{11})	(0, 1)	24	(t^{21}, t^9)	(1, 1)
7	(t^{14}, t^{20})	(0, 1)	16	(t^{24}, t^8)	(1, 0)	25	(t, t^{28})	(0, 0)
8	$(1, t^{27})$	(1, 0)	17	(t^{28}, t^2)	(0, 0)	26	$\mathcal{O}$	(0, 0)
9	(t^{28}, t^{30})	(0, 0)	18	$(1, t^6)$	(1, 1)			

Table 2.

Take $R = 2P$, then R has order 13. Choose $f(x) = x^2 - 12x - 11 \in \mathbb{Z}_{13}[x]$. Taking an initial state $Q = \mathcal{O}$, $P_0 = R$ and $P_1 = 2R$, the single point EC sequence $\underline{\mathbf{P}} = \{P_k\}$ is given by

$$P_k = 12P_{k-1} + 11P_{k-2} = a_kR, \; k = 2, 3, \ldots \tag{8}$$

where

$$a_k = 12a_{k-1} + 11a_{k-2},\ k = 2, 3, \ldots$$

and hence $\{\underline{\mathbf{a}}\} \in \mathbb{Z}_{13}(f)$. With the initial state $(a_0, a_1) = (1, 2)$, recurrence (8) gives the EC sequence

$$\underline{\mathbf{P}} = (R, 2R, 9R, \mathcal{O}, 8R, 5R, 5R, 11R, 5R, 12R, 4R, 11R, 7R, 10R, 2R, 4R, 5R, \mathcal{O}, 3R, 10R, 10R, 9R, 10R, 11R, \ldots)$$

and the binary bit sequence $\underline{\mathbf{u}}$ obtained from $\underline{\mathbf{P}}$ by applying the map τ is

$$\underline{\mathbf{u}} = (0,1,1,0,0,1,1,0,1,1,0,0,0,0,0,1,0,1,0,1,0,0,1,0,0,\ldots).$$

by looking up from Table 2. Note that the mapping of the x-coordinate resulted in the subsequence $(1111\ldots)$.

For general E and f, the periods and linear spans of EC bit sequences are not easy to determine. However, by a result from Chan and Games [3], we can obtain the linear span of single point EC bit sequences under certain conditions.

Let E be an elliptic curve over $\mathbb{F}_q$, $P \in E$ with order p, $f(x)$ be a primitive polynomial of degree n over $\mathbb{F}_p$, and $\underline{\mathbf{P}} = \{P_i\} \in G_P(f)$ where $P_i = a_i P$ where $\underline{\mathbf{a}} = \{a_k\} \in G_{\mathbb{F}_p}(f)$ is an m-sequence over $\mathbb{F}_p$ generated by $f(x)$. Let $\underline{\mathbf{s}}$ and $\underline{\mathbf{u}}$ be the EC bit sequences formed by mappings τ and δ respectively. I.e.,

$$u_i = \tau(P_i) = \tau(a_i P),\ i = 0, 1, \ldots \tag{9}$$

Let $d = \frac{p^n - 1}{p-1}$ and $\underline{\mathbf{u}}_j = \{u_{id+j}\}_{i \geq 0}$, $0 \leq j < d$. Denote $\mathrm{LS}(\underline{\mathbf{w}})$ to be the linear span of a periodic sequence $\underline{\mathbf{w}}$.

Theorem 3 *With the above notations and assumptions, the linear span of* $\underline{\mathbf{u}}$ *is given by*

$$\mathrm{LS}(\underline{\mathbf{u}}) = d \cdot \mathrm{LS}(\underline{\mathbf{u}}_j)$$

for any j *with* $\underline{\mathbf{u}}_j \neq 0$. *It is noted that all non-zero* $\underline{\mathbf{u}}_j$*'s are cyclic shifts of each other.*

In order to prove Theorem 3, we need a lemma from [3].

Lemma 2 (Chan and Games [3]) *With the notations above. let* $\rho : \mathbb{F}_p \to \mathbb{F}_2$ *be a function from* $\mathbb{F}_p$ *to* $\mathbb{F}_2$, *and a binary sequence* $\underline{\mathbf{u}} = \{b_i\}$ *obtained by* $b_i = \rho(a_i)$, $i = 0, 1, \ldots$.

Then, the linear span of $\underline{\mathbf{b}}$, $\mathrm{LS}(\underline{\mathbf{b}})$, *is given by*

$$\mathrm{LS}(\underline{\mathbf{b}}) = d \cdot \mathrm{LS}(\underline{\mathbf{b}}_j)$$

for any j *such that* $\underline{\mathbf{b}}_j = \{b_{id+j}\}_{i \geq 0} \neq \underline{0}$, $0 \leq j < d$. *It is noted that all non-zero* $\underline{\mathbf{b}}_j$*'s are shift equivalent.*

Proof: (of Theorem 3) We construct ρ as follows. Let η be a map from $\mathbb{F}_p$ to E defined by

$$\eta : c \to cP(\in E),\ \forall c \in \mathbb{F}_p, \tag{10}$$

and ρ be the composite of η and τ:

$$\rho = \tau \circ \eta : c \to \tau(\eta(c)) = \tau(cP),\ \forall c \in \mathbb{F}_p. \tag{11}$$

Thus ρ is a map from $\mathbb{F}_p$ to $\mathbb{F}_2$. Combining with (9), we have

$$u_i = \tau(a_i P) = \tau(\eta(a_i)) = \rho(a_i),\ i = 0, 1, \ldots$$

Applying Lemma 2, the assertion is established. □

Corollary 2 *The linear span of* $\underline{\mathbf{s}}$*,* $\mathrm{LS}(\underline{\mathbf{s}})$ *is bounded by*

$$\mathrm{LS}(\underline{\mathbf{s}}) \geq \frac{\mathrm{LS}(\underline{\mathbf{u}})}{2}.$$

Proof: Let $m(x)$ and $t(x)$ be the minimal polynomial of $\underline{\mathbf{s}}$ and $\underline{\mathbf{u}}$ respectively. Then $m(x^2)$ is a characteristic polynomial of $\underline{\mathbf{u}}$. Thus $t(x) \mid m(x^2)$. Hence

$$\mathrm{LS}(\underline{\mathbf{u}}) \leq 2\mathrm{LS}(\underline{\mathbf{s}}).$$

The result follows. □

Corollary 3 *The period of* $\underline{\mathbf{u}}$*,* $\mathrm{per}(\underline{\mathbf{u}})$*, is given by*

$$\mathrm{per}(\underline{\mathbf{u}}) = d \cdot \mathrm{per}(\underline{\mathbf{u}}_j)$$

for any $0 \leq j < d$ *such that* $\underline{\mathbf{u}}_j \neq 0$*.*

Proof: Note that $\underline{\mathbf{u}}$ can be considered as a binary sequence obtained from and m-sequence $\{a_i\}$ through the map $\rho = \tau \circ \eta$ defined by (10) and (11). According to the proof of Lemma 2 in [5], the result is immediate. □

Remark 3 *Theorem 3 and Corollary 3 show that the linear span and period of* $\underline{\mathbf{u}}$ *are determined by a shorter sequence* $\underline{\mathbf{u}}_j \neq 0$*. The results only depend on the order of the point* P *and the characteristic polynomial* $f(x)$ *and do not depend on* $\mathbb{F}_q$ *from which the coordinates of the points are taken. We will use the sequences obtained from equations (10) and (11) to explain this remarkable result. The study of linear span and period of* $\underline{\mathbf{u}}_j$ *is in progress.*

Example 3 (curve defined on $\mathbb{F}_{23}$*).* With the sequences in example 1, note that $f(x) = x^2 - 28x - 26$ is primitive in $\mathbb{Z}_{29}$. Let $d = \frac{29^2-1}{29-1} = 30$. Let

$\{\underline{\mathbf{a}}_j\} = \{a_{id+j}\}_{i\geq 0}$, and $\underline{\mathbf{P}}_j = \{P_{id+j}\}_{i\geq 0}$ for $0 \leq j < d$. Then we have

$$\begin{aligned}
\underline{\mathbf{a}}_0 &= (1,3,9,27,23,11,4,12,7,21,5,15,16,19,28,26,20,2,6,18,25,\\
&\quad 17,22,8,24,14,13,10,1,3,9,\ldots),\\
\underline{\mathbf{P}}_0 &= (P,3P,9P,27P,23P,11P,4P,12P,7P,21P,5P,15P,16P,19P,\\
&\quad 28P,26P,20P,2P,6P,18P,25P,17P,22P,8P,24P,14P,13P,10P,\\
&\quad P,3P,9P,\ldots),\\
\underline{\mathbf{u}}_0 &= (0,0,1,0,1,0,0,1,1,1,0,1,1,1,1,1,0,1,0,1,1,0,0,0,1,0,0,0,\\
&\quad 0,0,1,0,\ldots).
\end{aligned}$$

Using the Berlekamp-Massey algorithm [2,13], the linear span of $\underline{\mathbf{u}}_0$ is 15, and by directly checking, the period of $\underline{\mathbf{u}}_0$ is 28. Thus, by Theorem 3 and Corollary 3,

$$\mathrm{LS}(\underline{\mathbf{u}}) = d\cdot \mathrm{LS}(\underline{\mathbf{u}}_0) = 30\cdot \mathrm{LS}(\underline{\mathbf{u}}_0) = 450 \text{ and } \mathrm{per}(\underline{\mathbf{u}}) = 840.$$

Example 4 (curve defined on $\mathbb{F}_{25}$). With the sequences from example 2. Since $f(x) = x^2 - 12x - 11$ is primitive over $\mathbb{Z}_{13}$, therefore $d = \frac{13^2-1}{13-1} = 14$. We have

$$\begin{aligned}
\underline{\mathbf{a}}_0 &= (1,2,4,8,3,6,12,11,9,5,10,7,1,2,4,8,\ldots),\\
\underline{\mathbf{P}}_0 &= (R,2R,4R,8R,3R,6R,12R,11R,9R,5R,10R,7R,R,2R,\ldots),\\
\underline{\mathbf{u}}_0 &= (0,1,0,0,1,1,1,0,1,1,0,0,0,1,0,0\ldots).
\end{aligned}$$

Similarly, applying the Berlekamp-Massey algorithm, the linear span of $\underline{\mathbf{u}}_0$ is 7, by directly checking, the period of $\underline{\mathbf{u}}_0$ is 12. By Theorem 3 and Corollary 3, $\mathrm{LS}(\underline{\mathbf{u}}) = d\cdot \mathrm{LS}(\underline{\mathbf{u}}_0) = 98$ and $\mathrm{per}(\underline{\mathbf{u}}) = 168$.

Remark 4 *The proofs of Theorem 3 and Corollary 3 do not involve any properties of the map τ (from E to $\mathbb{F}_2$). Hence, the map τ can be replaced by any map from E to $\mathbb{F}_2$, and the results in Theorem 3 and Corollary 3 will still hold. Therefore, τ can be considered as a map from $\mathbb{F}_q$ to $\mathbb{F}_2$.*

Remark 5 *If $q = p$, that is, the curve defined over $\mathbb{F}_p$ has order p, in this case the discrete logarithm problem over such curves is solved by Samaev, Smart, and Satoh and Araki [17,18,16].*

4 Binary Sequences Obtained from Supersingular Elliptic Curves with $f(x) = x - 1$

For the case $n = 1$, we take $f(x) = x - 1, c_{-1} = 1$ and $Q = P_0$. Then we have

$$P_k = P_{k-1} + P_0,\ k = 1,2,\ldots.$$

Thus

$$P_k = (k+1)P_0,\ k = 0,1,\ldots \tag{12}$$

Let $\underline{u}$ and $\underline{s}$ be the sequences constructed as in Section 3 from $\underline{P}$ given by (12). In [5], Gong, Berson and Stinson studied $\underline{s}$ when E is a supersingular elliptic curve over $\mathbb{F}_{2^l}$ and $\underline{s}$ is referred to as an elliptic curve pseudorandom sequence. The $0-1$ distribution, period and linear span of $\underline{s}$ was also analysed in [5] when l is odd. The bound for the $0-1$ distribution of $\underline{s}$ where E is an elliptic curve over $\mathbb{F}_{2^l}$ was discussed in [9]. We have proved that the linear span of $\underline{u}$ in some cases is at least a quarter of the order of E, for which the proof is given in a separate paper [7].

In the following, we will discuss the $0-1$ distribution of $\underline{s}$ when l is even. First we introduce some notations. Let $H(\underline{\mathbf{w}})$ represent the Hamming weight of a binary vector. Let $\underline{\mathbf{w}} = (w_0, w_1, \ldots, w_{k-1})$, then the reverse of $\underline{\mathbf{w}}$ is $\overleftarrow{\underline{\mathbf{w}}} = (w_{k-1}, \ldots w_1, w_0)$. We also denote a periodic sequence with period t (not necessarily be the smallest period) as a vector of dimension t. Given the field $\mathbb{F}_{2^l}$ where l is even. Define

$$\mathrm{Te}(\alpha) = \alpha + \alpha^{2^2} + \alpha^{2^4} + \cdots + \alpha^{2^{l-2}}.$$

Let δ, ω be such that $\mathrm{Te}(\delta) \neq 0$ and $\mathrm{Tr}(\omega) = 1$. When l is even, there are 7 isomorphism classes [14]. Three of the 7 isomorphic classes of curves are:

1. $y^2 + y = x^3 + \delta x$,
2. $y^2 + y = x^3$,
3. $y^2 + y = x^3 + \omega$.

Let the order of the curve be $2v + 1$, and let

$$A = (s_0, s_2, \ldots, s_{2(v-1)}),\ B = (s_1, s_3, \ldots, s_{2v-1}),$$

and $\underline{s}' = (s_0, s_1, \ldots, s_{4v-1})$. Note that $s_{4v} = s_{4v+1} = 0$ since (s_{4v}, s_{4v+1}) is mapped from $\mathcal{O}$. We have $H(\underline{s}') = H(\underline{s})$. $\underline{s}'$ is the exact sequence discussed in [5].

Theorem 4 *$\underline{s}$ can be arranged into an array*

$$S = \begin{bmatrix} A & \overleftarrow{A} & 0 \\ B & \overleftarrow{B} & 0 \end{bmatrix}.$$

The weight of the $S_0 = (A, \overleftarrow{A})$ in all three cases is one of the five possible values: $2^{l-1} + \{0, \pm 2^{\frac{l}{2}+1},$ or $\pm 2^{\frac{l}{2}}\}$.

Proof: The negative of a point (x, y) on a supersingular curve is $(x, y + c_3)$. In all the three cases, $c_3 = 1$, hence $\mathrm{Tr}_2^{2^l}(y + c_3) = \mathrm{Tr}_2^{2^l}(y)$. Therefore we must have

$$S = \begin{bmatrix} A & \overleftarrow{A} & 0 \\ B & \overleftarrow{B} & 0 \end{bmatrix}.$$

For the second half of the statement, first consider the case $y^2+y=x^3+\delta x$. This curve has order 2^l+1. Let

$$n_{i,j}=|\{x\in\mathbb{F}_{2^l}:\mathrm{Tr}_2^{2^l}(x)=i,\mathrm{Tr}_2^{2^l}(x^3+\delta x)=j\}|.$$

We proceed to count the quantities of $n_{i,j}$ to find the weight distribution of the sequence. We have $n_{0,0}+n_{0,1}=|\{x\in\mathbb{F}_{2^l}:\mathrm{Tr}_2^{2^l}(x)=0\}|=2^{l-1}$. Similarly, $n_{1,0}+n_{1,1}=2^{l-1}$. Also,

$$\begin{aligned}n_{0,0}+n_{1,0}&=|\{x\in\mathbb{F}_{2^l}:\mathrm{Tr}_2^{2^l}(x^3+\delta x)=0\}|\\&=|\{x\in\mathbb{F}_{2^l}:y^2+y=x^3+\delta x\text{ has a solution}\}|\\&=(|E|-1)/2=2^{l-1}.\end{aligned}$$

and

$$\begin{aligned}&n_{0,0}+n_{1,1}-(n_{0,1}+n_{1,0})\\&=|\{x\in\mathbb{F}_{2^l}:\mathrm{Tr}_2^{2^l}(x+x^3+\delta x)=0\}|\\&\quad-|\{x\in\mathbb{F}_{2^l}:\mathrm{Tr}_2^{2^l}(x+x^3+\delta x)=1\}|\\&=2|\{x\in\mathbb{F}_{2^l}:\mathrm{Tr}_2^{2^l}(x+x^3+\delta x)=0\}|-|\{x\in\mathbb{F}_{2^l}\}|\\&=2|\{x\in\mathbb{F}_{2^l}:y^2+y=x^3+(\delta+1)x\text{ has a solution}\}|-2^l\\&=\begin{cases}0 & \text{if }\mathrm{Te}(\delta+1)\neq 0,\\ \pm 2^{\frac{l}{2}+1} & \text{otherwise.}\end{cases}\end{aligned}$$

When $\mathrm{Te}(\delta+1)\neq 0$, we have $(n_{0,0},n_{0,1},n_{1,0},n_{1,1})=(2^{l-2},2^{l-2},2^{l-2},2^{l-2})$, then $H(S_0)=2H(A)=2n_{1,0}=2^{l-1}$. If $\mathrm{Te}(\delta+1)=0$, then $H(S_0)=2H(A)=2n_{1,0}=2^{l-1}\pm 2^{\frac{l}{2}}$.

The cases for the curves $y^2+y=x^3$ and $y^2+y=x^3+\omega$ are similar, except now the curve orders are as follows:

$$\#E=\begin{cases}2^l+1-2^{\frac{l}{2}+1} & \text{if } m\equiv 0 \pmod 4,\\ 2^l+1+2^{\frac{l}{2}+1} & \text{if } m\equiv 2 \pmod 4.\end{cases}\quad\text{for } E: y^2+y=x^3,$$

$$\#E=\begin{cases}2^l+1+2^{\frac{l}{2}+1} & \text{if } m\equiv 0 \pmod 4,\\ 2^l+1-2^{\frac{l}{2}+1} & \text{if } m\equiv 2 \pmod 4.\end{cases}\quad\text{for } E: y^2+y=x^3+\omega.$$

Similarly, the weight of S_0 must be one of the five possible values $2^{l-1}+\{0,\pm 2^{\frac{l}{2}+1}, \text{ or } \pm 2^{\frac{l}{2}}\}$. □

References

1. P.H.T. Beelen, J.M. Doumen, Generating pseudo-random sequences using linear recurrency relations on elliptic curves, *Sixth International Conference on Finite Fields and Applications*, abstract, `http://www.iztapalapa.uam.mx/fq6/`, May 2001.

2. E. R. Berlekamp, *Algebraic coding theory*, McGraw-Hill, New York, 1968.
3. A.H. Chan and R.A. Games, On the Linear Span of Binary Sequences Obtained from Q-Ary M-Sequences, Q Odd, *IEEE Trans. Information Theory*, Vol 36, No. 3, 548-552, 1990.
4. Z. Dai, M. Huang, Linear Complexity and the Minimal Polynomial of Linear Recurring Sequences over $\mathbb{Z}_m$, *Systems Science and Mathematical Sciences*, Vol. 4, No. 1, 51-54, Feb. 1991.
5. G. Gong, T. Berson and D. Stinson, Elliptic Curve Pseudorandom Sequence Generator, *Proceedings of the Sixth Annual Workshop on Selected Areas in Cryptography*, LNCS 1758, Springer-Verlag, 34-48, 1999.
6. S.W. Golomb, *Shift Register Sequences*, Holden-Day, Inc., 1967. Revised edition, Aegean Park Press, 1982.
7. G. Gong, C. Lam, Randomness of Several Elliptic Curve Sequences, *preprint*, 2001.
8. S. Hallgren, Linear congruential generators over elliptic curves, CS 94-143, Dept. of Comp. Sci., Carnegie Mellon University, 1994.
9. E.E. Mahassni and I. Shparlinski, On the Uniformity of Distribution of Congruential Generators over Elliptic Curves, *preprint*, November 2000.
10. B. Kaliski, *A Pseudo-Random Bit Generator Based on Elliptic Logarithms*, Advances in Cryptology–Crypto '86, Lecture Notes in Computer Science, LNCS 263(1986), Springer-Verlag, 84-103.
11. N. Koblitz, Elliptic Curve Cryptosystems, *Mathematics of Computation (48)*, 203-209, 1987.
12. R. Lidl and H. Niederreiter, *Finite Fields*, Encyclopedia of Mathematics and its Applications, Volume 20, Addison-Wesley, 1983, Theorem 3.35, 97-98.
13. J. L. Massey, *Shift-register synthesis and BCH decoding*, IEEE Trans. Information Theory, January 1969.
14. A. J. Menezes, *Elliptic Curve Public Key Cryptosystems*, Kluwer Academic Publishers, 1993.
15. V. Miller, Uses of Elliptic Curves in Cryptography, *Advances in Cryptology – CRYPTO '85*, LNCS 218, Springer-Verlag, 417-426, 1986.
16. T. Satoh and K. Araki, Fermat quotients and the polynomial time discrete log algorithm for anomalous elliptic curves, *Commentarii Mathematici Universitatis Sancti Paul*, 47 (1998), 81-92.
17. I. Samaev, Evaluation of discrete logarithms in a group of p-torsion points of an elliptic curve in characteristic p, *Mathematics of Computation*, 67 (1998), 353-356.
18. N. Smart, The discrete logarithm problem on elliptic curves of trace one, *Journal of Cryptology*, 12 (1999), 193-196.
19. J. Solinas, Improved Algorithms for Arithmetic on Anomalous Binary Curves, *CACR Technical Reports*, CORR 99-46, University of Waterloo, 1999.
20. M. Ward, The characteristic number of a sequence of integers satisfying a linear recursion relation, Trans. Amer. Math. Soc., 33 (1931), 153-165.

On the Distinctness of Decimations of ℓ-Sequences

Mark Goresky[1], Andrew Klapper[2], and Ram Murty[3]

[1] The Institute for Advanced Study, School of Mathematics,
www.math.ias.edu/~goresky/.
Research partially supported by NSF grant #0002693.
[2] Dept. of Computer Science, University of Kentucky, Lexington, KY, 40506
and The Institute for Advanced Study, School of Mathematics.
e-mail: klapper@cs.uky.edu.
Research partially sponsored by NSF grant #9980429.
[3] Math. Dept., Queen's University, Kingston, Ontario, K7L 3N6, Canada
and The Institute for Advanced Study, School of Mathematics.
e-mail: murty@mast.queensu.ca.

Summary. Maximal length Feedback with Carry Shift Register sequences (or ℓ-sequences) have several remarkable statistical properties. Among them is the property that the arithmetic correlations between any two cyclically distinct decimations are precisely zero. It is open, however, whether all such pairs of decimations are indeed cyclically distinct. In this paper we show that a d-fold decimation of an ℓ-sequence with prime connection number q is cyclically distinct from the original ℓ-sequence if (1) $d = -1$, (2) $d = (q+1)/2$, for $q \equiv 1 \bmod 8$ or (3) $d < q/(c\log(q)^4)$ for a certain constant c.

1 Introduction

Let q be a prime integer such that 2 is primitive modulo q (thus q is a power of a prime). The class of sequences known as ℓ-sequences can be described in three ways [2]. They are the output sequences from maximal period feedback with carry shift registers (FCSRs) with connection number q. They are the 2-adic expansions of rational numbers r/q, where $\gcd(r, q) = 1$. And they are sequences $a_i = (A2^{-i} \bmod q) \bmod 2$, where $\gcd(A, q) = 1$. By $(x \bmod q) \bmod 2$ we mean first reduce x modulo q to a number between 0 and $q-1$, then reduce the result modulo 2. The period of such an ℓ-sequence is $\phi(q)$. In this paper we are concerned only with the case when q is prime.

These sequences are known to have several good statistical properties similar to those of m-sequences. In particular, they form families of sequences with remarkable *arithmetic autocorrelations*. That is, if $\mathbf{a} = a_0, a_1, \cdots$ and $\mathbf{b} = b_0, b_1, \cdots$ are two binary sequences with period L, their arithmetic correlation $C_{\mathbf{a},\mathbf{b}}(\tau)$ with shift τ is defined as follows. First shift $\mathbf{b}$ by τ positions. Next form the associated 2-adic numbers

$$\alpha = \sum_{i=0}^{\infty} a_i 2^i \quad \text{and} \quad \beta_\tau = \sum_{i=0}^{\infty} b_{i+\tau} 2^i .$$

Then find their difference $\gamma = \alpha - \beta_\tau$ as a 2-adic number. Finally, count the number of zeros minus the number of ones in one period of the periodic part of γ (which may only be eventually periodic). A pair of sequences has *ideal arithmetic correlations* if $C_{\mathbf{a},\mathbf{b}}(\tau) = 0$ for every τ.

Let $\mathbf{a}$ be an ℓ-sequence with prime connection number q, and let S be the set of d-fold decimations $\mathbf{a}^d$ of $\mathbf{a}$,

$$\mathbf{a}^d = a_0, a_d, a_{2d}, \cdots,$$

with $\gcd(q-1, d) = 1$.

Theorem 1. *(Klapper and Goresky [3]) Every pair of cyclically distinct sequences in S has ideal arithmetic correlations.*

On the basis of extensive experimental evidence (covering all primes less than 50,000), we made the following conjecture.

Conjecture 1 *If $q > 13$ is prime and* $\mathbf{a}$ *is an ℓ-sequence based on q, then every pair of decimations of* $\mathbf{a}$ *is cyclically distinct.*

The conjecture implies that the set of decimations of $\mathbf{a}$ is a family of $\phi(q-1)$ sequences with period $q-1$ and ideal arithmetic correlations. This result would be in stark contrast to the case of ordinary correlations, where there are well known upper bounds on the size of a family of sequences with bounded correlations. In this paper we report on progress toward proving Conjecture 1. We do not have a complete proof, but in many cases can show that decimations are distinct. By more careful search we have shown the conjecture is true for $d < 2{,}000{,}000$.

First note that $\mathbf{a}^d$ is cyclically distinct from $\mathbf{a}^e$ if and only if $\mathbf{a}^{de^{-1}}$ is cyclically distinct from $\mathbf{a}$ (where e^{-1} is computed modulo $q-1$). In this paper we show that for various d, the decimation $\mathbf{a}^d$ is cyclically distinct from $\mathbf{a}$. Throughout we assume $q > 13$. In particular, we show that if $\mathbf{a}$ is an ℓ-sequence with prime connection integer $q > 13$, then $\mathbf{a}^d$ is cyclically distinct from $\mathbf{a}$ if

1. $d = -1$ (equivalently, $d = q-2$);
2. $q \equiv 1 \bmod 8$ and $d = (q+1)/2$; or
3. $d \le \pi^4 q/(2^{18}(1+\ln(q-3))^4)$.

2 $d = -1$

Suppose that $d = -1$, so $\mathbf{a}^d$ is the reverse of $\mathbf{a}$. In this case we can use an earlier result on the distribution of patterns of consecutive bits in an ℓ-sequence. By an occurence of an s-bit sequence $C = c_0, c_1, \cdots, c_{s-1}$ in a periodic sequence $\mathbf{a}$ of period L we mean an index $0 \le i < L$ such that $c_j = a_{i+j}$ for $0 \le j < s$.

Theorem 2. *(Klapper and Goresky [2]) Let q be a power of a prime, and suppose that 2 is primitive modulo q. Let s be any nonnegative integer, and let C and D be s-bit sequences. Let* $\mathbf{a}$ *be an ℓ-sequence with connection integer q. Then the numbers of occurences of C and D in* $\mathbf{a}$ *differ by at most 2.*

It was also shown that the number of occurences of any bit pattern equals the number of occurences of its bitwise complement. Furthermore, an examination of the proof reveals that when q is prime the numbers of occurrences of C and D in $\mathbf{a}$ differ by at most 1. If $q = 2^t + e$ with $1 \le e < 2^t$, then this says that every t-bit pattern occurs either once or twice, and every $(t+1)$-bit pattern occurs either once or not at all. In fact we can say exactly which t-bit and $(t+1)$-bit patterns occur once and which occur twice. For any integer y, we denote by $[y]$ the reduced residue of y modulo 2^t. That is $y \equiv [y] \bmod 2^t$ and $0 \le [y] < 2^t$.

Lemma 1. *If $C = (c_0, c_1, \cdots, c_{t-1})$ is a t-bit pattern, and $c = \sum_{i=0}^{t-1} c_i 2^i$, then C occurs twice in* $\mathbf{a}$ *if and only if $[-ec] < e$.*

Proof: The cyclic shifts of $\mathbf{a}$ are the 2-adic expansions of the rational numbers $-x/q$ with $0 < x < q$ [2]. Thus the occurences of the t-bit pattern C occurs correspond to the integers x such that $c \equiv -x/q \bmod 2^t$ and $0 < x < q$. The first condition is equivalent to $qc \equiv -x \bmod 2^t$, which is equivalent to $ec \equiv -x \bmod 2^t$, which is equivalent to $-ec \equiv x \bmod 2^t$. That is $[-ec] = x$.

If $[-ec] < e$, then $2^t + [-ec] < 2^t + e = q$, so we have two occurences of C. Conversely, if $[-ec] \ge e$, then any integer congruent to $[-ec]$ modulo 2^t must either be greater than or equal to q or less than 0. Thus there is only one occurence of C. $\Box$.

For example, if C is the all one pattern of length t, then $c = 2^t - 1$, so $[-ec] = e$, so this C occurs once. Also, the all zero pattern of length t, occurs once.

Theorem 3. *The decimation* $\mathbf{a}^{-1}$ *(reversal) is cyclically distinct from* $\mathbf{a}$.

Proof: Suppose to the contrary that $\mathbf{a}^{-1}$ is a cyclic shift of $\mathbf{a}$. Then for some k we have $a_i = a_{k-i}$ for all i.

Consider the pattern of t ones. It occurs once and is its own reverse. If it occurs at index i, then its reverse occurs at index $k - i - t + 1$, so we must have $i = k - i - t + 1$ or $t = k - 2i + 1$. In particular, k and t have opposite parity. Suppose that C is a $(t+1)$-bit pattern that is equal to its reverse. We claim that C cannot occur in $\mathbf{a}$. We know that C can occur at most once. Thus if it occurs at position j, then its reverse occurs at position $k - j - t$, so $t = k - 2j$. This would imply that k and t have the same parity, which is false. It follows that any $t+1$-bit pattern that equals its own reverse cannot occur in $\mathbf{a}$.

Suppose that C is a t-bit pattern that equals its own reversal and occurs just once in $\mathbf{a}$, say at index i. Then its reverse occurs at index $k - i - t + 1$, so we must have $i = k - i - t + 1$. That is, $i \in \{(k-t+1)/2, (k-t+q)/2\}$. Thus there are at most two such bit patterns. By the comments preceeding the theorem, there are precisely two, the all zero and all one t-bit patterns.

Now suppose that C is t-bit pattern and b is a bit such that the $(t+1)$-bit pattern C, b equals its own reverse. Then C, b cannot occur in **a**. Let b' denote the complement of b. Then C, b' can occur at most once in **a**, so C can also occur at most once. In particular, by Lemma 1 if $c = c_0 + 2c_1 + \cdots + 2^{t-1}c_{t-1}$, then $[-ec] \geq e$. This gives us a crank to turn. We let

$$e = \sum_{i=0}^{t-1} e_i 2^i,$$

where $e_0 = 1$ and $e_i \in \{0, 1\}$ for $i = 1, \cdots, t-1$.

1. Let $C = 1, 0, 0, \cdots, 0$. Then C occurs once in **a** and $c = 1$, so $e \leq [-e] = 2^t - e$, so $e \leq 2^{t-1}$. In fact
$$e < 2^{t-1} \tag{1}$$
since q is odd.
2. Let $C = 0, 0, 1, 1, \cdots, 1, 0$. Thus C occurs once in **a** and $c = 2^{t-1} - 4$, so $e \leq [-(2^{t-1} - 4)e] = [2^{t-1} + 4e]$ since e is odd. We have $2^{t-1} + 4e < 2^{t-1} + 4 \cdot 2^{t-1} < 3 \cdot 2^t$ by equation (1). Suppose that $2^{t-1} + 4e \geq 2^{t+1}$. Then $[2^{t-1} + 4e] = 2^{t-1} + 4e - 2 \cdot 2^t \geq e$. It follows that $e \geq 2^{t-1}$, a contradiction. Therefore $2^{t-1} + 4e < 2^{t+1}$ so
$$e < \frac{3}{4} 2^{t-1}. \tag{2}$$
3. Let $C = 0, 1, \cdots, 1, 0$. Then C occurs twice in **a** and $c = 2^{t-1} - 2$. Thus $e > [-(2^{t-1} - 2)e] = [2^{t-1} + 2e]$ since e is odd. We have $2^{t-1} + 2e < 5 \cdot 2^{t-2} < 2 \cdot 2^t$ by equation (2). Thus $[2^{t-1} + 2e] \in \{2^{t-1} + 2e, 2^{t-1} + 2e - 2^t = 2e - 2^{t-1}\}$. If $[2^{t-1} + 2e] = 2^{t-1} + 2e$ then $e > 2^{t-1} + 2e$, which is impossible. Thus $[2^{t-1} + 2e] = 2e - 2^{t-1}$ and $2^{t-1} + 2e \geq 2^t$. That is, $e \geq 2^{t-2}$. Since e is odd, we have
$$e \geq 2^{t-2} + 1. \tag{3}$$
4. Let $C = 1, 0, 1, 1, \cdots, 1, 0, 1$. Then C occurs twice in **a** and $c = 2^t - 2^{t-2} - 3$. Thus $e > [-(2^t - 2^{t-2} - 3)e] = [3e + 2^{t-2}(1 + 2e_1)]$. Suppose that $e_1 = 1$. Then $3e + 2^{t-2}(1 + 2e_1) = 3e + 3 \cdot 2^{t-2} \geq 3(2^{t-2} + 1) + 3 \cdot 2^{t-2} > 2^t$ by equation (3). Also, $3e + 3 \cdot 2^{t-2} < (9/2)2^{t-2} + 3 \cdot 2^{t-2} = (15/2)2^{t-2} < 2 \cdot 2^t$ by equation (2). Thus $e > [3e + 3 \cdot 2^{t-2}] = 3e + 3 \cdot 2^{t-2} - 2^t = 3e - 2^{t-2}$, so $2^{t-2} > 2e$, which contradicts equation (3). Therefore $e_1 = 0$.
5. Let $C = 0, 0, 1, 1, \cdots, 1, 0, 0$. Then C occurs twice in **a** and $c = 2^{t-2} - 4$. Thus $e > [-(2^{t-2} - 4)e] = [4e - 2^{t-2}]$. We have $4e - 2^{t-2} \geq 4(2^{t-2} + 1) - 2^{t-2} \geq 0$ by equation (3). If $4e - 2^{t-2} < 2^t$, then $e > [4e - 2^{t-2}] = 4e - 2^{t-2}$.

This would imply that $2^{t-2} > 3e$, which contradicts equation (3). Thus $2^t \leq 4e - 2^{t-2}$ and so

$$e \geq \frac{5}{4}2^{t-2}. \tag{4}$$

6. Let $C = 0,0,0,1,1,\cdots,1,0,0$. Then C occurs once in **a** and $c = 2^{t-2} - 8$. Thus $e \leq [-(2^{t-2}-8)e] = [8e - 2^{t-2}]$. We have $8e - 2^{t-2} > 5\cdot 2^{t-1} - 2^{t-2} \geq 2 \cdot 2^t$ by equation (4). Moreover, $8e - 2^{t-2} < 3 \cdot 2^t$ by equation (2). Thus $e \leq [8e - 2^{t-2}] = 8e - 2^{t-2} - 2 \cdot 2^t = 8e - 7 \cdot 2^{t-2}$ and so

$$e > \frac{9}{7}2^{t-2}. \tag{5}$$

7. Let $C = 1,1,0,1,1,\cdots,1,0,1$. Then C occurs once in **a** and $c = 2^t - 2^{t-2} - 5$. Thus $e \leq [-(2^t - 2^{t-2} - 5)e] = [5e + 2^{t-2}]$. We have $5e + 2^{t-2} > (45/7)2^{t-2} + 2^{t-2} = (52/28)2^{t-2} > 2^t$ by equation (5). Suppose $5e + 2^{t-2} \geq 2 \cdot 2^t$. Then $e < [5e + 2^{t-2}] \leq 5e + 2^{t-2} - 2 \cdot 2^t = 5e - 7 \cdot 2^{t-2}$, so $7 \cdot 2^{t-2} < 4e$, which contradicts equation (2). Thus $5e + 2^{t-2} < 2 \cdot 2^t$, so

$$e < \frac{7}{5}2^{t-2}. \tag{6}$$

8. Let $C = 0,1,0,1,1,\cdots,1,0,1$. Then C occurs once in **a** and $c = 2^t - 2^{t-2} - 6$. Thus $e \leq [-(2^t - 2^{t-2} - 6)e] = [6e + 2^{t-2}]$. We have $6e + 2^{t-2} > (54/7)2^{t-2} + 2^{t-2} = (61/28)2^{t-2} > 2 \cdot 2^t$ by equation (5). Thus $e \leq [6e + 2^{t-2}] \leq 6e + 2^{t-2} - 2 \cdot 2^t = 6e - 7 \cdot 2^{t-2}$. It follows that

$$e > \frac{7}{5}2^{t-2}. \tag{7}$$

Equation (7) contradicts equation (6), proving the theorem if t is large enough that all the choices of C used are valid strings of length t. This holds as long as $t \geq 6$, so $q \geq 64$. For smaller values of q, the theorem can be verified by exhaustive search. □

3 Small *d*

More generally, by the third characterization of an ℓ-sequence, $\mathbf{a}^d$ is a cyclic permutation of **a** if and only if $(A2^{-id} \bmod q) \bmod 2 = (2^{-i} \bmod q) \bmod 2$ for every i, for some A. By the primitivity of 2 modulo q, this holds if and only if $(Ax^d \bmod q) \bmod 2 = (x \bmod q) \bmod 2$ for every x. That is, the map $x \mapsto Ax^d$ permutes the set E of even elements modulo q.

Using this point of view and deep results from analytic number theory [1] we obtain the following results.

Theorem 4. *For q sufficiently large, the decimation $\mathbf{a}^d$ is cyclically distinct from* **a** *if*

$$d \leq \frac{\pi^4 q}{2^{18}(1 + \ln(q-3))^4}.$$

The remainder of this section consists of a proof of this result. We define

$$f_E(x) = \begin{cases} 1 \text{ if } x \in E \\ 0 \text{ otherwise.} \end{cases}$$

We proceed using Fourier analysis of f_E. Let ζ be a primitive complex qth root of 1 and let

$$\hat{f_E}(b) = \frac{1}{q}\sum_{c=0}^{q-1} f_E(c)\zeta^{-bc}$$

be the cth Fourier coefficient of f_E. Thus by Fourier inversion,

$$f_E(a) = \sum_{b=0}^{q-1} \hat{f_E}(b)\zeta^{ba}.$$

Suppose as above that $x \mapsto Ax^d$ permutes E. Then

$$\sum_{x\in E} f_E(Ax^d) = \sum_{b=0}^{q-1} \hat{f_E}(b) \sum_{x\in E} \zeta^{bAx^d}.$$

The left hand side equals $|E| = (q+1)/2$. Let

$$S_b = \sum_{x\in E} \zeta^{bAx^d} = \sum_{x=0}^{(q-1)/2} \zeta^{bA2^d x^d}$$

If $b = 0$, then $\hat{f_E}(b) = (q+1)/(2q)$ and $S_b = |E| = (q+1)/2$. Thus

$$\frac{q^2-1}{4q} = |\sum_{b=1}^{q-1} \hat{f_E}(b)S_b| \leq (\sum_{b=1}^{q-1} |\hat{f_E}(b)|) \max_{b\neq 0} |S_b|.$$

Lemma 2. *We have*

$$\sum_{b=1}^{q-1} |\hat{f_E}(b)| \leq 1 + \frac{1}{2}\ln(\frac{q-3}{2}) = \frac{1+\ln(q-3)}{2}.$$

Proof: We have

$$\begin{aligned} \hat{f_E}(b) &= \frac{1}{q}\sum_{i=0}^{q-1} f_E(i)\zeta^{-bi} \\ &= \frac{1}{q}\sum_{i \text{ even}} \zeta^{-bi} \\ &= \frac{1}{q}\sum_{i=0}^{(q-1)/2} \zeta^{-2bi} \end{aligned}$$

$$= \frac{1}{q}\frac{(\zeta^{-2b})^{(q+1)/2} - 1}{\zeta^{-2b} - 1}$$

$$= \frac{1}{q}\frac{\zeta^{-b} - 1}{\zeta^{-2b} - 1}$$

$$= \frac{1}{q}\frac{\zeta^{b/2}}{\zeta^{-b/2} + \zeta^{b/2}}$$

$$= \frac{\zeta^{b/2}}{2q\cos(b\pi/q)}.$$

Also, for $0 \le x \le q/2$ we have $0 \le \cos(x\pi/q) \le 1 - 2x/q$. Thus

$$\sum_{b=1}^{q-1} |\hat{f}_E(b)| = \sum_{b=0}^{q-1} \frac{1}{2q|\cos(b\pi/q)|}$$

$$= \frac{1}{q}\sum_{b=0}^{(q-1)/2} \frac{1}{\cos(b\pi/q)}$$

$$\le \sum_{b=0}^{(q-1)/2} \frac{1}{q - 2b}$$

$$= \sum_{b=0}^{(q-1)/2} \frac{1}{2b+1}$$

$$\le 1 + \frac{1}{2}\int_1^{\frac{p-3}{2}} \frac{1}{x}dx$$

$$= 1 + \frac{1}{2}\ln\left(\frac{q-3}{2}\right),$$

which proves the lemma. □

Thus

$$\frac{q^2-1}{2q} \le (1 + \ln(q-3)) \max_{b\neq 0} |S_b|. \tag{8}$$

Furthermore, sums of the form S_b have been estimated by Davenport and Heilbronn [1]. Their results can be tightened by the following lemma which improves their first lemma. We let

$$S(a,b) = \sum_{i=0}^{q-1} \zeta^{ai^d + bi}.$$

Lemma 3. *For $b \neq 0$ and $d > 1$ we have*

$$\sum_{b=0}^{q-1} |S(1,b)|^4 \leq \frac{(d-1)q^4}{q-1} \leq dq^3$$

and

$$|S(a,b)|^4 \leq \frac{(d-1)q^4}{q-1} \leq dq^3.$$

Proof: Let $R(w,t)$ denote the number of solutions to the congruences

$$\begin{aligned} x + y &\equiv w \bmod q \\ x^d + y^d &\equiv t \bmod q. \end{aligned}$$

By solving for y using the first equation, we can reduce this to a single equation of degree $d-1$ (since d is odd). Such an equation has at most $d-1$ solutions unless $w = t = 0$, when it has q solutions. Also, $R(w,t) = 0$ if one but not both of w and t is zero. Thus we have

$$\begin{aligned} \sum_{w=0}^{q-1}\sum_{t=0}^{q-1} R^2(w,t) &\leq (d-1)\sum_{w=1}^{q-1}\sum_{t=1}^{q-1} R(w,t) + q^2 \\ &= dq^2. \end{aligned}$$

Therefore

$$\begin{aligned} &\sum_{a=0}^{q-1}\sum_{b=0}^{q-1} |S(a,b)|^4 \\ &= \sum_{a=0}^{q-1}\sum_{b=0}^{q-1}\sum_{x_1=0}^{p-1}\sum_{x_2=0}^{p-1}\sum_{x_3=0}^{p-1}\sum_{x_4=0}^{p-1} \zeta^{a(x_1^d+x_2^d-x_3^d-x_4^d)+b(x_1+x_2-x_3-x_4)} \\ &= \sum_{\substack{x_1,x_2,x_3,x_4 \in \mathbf{Z}/(q) \\ x_1+x_2=x_3+x_4 \\ x_1^d+x_2^d=x_3^d+x_4^d}} q^2 \\ &= q^2 \sum_{w=0}^{q-1}\sum_{t=0}^{q-1} R^2(w,t) \\ &\leq dq^4. \end{aligned}$$

Furthermore, we do not need the terms for which $a = 0$. Arguing similarly we can show that

$$\sum_{b=0}^{q-1} |S(0,b)|^4 = q^4.$$

Thus

$$\sum_{a=1}^{q-1}\sum_{b=0}^{q-1}|S(a,b)|^4 \le dq^4 - q^4 = (d-1)q^4 \le d(q-1)q^3.$$

For any $\lambda \neq 0$, we have $S(1,b) = S(\lambda^d, \lambda b)$. Since the mapping $x \mapsto x^d$ is a permutation, we have

$$\begin{aligned}\sum_{b=0}^{q-1}|S(1,b)|^4 &= \frac{1}{q-1}\sum_{\lambda=1}^{q-1}|S(\lambda^d,\lambda b)|^4\\ &\le \frac{1}{q-1}\sum_{u=0}^{q-1}\sum_{v=0}^{q-1}|S(u,v)|^4\\ &\le \frac{(d-1)q^4}{q-1}\\ &\le dq^3.\end{aligned}$$

It follows also that for $a \neq 0$,

$$|S(a,b)|^4 \le \frac{(d-1)q^4}{q-1} \le dq^3.$$

The lemma follows. □

Davenport and Heilbronn gave estimates on certain exponential sums [1]. If we let

$$\sum_{x=0}^{q-1}\zeta^{f(x)+nx} = F(n),$$

where n is an integer, then their Lemma 4 says that for any m,

$$\sum_{x=0}^{m}\zeta^{f(x)} = \frac{m}{q}F(0) + O\left(\sum_{n=1}^{q-1}\frac{1}{n}(|F(n)| + F(-n)|)\right) + O(\ln(q)).$$

In our case $m = (q-1)/2$, $f(x) = Ax^d$, $F(0) = 0$ and $F(n) = S(1,n)$. By carefully examining Davenport and Heilbronn's proof, one sees that the constant on the first big-O is $2/\pi$ and the second big-O can be replaced by $4\ln(q) + 4$.

Recall that Hölder's inequality says that if for real vectors $x = (x_1, \cdots, x_r)$ and $y = (y_1, \cdots, y_r)$ and positive real numbers a and b with $(1/a) + (1/b) = 1$ we have

$$|\sum_{i=1}^{r} x_i y_i| \le (\sum_{i=1}^{r} x_i^a)^{1/a}(\sum_{i=1}^{r} y_i^b)^{1/b}. \tag{9}$$

We shall use this to obtain estimates on the sum $\sum_{n=1}^{q-1}(1/n)|S(1,n)|$.

Lemma 4. *Let b be any real number strictly greater than one. Then*

$$\sum_{n=1}^{r} \frac{1}{n^b} \leq \frac{b}{b-1}.$$

Proof: We have

$$\begin{aligned}\sum_{n=1}^{r} \frac{1}{n^b} &\leq 1 + \int_1^r \frac{1}{x^b}dx \\ &= 1 + \frac{1}{b-1} - \frac{1}{(b-1)r^{b-1}} \\ &\leq \frac{b}{b-1}\end{aligned}$$

proving the lemma. □

We use equation (9) with $x_n = |S(1,n)|$, $y_n = 1/n$, $a = 4$, and $b = 4/3$, so $b/(b-1) = 4$.

$$\begin{aligned}\sum_{n=1}^{q-1} \frac{1}{n}|S(1,n)| &\leq (\sum_{n=1}^{q-1} |S(1,n)|^4)^{1/4}(\sum_{n=1}^{q-1} \frac{1}{n^{4/3}})^{3/4} \\ &\leq \left(\frac{(d-1)q^4}{q-1}\right)^{1/4} \cdot 4^{3/4} \\ &= 2^{3/2}\frac{(d-1)^{1/4}q}{(q-1)^{1/4}},\end{aligned}$$

where the second line follows from Lemma 3 and Lemma 4. The same bound applies to the sum using $S(1,-n)$ in place of $S(1,n)$.

Therefore

$$\frac{q^2-1}{2q} \leq (1+\ln(q-3))\left(\frac{2^{7/2}(d-1)^{1/4}q}{\pi(q-1)^{1/4}} + \ln(q) + 4\right).$$

It can be shown that this implies that

$$\frac{q}{2} \leq (1+\ln(q-3))\left(\frac{2^{7/2}d^{1/4}q^{3/4}}{\pi}\right).$$

Theorem 4 follows from this. □

4 $d = (q+1)/2$

Now suppose $d = (q+1)/2$. Since d must be relatively prime to $q-1$, this only occurs if $(q+1)/2$ is odd, so $q \equiv 1 \bmod 4$ and $(q-1)/2$ is even.

Lemma 5. *Let c be any integer. If x is a square modulo q, then* $cx^d \equiv cx \bmod q$. *Otherwise,* $cx^d \equiv -cx \bmod q$.

Proof: Let α be primitive mod q. Then $-1 \equiv \alpha^{(q-1)/2} \equiv (\alpha^{(q-1)/4})^2 \bmod q$ is a square. Suppose $x \equiv u^2 \bmod q$. Then $cx^d \equiv cu^{(q+1)} \equiv cu^2 \equiv cx \bmod q$. Suppose x is not a square. Then $x \equiv \alpha u^2 \bmod q$ for some u. Therefore $cx^d \equiv c\alpha^d u^2 \equiv c\alpha^{(q-1)/2}x \equiv -cx \bmod q$. □

The following definition will be useful in analyzing parity.

Definition 1. Let $\sigma(x) = 1$ if x is even and $\sigma(x) = -1$ if x is odd.

If x is an integer, then by $x \bmod q$ we mean the reduced residue of x modulo q. That is, $x \bmod q$ is the integer between 0 and $q-1$ that is congruent to x modulo q.

Lemma 6. *If x is an arbitrary positive integer, then $\sigma(x) = \sigma(x \bmod q) \cdot (-1)^{\lfloor x/q \rfloor}$ and $\sigma((-x) \bmod q) = -\sigma(x)$*

Proof: The first statement follows from the fact that subtracting or adding q to an integer reverses its parity. Next suppose $0 \le x < q$. Then $(-x \bmod q) = q - x$ which proves the second statement for such x. The general case follows from this case and the first statement. □

Theorem 5. *Suppose that $q \equiv 1 \bmod 8$. Then the decimation $\mathbf{a}^{(q+1)/2}$ is cyclically distinct from* $\mathbf{a}$.

Proof: Suppose to the contrary that $x \mapsto cx^d \bmod q$ permutes $\{0, 2, \cdots, q-1\}$. Let $\mathrm{J}(x,q)$ be the Jacobi symbol of x over q. Then Lemma 5 says that for all $x \in \{0, 1, \cdots, q-1\}$,

$$\sigma(x) = \mathrm{J}(x,q)\sigma(cx \bmod q).$$

By multiplicativity we have

$$\begin{aligned}
\sigma(x) &= \mathrm{J}(x,q)\sigma(cx \bmod q) \\
&= \mathrm{J}(x,q)\mathrm{J}(cx,q)\sigma(c^2x \bmod q) \\
&= \mathrm{J}(cx^2,q)\sigma(c^2x \bmod q) \\
&= \mathrm{J}(c,q)\sigma(c^2x \bmod q).
\end{aligned}$$

Let $e \equiv c^2 \bmod q$. We consider two cases.

1. Suppose c is a square, so $\mathrm{J}(c,q) = 1$ and for all x we have $\sigma(x) = \sigma(ex \bmod q)$. We claim that $e = 1$. To see this we prove by induction on a that for $a = 1, 2, ..., q-1$ we have $e < q/a$. Note first that e must be odd because $\sigma(1) = \sigma(e)$. For the base case, let $a = 1$. We have $e < q = q/a$. For the induction case, suppose $e < q/(a-1)$. Then $ae < q + q/(a-1) \le 2q$, so $ae \bmod q$ is ae or $ae - q$. But $\sigma(ae \bmod q) = \sigma(a)$ which implies that $ae \bmod q$ is ae (remember, e is odd). That is, $ae < q$ and $e < q/a$, completing the induction. In particular, $e < q/q - 1$, so $e \le 1$, and therefore $e = 1$.
   This means $c = 1$ or $c = -1$. If $c = 1$, then for all x we have $\sigma(x) = \mathrm{J}(x,q)\sigma(x)$, which implies $\mathrm{J}(x,q) = 1$ for all x. This is impossible. If $c = -1$, then similarly $\mathrm{J}(x,q) = -1$ for all x. This is also impossible.

2. Suppose c is not a square. Then for all x we have $\sigma(x) = -\sigma(ex \bmod q) = \sigma((-ex) \bmod q) = \sigma((q-e)x \bmod q))$. The same argument as in the previous case shows that $q - e = 1$. But $q - e \equiv -e \bmod q$, so $e = -1$. That is, $c^2 = -1$, so $c = \pm 2^{(q-1)/4}$. Therefore c is a square, since $(q-1)/4$ is even. But this contradicts the assumption on c. □

5 Conclusions

We have shown that many decimations of an ℓ-sequence **a** are cyclically distinct from **a**. There is experimental evidence that all such decimations are cyclically distinct. We leave the complete proof as an open problem.

References

1. Davenport, H., and Heilbronn, H. (1936) On and Exponential Sum. Proc. London Math. Soc. **41**, 449-453.
2. Klapper, A. and Goresky, M. (1997) Feedback Shift Registers, Combiners with Memory, and 2-Adic Span. Jo. Crypt. **10**, 111-147.
3. Klapper, A. and Goresky, M. (1997) Arithmetic Cross-Correlations of FCSR Sequences, IEEE Trans. Info. Thy. **43**, 1342-1346.

On Binary Sequences of Period $n = p^m - 1$ with Optimal Autocorrelation *

Tor Helleseth[1] and Kyeongcheol Yang[2]

[1] Dept. of Informatics, University of Bergen, Høyteknologisenteret, N-5020 Bergen, Norway

[2] Dept. of Electronic and Electrical Engineering, Pohang University of Science and Technology, Pohang, Kyungbuk 790-784, Korea

Summary. Binary sequences of period $n = p^m - 1$ for an odd prime p are introduced in [4] by taking the characteristic sequence of the image set of the polynomial $(z+1)^d + az^d + b$ over the finite field F_{p^m} of p^m elements. It was shown in [4] that they are (almost) balanced and have optimal autocorrelation in the case where the polynomial can be transformed into the form $z^2 - c$. In this paper, we show that the sequences are (almost) balanced and have optimal autocorrelation in the case of $d = (p^m+1)/2$, $a = (-1)^{d-1}$ and $b = \pm 1$. Furthermore, we show that they are equivalent to the Lempel-Cohn-Eastman sequence in [2] in the balanced case. We also give a direct proof of the autocorrelation property of the Lempel-Cohn-Eastman sequence and discuss its linear complexity.

1 Introduction

Given a binary sequence $\{s(t)\}$ of period n, let D be the difference between the number of 1's and that of 0's in its period. $\{s(t)\}$ is said to be *balanced* if D is zero when n is even, and if D is one when n is odd. It is said to be *almost balanced* if D is 2. The number D will be referred to as the discrepancy of the sequence $\{s(t)\}$.

The periodic autocorrelation function $R(\tau)$ of $\{s(t)\}$ is defined by

$$R(\tau) = \sum_{t=0}^{n-1} (-1)^{s(t)+s(t+\tau)}$$

where the sum $t + \tau$ is computed mod n. Clearly, $R(0) = n$. It is also easily checked that $R(\tau) = n \pmod 4$ and

$$\sum_{\tau=0}^{n-1} R(\tau) = D^2.$$

* This work was supported in part by the Norwegian Research Council and by the BK21 Program of the Ministry of Education of Korea and the Com2Mac of Postech.

The sequence $\{s(t)\}$ of even period n is said to have *optimal autocorrelation* if

$$R(\tau) = \begin{cases} 0 \text{ or } -4 & \text{if } n = 0 \pmod 4, \\ 2 \text{ or } -2 & \text{if } n = 2 \pmod 4 \end{cases}$$

for any $\tau \neq 0 \pmod n$.

Let $\{s_1(t)\}$ and $\{s_2(t)\}$ be two binary sequences of period n. If there exists a constant τ such that $s_2(t) = s_1(t+\tau)$ for any t, then $\{s_2(t)\}$ is called a *cyclic shift* of $\{s_1(t)\}$. If there is an r with $(r,n) = 1$ such that $s_2(t) = s_1(rt)$ for all t, then $\{s_2(t)\}$ is called the *r-decimation* of $\{s_1(t)\}$. If $s_2(t) = s_1(t) + 1 \pmod 2$ for all t, then $\{s_2(t)\}$ is called the *complement* of $\{s_1(t)\}$. If $\{s_2(t)\}$ is a cyclic shift of any of decimations or complement of $\{s_1(t)\}$ or their combination, then they are said to be *equivalent*. Two equivalent sequences have the same distribution of autocorrelation and the same discrepancy.

For an odd prime p, let α be a primitive element of the finite field F of p^m elements and $F^* = F\backslash\{0\}$. For a nonempty S in F^*, let $\{s(t)\}$ be the sequence of length $n = p^m - 1$, defined by

$$s(t) = \begin{cases} 1 & \text{if } \alpha^t \in S, \\ 0 & \text{otherwise.} \end{cases}$$

The sequence $\{s(t)\}$ has period $n = p^m - 1$ and is called the *characteristic sequence* of S in F^*.

Binary sequences of period $n = p^m - 1$ for an odd prime p are introduced in [4] by taking the characteristic sequence of the image set of the polynomial $(z+1)^d + az^d + b$ over F. It was shown in [4] that they are (almost) balanced and have optimal autocorrelation in the case where the polynomial can be transformed into the form $z^2 - c$.

In this paper, we show that the sequences are (almost) balanced and have optimal autocorrelation in the case of $d = (p^m + 1)/2$, $a = (-1)^{d-1}$ and $b = \pm 1$. Furthermore, we show that they are equivalent to the Lempel-Cohn-Eastman sequence in [2]. We also give a direct proof of the autocorrelation property of the Lempel-Cohn-Eastman sequence and discuss its linear complexity.

2 Cyclotomic Numbers and Quadratic Character

From now on, we let F be the finite field of p^m elements and $F^* = F\backslash\{0\}$ and let α be a primitive element of F. For $i = 0$ and 1, let

$$C_i = \{\alpha^{2s+i} \mid s = 0, 1, \dots, (p^m - 1)/2 - 1\}.$$

In other words, C_0 (C_1, resp.) is the set of squares (non-squares, resp.) in F^*. For fixed i and j, the cyclotomic number (i,j) is defined to be the number of solutions to the equation

$$1 + z_i = z_j$$

where $z_i \in C_i$, $z_j \in C_j$. From the theory of cyclotomic numbers (cf. [6]), it is easy to determine (i, j) as in the following lemma.

Lemma 1. (Lemma 6, [6]) *The cyclotomic numbers are given as follows:*

(a) If $n \equiv 0 \pmod 4$, *then* $(0,0) = (p^m - 5)/4$; $(0,1) = (1,0) = (1,1) = (p^m - 1)/4$.

(b) If $n \equiv 2 \pmod 4$, *then* $(0,0) = (1,0) = (1,1) = (p^m - 3)/4$; $(0,1) = (p^m + 1)/4$.

Let I be the indicator function such that $I(0) = 1$ and $I(x) = 0$ for $x \neq 0$. Let χ be the quadratic character of F, defined by

$$\chi(z) = \begin{cases} +1 & \text{if } z \in C_0, \\ -1 & \text{if } z \in C_1, \\ 0 & \text{if } z = 0. \end{cases}$$

The sum $\Gamma(a) = \sum_{z \in F} \chi(az^2 + 1)$ can be easily determined using the theory of cyclotomic numbers.

Lemma 2. *For* $a \in F$,

$$\Gamma(a) = \begin{cases} p^m & \textit{if } a = 0, \\ -\chi(a) & \textit{otherwise.} \end{cases}$$

Proof. Clearly, $\Gamma(0) = p^m$. Assume that $a \neq 0$. First, we define the three numbers n_+, n_0, and n_- by

$$\begin{aligned} n_+ &= |\{z \in F \mid \chi(az^2 + 1) = +1\}|, \\ n_0 &= |\{z \in F \mid \chi(az^2 + 1) = 0\}|, \\ n_- &= |\{z \in F \mid \chi(az^2 + 1) = -1\}|. \end{aligned}$$

Note that $n_+ + n_0 + n_- = p^m$.

Case $a \in C_0$: Note that $z = 0$ is a solution to $\chi(az^2+1) = +1$. When z runs through F^*, az^2 takes each element in C_0 twice. Therefore, $n_+ = 1 + 2(0,0)$. Note that $\chi(az^2 + 1) = 0$ if and only if $az^2 + 1 = 0$ and that $az^2 + 1 = 0$ has exactly two solutions in F if and only if $-1 \in C_0$. Hence,

$$\begin{aligned} n_0 &= \begin{cases} 2 & \text{if } n = 0 \pmod 4, \\ 0 & \text{if } n = 2 \pmod 4 \end{cases} \\ &= 1 + (-1)^{n/2}. \end{aligned}$$

Since $n_- = p^m - (n_+ + n_0)$, we have

$$\begin{aligned} \Gamma(a) &= n_+ \cdot (+1) + n_0 \cdot 0 + n_- \cdot (-1) \\ &= 4(0,0) + 3 + (-1)^{n/2} - p^m \\ &= -1 \end{aligned}$$

and so $\Gamma(a) = -\chi(a)$.

Case $a \in C_1$: Similarly, it is easily shown that $n_+ = 2(1,0) + 1$ and $n_0 = 1 - (-1)^{n/2}$. Therefore, we have $\Gamma(a) = +1 = -\chi(a)$. □

More generally (cf. [5]),

Lemma 3. *For a, b and $c \in F$,*

$$\sum_{z\in F} \chi(az^2 + bz + c) = \begin{cases} I(b)\chi(c)p^m & \text{if } a = 0, \\ \chi(a)(p^m I(b^2 - 4ac) - 1) & \text{otherwise.} \end{cases}$$

3 Lempel-Cohn-Eastman Sequences

Let $S = \{\alpha^{2i+1} - 1 \mid i = 0, 1, \ldots, (p^m - 1)/2 - 1\}$. The characteristic sequence $\{s_L(t)\}$ of period $n = p^m - 1$ is defined by

$$s_L(t) = \begin{cases} 1 & \text{if } \alpha^t \in S, \\ 0 & \text{otherwise.} \end{cases}$$

This sequence is referred to as the Lempel-Cohn-Eastman Sequence. In [2], it was shown that $\{s_L(t)\}$ is balanced and has optimal autocorrelation.

Using the indicator function I and the quadratic character χ, $\{s_L(t)\}$ can be expressed in a closed form. First, we note that

$$\begin{aligned} s_L(t) = 1 &\Leftrightarrow \alpha^t = \alpha^{2i+1} - 1 \text{ for some } i \\ &\Leftrightarrow \chi(\alpha^t + 1) = -1 \text{ and } \alpha^t + 1 \neq 0. \end{aligned}$$

It is also easily checked that

$$1 + \chi(\alpha^t + 1) = 2(1 - s_L(t)) - I(\alpha^t + 1).$$

Therefore, we have

$$s_L(t) = \frac{1}{2}\left(1 - I(\alpha^t + 1) - \chi(\alpha^t + 1)\right). \tag{1}$$

It is possible to give a direct proof on the autocorrelation properties of $\{s_L(t)\}$ in the following, which is different from that in [2].

Theorem 4. *The autocorrelation of $\{s_L(t)\}$ is given as follows: If $n = 0 \pmod 4$, then*

$$R_L(\tau) = \begin{cases} -4 & \text{if } \chi(\alpha^\tau) = 1 \text{ and } \chi(1 - \alpha^\tau) = -1, \\ 0 & \text{otherwise.} \end{cases}$$

If $n = 2 \pmod 4$, then

$$R_L(\tau) = \begin{cases} 2 & \text{if } \chi(\alpha^\tau) = 1 \text{ and } \chi(1 - \alpha^\tau) = 1, \\ -2 & \text{otherwise.} \end{cases}$$

Proof. Since the ± 1 version of $\{s_L(t)\}$ is given by

$$(-1)^{s_L(t)} = 1 - 2s_L(t) = I(\alpha^t + 1) + \chi(\alpha^t + 1),$$

we have

$$\begin{aligned} R_L(\tau) &= \sum_{t=0}^{n-1} \left(I(\alpha^{t+\tau}+1) + \chi(\alpha^{t+\tau}+1)\right)\left(I(\alpha^t+1) + \chi(\alpha^t+1)\right) \\ &= \chi(1-\alpha^\tau) + \chi(1-\alpha^{-\tau}) + \sum_t \chi((\alpha^t+1)(\alpha^{t+\tau}+1)) \end{aligned}$$

by the properties of I and χ. For simple notation, let $z = \alpha^t$ and $a = \alpha^\tau$. Then the last term leads to

$$\begin{aligned} \sum_t \chi((\alpha^t+1)(\alpha^{t+\tau}+1)) &= \sum_{z\in F^*} \chi(az^2 + (a+1)z + 1) \\ &= \sum_{z\in F} \chi(az^2 + (a+1)z + 1) - \chi(1) \\ &= -\chi(a) - \chi(1) \end{aligned}$$

by Lemma 3. Therefore we have

$$\begin{aligned} R_L(\tau) &= \chi(1-a) + \chi(1-a^{-1}) - \chi(a) - \chi(1) \\ &= \chi(1-a)\left(1 + (-1)^{n/2}\chi(a)\right) - \chi(a) - \chi(1) \end{aligned}$$

which leads to the theorem. □

4 Sequences from the Image sets of Some Polynomials

In [4], it was shown that polynomials introduced in [3] and [1] can be generalized to generate binary sequences of period $p^m - 1$ with optimal autocorrelation for any prime p and an integer m. For $a, b \in F$ and a positive integer d, consider the subset of F^* given by

$$\{x \mid x = (z+1)^d + az^d + b, z \in F\}\backslash\{0\} \tag{2}$$

and its characteristic sequence $\{s(t)\}$. For $d = 2, 3, 4$ or $d = p^m - p^{m-1} - 1$, there are some combinations of p, a, and b such that the polynomial $(z+1)^d + az^d + b$ can be transformed ino the form $z^2 - c$. In these cases, the corresponding characteristic sequences are (almost) balanced binary sequences with optimal autocorrelation. In particular, they are all equivalent to the Lempel-Cohn-Eastman sequence in the balanced case.

For $d = (p^m + 1)/2$, $a = (-1)^{d-1}$ and $b = \pm 1$, it was shown in [4] that the corresponding characteristic sequence is balanced when $n = 0 \pmod 4$ and is almost balanced when $n = 2 \pmod 4$. Here we will give an expression of $\{s(t)\}$ and its derived sequences in a closed form and show that they have optimal autocorrelation.

For any $z \in F^*$, we have $z^{d-1} = \chi(z)$ since $d = (p^m-1)/2+1$. Therefore, the polynomial $f(z) = (z+1)^d+(-1)^{d-1}z^d+b$ over F can be simply expressed as

$$f(z) = (\chi(z+1) + \chi(-z))z + \chi(z+1) + b. \tag{3}$$

Lemma 5. *Let S_+ and S_- be the image sets in F^* defined by $f(z)$ with $b = +1$ and $b = -1$ in (3), respectively. Then*

$$S_+ = \{2\} \cup \{z \mid z/2 \in C_0, 1 - z/2 \in C_0\} \cup \{z \mid z/2 \in C_1, 1 - z/2 \in C_1\},$$
$$S_- = \{-2\} \cup \{z \mid z/2 \in C_0, 1 + z/2 \in C_0\} \cup \{z \mid z/2 \in C_1, 1 + z/2 \in C_1\}.$$

Furthermore,

$$|S_+| = |S_-| = \begin{cases} n/2 & \text{if } n = 0 \pmod 4, \\ n/2+1 & \text{if } n = 2 \pmod 4. \end{cases}$$

Proof. First, we consider S_+. Since $-1 \in C_0$ for $n = 0 \pmod 4$, we have

$$\begin{aligned} S_+ &= \{2\} \cup \{2z+2 \mid z \in C_0, 1+z \in C_0\} \cup \{-2z \mid z \in C_1, 1+z \in C_1\} \\ &= \{2\} \cup \{z \mid z/2 \in C_0, 1 - z/2 \in C_0\} \cup \{z \mid z/2 \in C_1, 1 - z/2 \in C_1\} \end{aligned}$$

and $|S_+| = 1 + (0,0) + (1,1) = (p^m-1)/2$. For $n = 2 \pmod 4$, we have $-1 \in C_1$. Therefore,

$$\begin{aligned} S_+ &= \{2\} \cup \{2z+2 \mid z \in C_1, 1+z \in C_0\} \cup \{-2z \mid z \in C_0, 1+z \in C_1\} \\ &= \{2\} \cup \{z \mid z/2 \in C_0, 1 - z/2 \in C_0\} \cup \{z \mid z/2 \in C_1, 1 - z/2 \in C_1\} \end{aligned}$$

and $|S_+| = 1 + (1,0) + (0,1) = (p^m+1)/2$. Similar works hold for S_-. □

Theorem 6. *Let $\{s_+(t)\}$ and $\{s_-(t)\}$ be the characteristic sequences of S_+ and S_-, respectively. Then*

$$s_+(t) = \frac{1}{2}\left(1 + I(1+\alpha^{-(t-\gamma)}) + (-1)^{n/2}\chi(1+\alpha^{-(t-\gamma)})\right),$$
$$s_-(t) = \frac{1}{2}\left(1 + I(1+\alpha^{-(t-\lambda)}) + (-1)^{n/2}\chi(1+\alpha^{-(t-\lambda)})\right)$$

where $\alpha^\gamma = -2$ and $\alpha^\lambda = 2$.

Proof. First, consider $s_+(t)$. Note that $s_+(t) = 1$ if and only if $\alpha^t \in S_+$. Then it can be eaily checked that

$$\begin{aligned} &(1+\chi(\alpha^t/2))\,(1+\chi(1-\alpha^t/2)) + (1-\chi(\alpha^t/2))\,(1-\chi(1-\alpha^t/2)) \\ &\qquad = \begin{cases} 4 & \text{if } \alpha^t \in S(1)\backslash\{2\}, \\ 2 & \text{if } \alpha^t = 2, \\ 0 & \text{otherwise} \end{cases} \end{aligned}$$

and this leads to

$$2+2\chi((1-\alpha^t/2)\alpha^t/2) = 4s_+(t) - 2I(\alpha^t - 2).$$

Note that $I(\alpha^t - 2) = I(1 - 2\alpha^t{-}t)$ and $\chi((1-\alpha^t/2)\alpha^t/2) = (-1)^{n/2}\chi(1 - 2\alpha^{-t})$. Therefore,

$$s_+(t) = \frac{1}{2}\left(1 + I(1-2\alpha^{-t}) + (-1)^{n/2}\chi(1-2\alpha^{-t})\right)$$

Similar works hold for the case $\{s_-(t)\}$. □

Corollary 7. *For $n = 0 \pmod 4$, the sequences $\{s_+(t)\}$ and $\{s_-(t)\}$ are equivalent to the Lempel-Cohn-Eastman sequence $\{s_L(t)\}$. More precisely,*

$$s_+(t) = 1 - s_L(\gamma - t),$$
$$s_-(t) = 1 - s_L(\lambda - t)$$

where $\alpha^\gamma = -2$ and $\alpha^\lambda = 2$.

Proof. It follows from and (1) and Theorem 6. □

Corollary 8. *For $n = 2 \pmod 4$, the sequences $\{s_+(t)\}$ and $\{s_-(t)\}$ are almost balanced and have optimal autocorrelation given by*

$$R_{s_+}(\tau) = R_{s_-}(\tau) = \begin{cases} 2 & \text{if } \chi(\alpha^\tau) = -1 \text{ and } \chi(1-\alpha^\tau) = -1, \\ -2 & \text{otherwise.} \end{cases}$$

Proof. The sequences $\{s_+(t)\}$ and $\{s_-(t)\}$ are almost balanced in the case of $n = 2 \pmod 4$ by Lemma 5. By Theorem 9 and Lemma 3, it can be easily checked that

$$R_{s_+}(\tau) = R_{s_-}(\tau) = \chi(a)\chi(1-a) - \chi(1-a) - \chi(a) - 1$$

where $a = \alpha^\tau$. □

Theorem 9. *Let $S_+^* = S_+\backslash\{2\}$ and $S_-^* = S_-\backslash\{-2\}$. Let $\{s_+^*(t)\}$ and $\{s_-^*(t)\}$ be their characteristic sequences, respectively. Then*

$$s_+^*(t) = \frac{1}{2}\left(1 - I(1+\alpha^{-(t-\gamma)}) + (-1)^{n/2}\chi(1+\alpha^{-(t-\gamma)})\right),$$
$$s_-^*(t) = \frac{1}{2}\left(1 - I(1+\alpha^{-(t-\lambda)}) + (-1)^{n/2}\chi(1+\alpha^{-(t-\lambda)})\right)$$

where $\alpha^\gamma = -2$ and $\alpha^\lambda = 2$.

Proof. In the case of S_+^*, it can be easily checked that

$$\left(1+\chi(\alpha^t/2)\right)\left(1+\chi(1-\alpha^t/2)\right) + \left(1-\chi(\alpha^t/2)\right)\left(1-\chi(1-\alpha^t/2)\right)$$
$$= 4s_+^*(t) + 2I(\alpha^t - 2).$$

The case S_+^* is also similar. □

Corollary 10. *For $n = 0 \pmod 4$, the sequences $\{s_+^*(t)\}$ and $\{s_-^*(t)\}$ are almost balanced and have optimal autocorrelation given by*

$$R_{s_+^*}(\tau) = R_{s_-^*}(\tau) = \begin{cases} -4 & \text{if } \chi(\alpha^\tau) = 1 \text{ and } \chi(1-\alpha^\tau) = 1, \\ 0 & \text{otherwise.} \end{cases}$$

Proof. The sequences $\{s_+^*(t)\}$ and $\{s_-^*(t)\}$ are almost balanced in the case of $n = 0 \pmod 4$ by Lemma 5. By Theorem 9 and Lemma 3, it can be easily checked that

$$R_{s_+}(\tau) = R_{s_-}(\tau) = -(\chi(a)+1)(\chi(1-a)+1)$$

where $a = \alpha^\tau$. □

Corollary 11. *For $n = 2 \pmod 4$, the sequences $\{s_+^*(t)\}$ and $\{s_-^*(t)\}$ are equivalent to the Lempel-Cohn-Eastman sequence $\{s_L(t)\}$. More precisely,*

$$s_+^*(t) = s_L(\gamma - t),$$
$$s_-^*(t) = s_L(\lambda - t)$$

where $\alpha^\gamma = -2$ and $\alpha^\lambda = 2$.

Proof. It follows from (1) and Theorem 9 . □

5 Linear Complexity of Lempel-Cohn-Eastman Sequences

Any binary sequence $\{s(t)\}$ of period n satisfies a linear recursion, say,

$$s(t+m) = a_{m-1}s(t+m-1) + \cdots + a_1 s(t+1) + a_0 s(t) \tag{4}$$

for all t, where $a_i \in \{0,1\}$ and $a_0 \neq 0$. Here, m is called the degree of the linear recusion. There are an infinite number of linear recursions of $\{s(t)\}$, including $s(t+n) = s(t)$ for all t. The degree of the minimum-degree linear recursion is called the ***linear complexity*** or ***linear span*** of $\{s(t)\}$.

For convenience, assume that (4) is the minimum-degree linear recursion of $\{s(t)\}$. Then the polynomial $a(x) = x^m + a_{m-1}x^{m-1} + \ldots + a_0$ is called the ***characteristic polynomial*** of $\{s(t)\}$. If we write $G(x) = \sum_{t=0}^{\infty} s(t)x^t$, it is easily checked that

$$G(x) = \frac{\sum_{t=0}^{n-1} s(t)x^t}{x^n + 1}$$

and $G(x)$ can be reduced to $b(x)/a(x)$, where $b(x)$ is a polynomial of degree $< m$ and relatively prime to $a(x)$.

Theorem 12. *If $n = 0 \pmod 4$, the linear complexity of $\{s_L(t)\}$ is at most $n-1$.*

Proof. Since $\{s_L(t)\}$ is balanced, it has $n/2$ 1's in one period. This implies that

$$\sum_{t=0}^{n-1} s_L(t)x^t$$

is divisible by $x+1$. On the other hand, x^n+1 has $x+1$ as its factor. Therefore, the degree of the charcteristic polynomial is at most $n-1$. □

Theorem 13. *If $n = 2p$ for an odd prime p and 2 is a primitive element mod p, then the linear complexity of $\{s_L(t)\}$ is n.*

Proof. Clearly, $z^n+1=(z^{n/2}+1)^2$. If we let $C(x)=\sum_{t=0}^{n-1} s_L(t)x^t$, it is sufficient to show that $C(\beta)\neq 0$ for any β such that $\beta^{n/2}=1$.

$$\begin{aligned} C(\beta) &= \sum_{t=0}^{n/2-1} (s_L(t)+s_L(t+n/2))\beta^t \\ &= \frac{1-\chi(2)}{2} + \sum_{t=0}^{n/2-1} \frac{2-\chi(1+\alpha^t)-\chi(1-\alpha^t)}{2}\beta^t \end{aligned}$$

Note that the minimal poynomial of β over F_2 is $x^{p-1}+x^{p-2}+\cdots+x+1$. Therefore, $C(\beta)=0$ implies that $\chi(2)=-1$ and

$$\chi(1+\alpha^t)+\chi(1-\alpha^t)=0$$

for any t, $1\leq t\leq n/2-1$, which is impossible. □

References

1. H. Dobbertin, "Kasami power functions, permutation polynomials and cyclic difference sets," *Proceedings of Difference Sets, Sequences and their Correlation Properties*, NATO Advanced Study Institute Workshop, Bad Windsheim, Germany, Aug. 3-14, 1998.
2. A. Lempel, M. Cohn, and W. L. Eastman, "A class of binary sequences with optimal autocorrelation properties," *IEEE Trans. on Inform. Theory*, vol. 23, no.1, pp. 38-42, Jan. 1977.
3. J. -S. No, H. Chung, and M. -S. Yun, "Binary pseudorandom sequences of period 2^m-1 with ideal autocorrelation generated by the polynomial $z^d+(z+1)^d$," *IEEE Trans. on Inform. Theory*, vol. 44, no. 2, pp. 1278-1282, May 1999.
4. J. -S. No, H. Chung, H. -Y. Song, K. Yang, J. -D. Lee and T. Helleseth, "New construction for binary sequences of period p^m-1 with optimal autocorrelation," to appear in *IEEE Trans. on Inform. Theory*.
5. R. Lidl and H. Niederreiter, *Finite Fields*, Encyclopedia of Mathematics and Its Applications, Addison-Wesley, 1983.
6. T. Storer, *Cyclotomy and Difference Sets*, Lecture Notes in Advanced Mathematics, Markham Publishing Company, Chicago, 1967.

On the Profile of the k-Error Linear Complexity and the Zero Sum Property for Sequences over $GF(p^m)$ with Period p^n

Takayasu Kaida[1], Satoshi Uehara[2], and Kyoki Imamura[3]

[1] Yatsushiro National College of Technology, Kumamoto 866-8501, JAPAN
e-mail: `kaida@as.yatsushiro-nct.ac.jp`
[2] The University of Kitakyushu, Wakamatsu, Kitakyushu 808-0135, JAPAN
e-mail: `uehara@env.kitakyu-u.ac.jp`
[3] Kyushu Institute of Technology, Iizuka, Fukuoka 820-8502, JAPAN
e-mail: `imamura@cse.kyutech.ac.jp`

Summary. The k-error linear complexity of periodic sequences is a natural generalization of the linear complexity which is one of important measures for pseudo-random sequences. In this paper, we give a relation between the minimum decrease and the zero sum property for sequences over $GF(p^m)$ with period p^n, where p is a prime. Moreover the parity of the decrease set in case of binary sequences with period 2^n is shown.

Keywords – k-error linear complexity(k-LC), zero sum property(ZSP), minimum decrease, sequence over $GF(p^m)$ with period p^n

1 Introduction

The linear complexity (LC) is one of important measures for pseudo-random sequences in the fields of cryptography and computer science, and so on. The LC of a sequence is the length of the shortest linear feedback shift register (LFSR) that generates the sequence [1]. We can determine such an LFSR from the consecutive sequence values of length $2LC$ [1]. Therefore it is desirable for pseudo-random sequence to have a large LC. The LC was generalized to the k-error LC (k-LC) by M. Stamp and C.F. Martin [2], which is defined as the minimum LC value among sequences obtained by changing k or fewer symbols within one period. A fast algorithm for computing the k-LC of binary sequences with period 2^n was also given by them [2]. (The sphere complexity was defined by C. Ding, G. Xiao and W. Shan earlier than the k-LC, and it is essentially the same as the k-LC, but not completely [3].) Later two fast algorithms for computing the k-LC of sequences over $GF(p^m)$ with period p^n (p: a prime) were proposed by authors in 1999 [4,5]. The LC of a sequence over $GF(p^m)$ with period p^n obtained from a sequence over $GF(p^m)$ with period $p^n - 1$ by inserting one symbol within one period was shown in [6]. Their results depend on whether the original sequence with period $p^n - 1$ has the zero sum property (ZSP) or not, where a periodic sequence is said to have the ZSP if the sum of all the symbols within one period is equal to zero over $GF(p^m)$.

Let S be a sequence over $GF(p^m)$ with period p^n. We will write the LC of S as $LC(S)$ and the k-LC of S as $LC_k(S)$. From the definition of the k-LC we have $LC(S) = LC_0(S) \geq LC_1(S) \geq LC_2(S) \geq \cdots \geq LC_{p^n}(S) = 0$. Recently K. Kurosawa, F. Sato, T. Sakata and W. Kishimoto gave a formula for the smallest k value such that the k-LC is strictly less than the LC of the original sequence over $GF(p^m)$ with period p^n [7].

In this paper we will first prove that for a sequence S over $GF(p^m)$ with period p^n we have (i) $LC(S) > LC_1(S)$ if S does not have the ZSP and (ii) $LC(S) = LC_1(S)$ if S has the ZSP. Secondly we will prove that in case of binary sequences with period 2^n, the necessary condition for $LC_{k-1} > LC_k(S)$ is that (i) k is odd if S does not have the ZSP and (ii) k is even if S has the ZSP.

2 Linear Complexity and k-Error Linear Complexity

Let p be a prime and m a positive integer. For an infinite sequence $S = (s_0, s_1, \cdots)$ over $K = GF(p^m)$, a polynomial $f(x) = \sum_{i=0}^{L} f_i x^i$ over K with $f_L \neq 0$ and $f_0 = 1$ is called the generator polynomial of the sequence S if

$$\sum_{j=0}^{L} f_j s_{L+i-j} = 0 \tag{1}$$

for any $i \geq 0$. The linear complexity (LC) of a sequence S, denoted by $LC(S)$, is defined by the minimum degree in generator polynomials of sequence S. A sequence S has the period N, if there is a positive integer N such that $s_i = s_{N+i}$ for any $i \geq 0$. In general if N is a period of sequence S, then all multiples of N are also the periods of sequence S. We define the period of sequence S as the minimum value of its periods and we consider sequences with such a period N in this paper. For a sequence S over K with period N, we have $LC(S) = \deg f(x) = N - \deg h(x)$ if

$$\sum_{i=0}^{\infty} s_i x^i = \frac{s(x)}{1 - x^N} = \frac{g(x)}{f(x)}, \quad \gcd(g(x), f(x)) = 1, \tag{2}$$

where $s(x) = \sum_{i=0}^{N-1} s_i x^i$, $g(x), f(x) \in K[x]$ and $h(x) = \gcd(1 - x^N, s(x))$.

Definition 1: The k-error linear complexity (k-LC) of a sequence S over K with period N is defined by

$$LC_k(S) = \min\{LC(S + E) \mid W(E^N) \leq k\}, \tag{3}$$

where sequence $E = (e_0, e_1, \cdots)$ over K has period N or a divisor of N, and $W(E^N)$ means the Hamming weight of $E^N = (e_0, e_1, \cdots, e_{N-1})$ with length

N, and we consider periodic sequence $S+E=(s_0+e_0, s_1+e_1, \cdots)$ over K. □

It is obvious that the k-LC is a generalization of the LC, since $LC(S) = LC_0(S)$. Moreover we have $LC_k(S) \geq LC_{k'}(S)$ for $k < k'$ from the definition of the k-LC.

Remark 1: C. Ding et al. studied the following complexities similar to the k-LC. For a sequence S with period N the weight complexity and the sphere complexity are defined as

$$WC_k(S) = \min\{LC(S+E) \mid W(E^N) = k\} \tag{4}$$

and

$$SC_k(S) = \min\{LC(S+E) \mid 1 \leq W(E^N) \leq k\}, \tag{5}$$

respectively. In general $SC_k(S) \leq LC(S)$ does not hold, since $W(E^N) = 0$ is missing in definition (5). Hence they introduced the stability of a sequence S with period N by

$$T_k(S) = \mid LC(S) - SC_k(S) \mid, \tag{6}$$

where $\mid t \mid$ is the absolute value of an integer t. □

3 Decrease Set and Zero Sum Property

In this section we will prove two theorems after briefly reviewing previous results necessary to prove our main theorems.

Definition 2: For a sequence S with period N, the decrease set of S is defined by

$$DS(S) = \{k \mid LC_k(S) < LC_{k-1}(S), 1 \leq k \leq N\}. \tag{7}$$

The minimum decrease of a sequence S is defined by

$$MD(S) = \min DS(S) = \min\{k \mid LC_k(S) < LC_{k-1}(S), 1 \leq k \leq N\}. \tag{8}$$

□

The zero sum property defined in the following Definition 3 is useful for a simple classification of periodic sequences.

Definition 3: A sequence S over K with period N is said to have the zero sum property (ZSP) if $\sum_{i=0}^{N-1} s_i = 0$ over K. □

We consider a sequence S over $K = GF(p^m)$ with period $N = p^n$, where n is a positive integer. In this case we have $1 - x^N = (1-x)^N$ and

$$LC(S) = N - t, \tag{9}$$

where $(1-x)^t = \gcd(1-x^N, s(x))$. Therefore we need only to calculate the multiplicity of $s(x)$ at $x=1$ for the LC of a sequence S over $GF(p^m)$ with period p^n. The following two lemmas are known about the minimum decrease of sequences [7]. We will give their proofs including the case of non-binary sequences, since they were given simply after showing for binary sequences in [7] .

Lemma 1 [7]: Let S be a sequence over $K = GF(p^m)$ with period $N = p^n$. We have

$$LC(S+E) < LC(S) \tag{10}$$

if and only if

$$e(x) = (1-x)^t e_1(x) \tag{11}$$

for some $e(x) \in K[x]$ such that $e_1(1) = -s_1(1) \neq 0$, where $s(x) = \sum_{i=0}^{N-1} s_i x^i = (1-x)^t s_1(x)$ with $s_1(1) \neq 0$ and $e(x) = \sum_{i=0}^{N-1} e_i x^i$ for a sequence $E = (e_0, e_1, \cdots)$ with period N or a divisor of N.

Proof: Let $\gcd(1-x^N, s(x)+e(x)) = (1-x)^z$. There is some $e(x)$ such that $z > t$ if and only if (10) holds. Then we show that $z > t$ if and only if (11) holds. We set

$$e(x) = (1-x)^c e_1(x) \tag{12}$$

for some $e_1(1) = -s_1(1) \neq 0$.

Firstly we suppose $c > t$ then

$$s(x) + e(x) = (1-x)^t (s_1(x) + (1-x)^{c-t} e_1(x)). \tag{13}$$

Therefore $z = t$ from $\gcd((1-x)^N, s(x)+e(x)) = (1-x)^t$. If $c < t$ then $z = c < t$ similarly. Hence it is shown that (11) is induced from $z > t$.

Finally we suppose $c = t$. Then

$$s(x) + e(x) = (1-x)^t (s_1(x) + e_1(x)). \tag{14}$$

We have $s(x) + e(x) = (1-x)^{t+1} r(x)$ for some $r(x)$ from $e_1(1) = -s_1(1)$. Therefore $z \geq t+1 > t$ from

$$s(x) + e(x) = (1-x)^{t+1} r(x). \tag{15}$$

Hence it is shown that $z > t$ is induced from (11). □

The following two properties shown in [9] are used in Lemma 2

Property 1 [8, Lemma 1]: Let c be a non-zero element in $GF(p^m)$ and $t =$

$\sum_{i=0}^{m-1} t_i p^i$ be a non-negative integer such that $0 \le t_i \le p-1$ for $0 \le i < m$. We have

$$PW((x-c)^t) = \prod_{i=0}^{m-1} (t_i + 1), \tag{16}$$

where $PW(f(x))$ is the number of non-zero coefficients of a polynomial $f(x)$. □

Property 2 [8, Theorem 6.2]: For a polynomial $f(x)$ over $GF(p^m)$, any non-zero element c in $GF(p^m)$ and any non-negative integer t, we have

$$PW(f(x)(x-c)^t) \ge PW((x-c)^t). \tag{17}$$

□

Lemma 2 [7]: Let S be a sequence over $GF(p^m)$ with period $N = p^n$. We have

$$MD(S) = \text{prod}(N - LC(S)), \tag{18}$$

where $\text{prod}(t) = \prod_{i=0}^{m-1}(t_i+1)$ for a non-negative integer $t = \sum_{i=0}^{m-1} t_i p^i$ such that $0 \le t_i \le p-1$ for $0 \le i < m$.

Proof: From Lemma 1, we have

$$MD(S) = \min\{PW((1-x)^t e_1(x)) \mid e_1(1) = -s_1(1)\}. \tag{19}$$

On the other hand, let $e_1(x) = -s_1(1) \ne 0$ then

$$PW((1-x)^t e_1(x)) = PW(-s_1(1)(1-x)^t) = \prod_{i=0}^{m-1}(t_i+1) \tag{20}$$

from Property 1 and Property 2. Finally we take $t = N - LC(S)$ from $LC(S) = N - t$. □

The following lemma is a slightly modified version of the result given in [6].

Lemma 3: Let S be a sequence over $K = GF(p^m)$ with period $N = p^n$. Let sequence $D = (d_0, d_1, \cdots)$ be a periodic sequence with period $N-1$ obtained from S by $d_i = s_{i+1} - s_0$ for $0 \le i \le N-2$. Then we have

$$\begin{cases} LC(S) = N & \text{if} \quad S \text{ dose not have the ZSP}, \\ N - LC(D) \le LC(S) \le N-1 & \text{if} \quad S \text{ has the ZSP}. \end{cases} \tag{21}$$

Proof: If S does not have the ZSP, then we have $s(1) \ne 0$, $t = 0$ and$LC(S) = N$. If S has the ZSP, then we have $t \ge 1$ and $LC(S) < N$.

We will prove $N - LC(D) \leq LC(S)$. Let $U = (1, 1, \cdots, 1)$ be an all one infinite sequence and

$$u(x) = \sum_{i=0}^{N-1} x^i = (1 - x^N)/(1 - x) = (1 - x)^{N-1}.$$

Consider the sequence $\hat{S} = S - s_0 U = (\hat{s}_0, \hat{s}_1, \cdots)$ and $\hat{s}(x) = \sum_{i=0}^{N-1} \hat{s}_i x^i$. Since $\hat{s}(x) = s(x) - s_0(1 - x)^{N-1}$, $(1 - x)^t = \gcd((1 - x)^N, s(x))$ is equal to $\gcd((1 - x)^N, \hat{s}(x))$ if $t < N$, where $t = N$ means that S is a zero sequence.

The sequence $\hat{S}$ has the ZSP if and only if S has the ZSP. Therefore without loss of generality we can assume that $s_0 = 0$ and $s(x) = xd(x)$. Since S and D have the ZSP, we have $s(x) = xd(x) = x(1 - x)^t d_0(x)$ with $d_0(1) \neq 0$. Therefore we have (9).

Let $h(x) = \gcd(1 - x^{N-1}, d(x))$. Since $1 - x^{N-1}$ has only simple zero at $x - 1$, we have $h(x) = (1 - x)h_0(x)$, where $h_0(x) = \gcd(1 - x^{N-1}, d_0(x))$, $h_0(1) \neq 0$ and

$$\deg h_0(x) \leq d_0(x) = \deg d(x) - t \leq N - 2 - t. \tag{22}$$

Substitution of (9) and (22) into $LC(D) = N - 1 - \deg h(x) = N - 2 - \deg h_0(x)$ gives $N - LC(D) \leq LC(S)$. □

Remark 2: Lemma 3 is a generalization of the following theorem in [9, Theorem 3]: For a binary sequence S with period 2^n, we have $LC(S) = 2^n$ if and only if the weight of one period $(s_0, s_1, \cdots, s_{2^n})$ is odd. □

From above discussion we show next theorem about the minimum decrease of sequences over $GF(p^m)$ with period p^n.

Theorem 1: For a sequence S over $GF(p^m)$ with period p^n, we have

$$\begin{cases} MD(S) = 1 & \text{if} \quad S \text{ does not have the ZSP,} \\ MD(S) \geq 2 & \text{if} \quad S \text{ has the ZSP.} \end{cases} \tag{23}$$

Proof: If S does not have the ZSP then $LC(S) = N$ from Lemma 3. Hence we have $MD(S) = \text{prod}(0) = 1$ from Lemma 2. If S has the ZSP then we have $LC(S) < N$ from Lemma 3 and $N - LC(S) \geq 1$. Hence we have $MD(S) = \text{prod}(N - LC(S)) \geq 2$. □

Finally we consider the case of $p = 2$ and $m = 1$, i.e., binary sequences with period 2^n. Next Theorem gives the parity of elements in the decrease set of binary sequences with period 2^n.

Theorem 2: For a binary sequence S with period 2^n, all elements in the $DS(S)$ are odd if a binary sequence S does not have the ZSP, or all elements in the $DS(S)$ are even if a binary sequence S has the ZSP.

Proof: We use the weight complexity defined by (4), i.e.,

$$WC_k(S) = \min\{LC(S+E) \mid W(E^N) = k\}.$$

In case that a binary sequence S does not have the ZSP, then the sequence $S+E$ does not have the ZSP and $LC(S+E) = N$ from Lemma 3 if $k = W(E^N)$ is even. So we have $LC_k(S) = LC_{k-1}(S)$ for even k if S does not have the ZSP.

The proof of the case when S has the ZSP is similar. □

4 Examples

In this section we show some numerical examples of sequences over $GF(2)$ with period $N = 16$ and $N = 64$, respectively. We make a sequence S from an M-sequence $M = (m_0, m_1, \cdots)$ with period $N-1$

$$M^{N-1} = \begin{cases} (011110101100100) & \text{in Fig.1 and Fig.2,} \\ (011111101010110011011 & \\ \quad 101101001001110001011 & \\ \quad 110010100011000010000) & \text{in fig.3 and Fig.4} \end{cases} \tag{24}$$

generated by polynomial

$$f(x) = \begin{cases} x^4 + x + 1 & \text{in Fig.1 and Fig.2,} \\ x^6 + x + 1 & \text{in Fig.3 and Fig.4} \end{cases} \tag{25}$$

by one-symbol ($b = 0$ or 1) insertion at each position i $(0 \le i \le N-1)$, where the inserted position i means that we put symbol b at between m_{i-1} and m_i of an original sequence M, i.e., one period of a sequence S is given by

$$s_j = \begin{cases} m_j & \text{if } 0 \le j < i, \\ b & \text{if } j = i, \\ m_{j+1} & \text{if } i < j < N. \end{cases} \tag{26}$$

As the inserted symbol b, it is shown in Fig.1 and Fig.3 in the case of $b = 0$ and it is shown in Fig.2 and Fig.4 in the case of $b = 1$. From these figures two theorems can be confirmed, i.e., we have $LC_1(S) = LC(S)$ in the case of $b = 0$ satisfying the ZSP (Fig.1 and Fig.3) and $LC_1(S) < LC(S)$ in the case of $b = 1$ not satisfying the ZSP (Fig.2 and Fig.4).

5 Conclusion

In this paper we discussed the relation between the linear complexity (LC) of a sequence S over $GF(p^m)$ with period p^n and the k-error LC (k-LC) of S. Firstly we showed that $LC(S) > LC_1(S)$ if S does not have the zero sum property (ZSP) and $LC(S) = LC_1(S)$ if S has the ZSP, where S is said to have the ZSP if the sum of all the symbols within one period of S is equal to zero. Secondly we showed that in case of a binary sequence S with period 2^n the necessary condition for $LC_{k-1}(S) < LC_k(S)$ is that (i) k is odd if S does not have the ZSP and (ii) k is even if S has the ZSP.

Further works are desirable about the value of k which gives $LC_{k-1}(S) > LC_k(S)$ and the decrease value $LC_{k-1}(S) - LC_k(S)$. The study for sequences with arbitrary period not restricted to the form of p^n is of course desirable.

Acknowledgments

The authors would like to thank Dr. Kenny Paterson for helpful discussion and bringing our attention to the paper [9] during SETA'01 at Bergen.

References

1. Massey, J.L. (1969) Shift register synthesis and BCH decoding. IEEE Trans. on Information Theory. **IT-15**, 122–127
2. Stamp M., Martin C.F. (1993) An algorithm for the k-error linear complexity of binary sequences with period 2^n. IEEE Trans. on Information Theory. **39**, 1398–1401
3. Ding C., Xiao G., Shan W. (1991) The Stability Theory of Stream Ciphers. Lecture Notes in Compute Science, **561**, Springer-Verlag
4. Kaida T., Uehara S., Imamura K. (1999) An algorithm for the k-error linear complexity of sequences over $GF(p^m)$ with period p^n, p a prime. Information and Computation. **151**, Academic Press, 134–147
5. Kaida T., Uehara S., Imamura K. (1999) A new algorithm for the k-error linear complexity of sequences over $GF(p^m)$ with period p^n. Sequences and their Applications. Springer-Verlag, 284–296
6. Uehara S., Imamura K. (1996) Linear complexity of periodic sequences obtained from $GF(q)$ sequences with period $q^n - 1$ by one-symbol insertion. IEICE Trans. on Fundamentals. **79-A**, 1739–1740
7. Kurosawa K., Sato F., Sakata T., Kishimoto W. (2000) A relationship between linear complexity and k-error linear complexity. IEEE Trans. on Information Theory. **46**, 694–698
8. Massey J.L., Costello D.J., Justesen J. (1973) Polynomial weight and code construction, IEEE Trans, on Information Theory. **IT-19**, 101–110
9. Chan A.H., Games R.A., Key E.L. (1982) On the complexities of de Bruijn sequences, Journal of Combinatorial Theory, Series A, **33**, 233–246

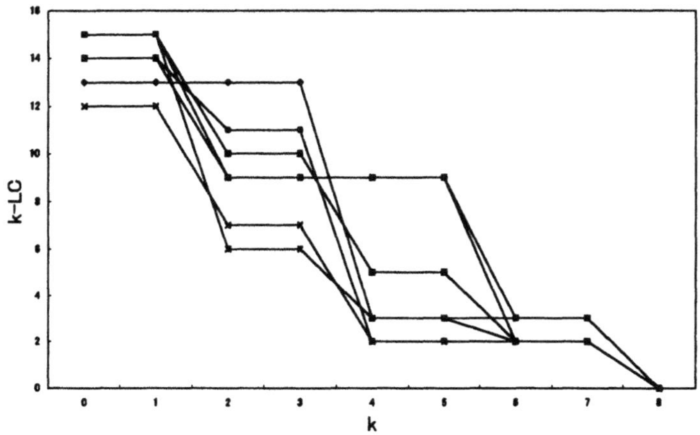

Fig. 1. The profile of k-LC for binary sequence ($N = 16, b = 0$)

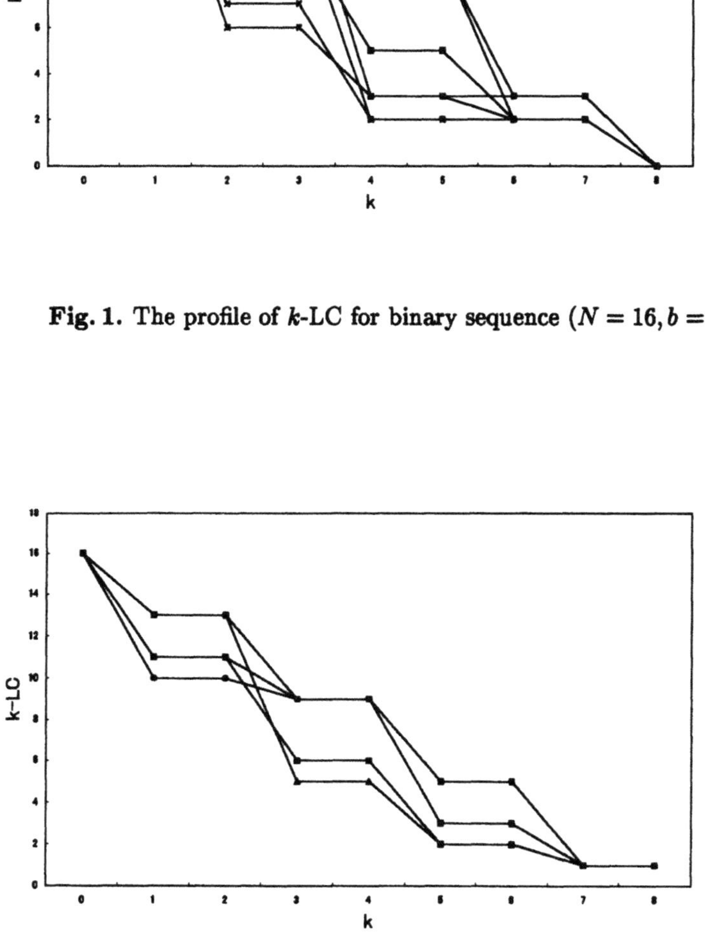

Fig. 2. The profile of k-LC for binary sequence ($N = 16, b = 1$)

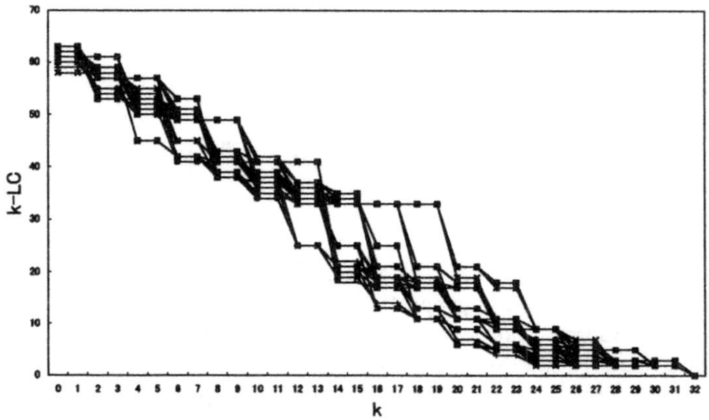

Fig. 3. The profile of k-LC for binary sequence ($N = 64, b = 0$)

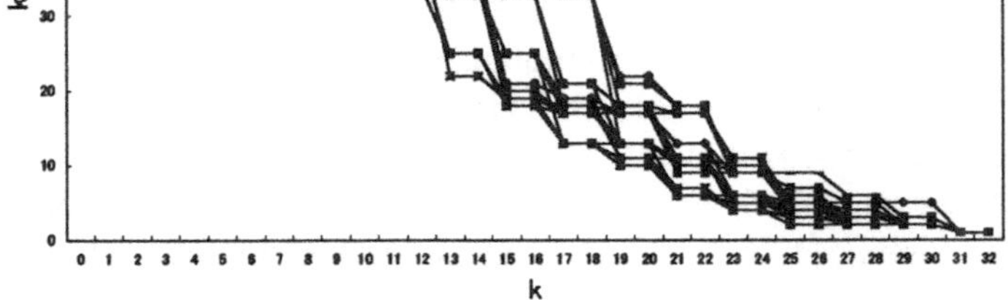

Fig. 4. The profile of k-LC for binary sequence ($N = 64, b = 1$)

Constant Sum Implies Statistical Independence of Chaotic Sequences

Tohru Kohda

Department of Computer Science and Communication Engineering,
Kyushu University, Fukuoka 812-8581, Japan

Summary. We have recently given a simple sufficient condition for a class of ergodic maps with some symmetric properties to produce a sequence of independent and identically distributed binary random variables which consists of two simple symmetric poperties: the *equidistributivity property* and *constant summation property*. The objective of this note is to show that the latter property results when the requirement of the former is replaced by an idependence requirement of real-valued random variables, as functions of a chaotic orbit.

1 Introduction

Several computational applications requiring random numbers are expanding [1–3]. These include computer programming, stochastic simulation, Monte Carlo technique, simulated annealing, and so on. In particular, a sequence of independent and identically distributed (i.i.d.) binary random variables [4–6] has found significant applications in modern digital communication systems, such as in spread spectrum (SS) communication systems [7] or cryptosystems [8]. For such binary sequences, LFSR (linear feedback shift register) sequences have often been used [9].

It is well known from probability theory [4–6] that coin tossing has always been considered as the symbol of randomness. Randomness itself, however, is very difficult to define. In fact, throwing a die may be seen as both a deterministic and a random event. Both of the dyadic map and its associated binary function, called the Rademacher function, have been considered to realize coin tossing [4–6]. Thus, it is an important problem whether any other maps and their associated binary functions than the dyadic map and Rademacher function, respectively can realize coin tosing or not.

We have recently given several methods for generating a binary sequence based on chaotic dynamics [10,11] and shown that a class of ergodic maps with *equidistributivity property* (or briefly *EDP*) can generate a sequence of i.i.d. binary random variables if their associated binary functions satisfy the *constant summation property* (or briefly *CSP*) [12,13] . Furthermore, we have also given a simple method for generating a sequence of i.i.d. $p \geq 2$-ary random variables [14,15]. Note that maps with the EDP includes a class of the well-known maps, *e.g.*, dyadic map, logistic map, tent map, R-adic map, and Chebyshev map. This suggests that we can deduce a great deal of information about pseudorandom number generation based on ergodic maps from the well-known facts in chaos theory. In fact, close relationship between binary sequence generation and ergodic theory has recently been reported [16].

On the other hand, many communication systems based on chaotic phenomena [17–20] have recently been proposed. Most of these systems, however, have been based on a chaotic real-valued orbit itself rather than its binary version (namely, *code level*). Furthermore, sequences of random numbers generated by a Markov chain are discussed as candidates of spreading sequences for asynchronous CDMA [21,22], running-key sequences in the chaotic encryption systems [23] as well as real-valued spreading sequence [24,25].

There is another question: whether the above two conditions play an important role in the discussion of statistical independence of a real-valued sequence or not. In physical systems, sequences whose autocorrelation exponentially decays are often discussed. As well-known, independence implies not only that (linear-) autocorrelation is 0 but also that correlation coefficient between the N_1-th power and the N_2-th power sequence is 0 for any N_1 and N_2 (as discussed in section 4). Recently, we have shown that the Chebyshev polynomial map of degree p, $T_p(\omega)$, has the p-th *order correlated* property and the $\lfloor \log_p N \rfloor$-*delay correlated* property of a real-valued trajectory [26]. Furthermore, we also have shown that a rational function version of $T_p(\omega)$, *i.e., Jacobian elliptic Chebyshev rational maps* can easily generate a sequence of i.i.d. p–ary random variables [27]. This paper is intended to discuss a gap between CSP and statistical independence of a real-valued trajectory generated by ergodic map with the EDP.

2 Bernoulli Shift and Rademacher Function

Let us start with reviewing fundamental subjects from the well-known textbooks of elementary probability theory [4–6].

Let ω be a point drawn at random from the unit interval $(0,1)$. With each ω associate its nonterminating binary expansion

$$\omega = \sum_{k=1}^{\infty} \frac{d_k(\omega)}{2^k} = 0.d_1(\omega)d_2(\omega)\cdots, \text{ where } d_i(\omega) \in \{0,1\}. \tag{1}$$

Imagine now a coin with faces labeled 1 and 0 instead of the usual heads (H) and tails (T). If ω is drawn at random, $\{d_k(\omega)\}_{k=1}^{\infty}$ behaves as if it resulted from an infinite sequence of tosses of a fair coin.

On the other hand, define the map $\tau_B(\cdot)$, referred to as the *Bernoulli shift* or *dyadic map* by

$$\tau_B(\omega) = 2\omega, \quad \mod 1 \tag{2}$$

which gives

$$\tau_B(\omega) = 0.d_2(\omega)d_3(\omega)\cdots. \tag{3}$$

This implies that $\tau_B(\cdot)$ shifts the digits one place to the left. Namely,

$$d_k(\tau_B(\omega)) = d_{k+1}(\omega), \quad \text{for } k \geq 1 \tag{4}$$

The functions $d_k(\omega), k = 1, 2, \cdots$ are often replaced by the *Rademacher functions* [4,6] defined as, $r_k(\omega) = 1 - 2d_k(\omega), \quad k = 1, 2, \cdots$. Since the functions $r_k(\omega)$ and $d_k(\omega)$, $k = 1, 2, \cdots$ furnish us with a *model* of independent tosses of a "fair" coin, they are ***independent random variables.*** Figure 1 shows the binary functions $d_k(\omega)$ which define $r_k(\omega)$.

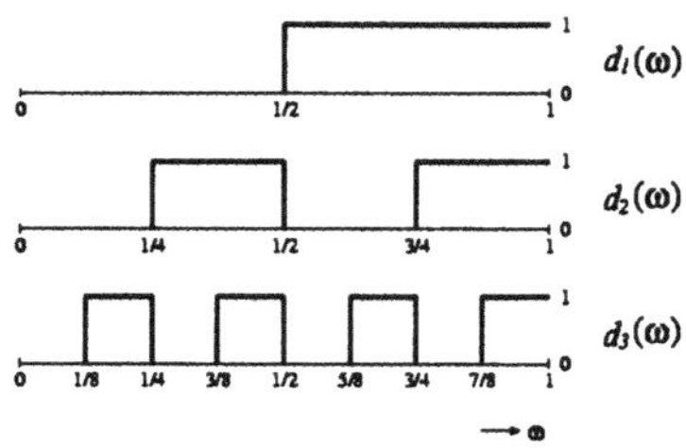

Fig. 1. Binary functions $d_k(\omega)$ which define Rademacher functions $r_k(\omega)$.

3 EDP and CSP

In the application of chaotic dynamics to cryptosystem, the dyadic map is impracticable with the help of a computer with its limited accuracy in the sense that the period of its chaotic orbit is very short. Of cruicial importance is the question whether any other maps than the dyadic map can realize coin tosing or not.

To discuss this problem, let us start with giving three simple methods [10,11] for obtaining binary sequences from chaotic real-valued sequences $\{\omega_n\}_{n=0}^{\infty}$, generated by an ergodic map $\tau(\cdot) : I = [d, e] \to I$

$$\omega_{n+1} = \tau(\omega_n), \qquad \omega_n = \tau^n(\omega_0), \omega_0 \in I, \, n = 1, 2, \cdots. \tag{5}$$

Method-1: We define a threshold function $\Theta_t(\omega)$ as

$$\Theta_t(\omega) = \begin{cases} 0 \text{ for } \omega < t \\ 1 \text{ for } \omega \geq t \end{cases} \tag{6}$$

and define its complementary function $\overline{\Theta}_t(\omega) = 1 - \Theta_t(\omega)$. Using these functions, we can obtain a binary sequence $\{\Theta_t(\tau^n(\omega))\}_{n=0}^{\infty}$, which is referred to as a *chaotic threshold sequence.*

Method-2: We write the value of ω $(|\omega| \leq 1)$ in a binary representation:

$$|\omega| = 0.A_1(\omega)A_2(\omega) \cdots A_i(\omega) \cdots, \text{ where } A_i(\omega) \in \{0, 1\}. \tag{7}$$

The i-th bit $A_i(\omega)$ can be expressed as

$$A_i(\omega) = \sum_{r=1}^{2^i} (-1)^{r-1} \left\{ \Theta_{\frac{r}{2^i}}(\omega) + \overline{\Theta}_{-\frac{r}{2^i}}(\omega) \right\}. \tag{8}$$

Thus we can obtain a binary sequence $\{A_i(\omega_n)\}_{n=0}^{\infty}$ which we call a ***chaotic bit sequence***. Since $\Theta_t(\omega)$ can be regarded as a Boolean function whose variable, seed ω, is not binary but real-valued, $A_i(\omega)$ can be rewritten by

$$A_i(\omega) = \bigoplus_{r=1}^{2^i} \left\{ \Theta_{-\frac{r}{2^i}}(\omega) \oplus \Theta_{\frac{r}{2^i}}(\omega) \right\} \tag{9}$$

where $\oplus$ denotes modulo 2 addition.

Furthermore, for any map $\tau(\cdot)$ defined on the interval $[d, e]$, we can give a generalized version of the second method as follows [12] .
Method-3: We define a binary function with the set of thresholds, $T = \{t_r\}_{r=0}^{N}$, defined by

$$C_T(\omega) = \bigoplus_{r=0}^{N} \Theta_{t_r}(\omega) = \sum_{r=0}^{N} (-1)^r \Theta_{t_r}(\omega). \tag{10}$$

This function $C_T(\omega)$ gives the following two simple examples.

Example 1 of $C_T(\omega)$: $C_T(\omega) = \Theta_{t_0}(\omega)$, when $N = 0$.

Example 2 of $C_T(\omega)$: The value of $\dfrac{\omega - d}{e - d} \in [0, 1]$ has

$$\frac{\omega - d}{e - d} = 0.B_1(\omega)B_2(\omega)\cdots B_i(\omega)\cdots \quad \text{where } B_i(\omega) \in \{0, 1\} \tag{11}$$

which implies that

$$C_T(\omega) = B_i(\omega), \text{ when } N = 2^i, t_r = (e - d)\frac{r+1}{2^i} + d. \tag{12}$$

Note that if the interval $I = [0, 1]$, then $A_i(\omega) = B_i(\omega)$. Thus each of $\{A_i(\tau^n(\omega))\}_{n=0}^{\infty}$ and $\{B_i(\tau^n(\omega))\}_{n=0}^{\infty}$ is referred to as a ***chaotic bit sequence***. Note that $B_i(\omega) = d_i(\omega)$ when $\tau(\cdot) = \tau_B(\cdot)$.

Thus, it is an important problem whether any other maps and its associated binary functions than the dyadic map and the Rademacher function, respectively can realize a sequence of i.i.d. binary random variables or not. Recently we have replied in the affirmative: There is a wide class of ergodic maps with the equidistributivity property (EDP) and their associative binary functions with the constant summation property (CSP) [12,13].

To review this, we begin by introducing a piecewise monotonic onto ergodic map $\tau(\cdot) : I = [d, e] \to I$ that satisfies the following properties:

i) there is a partitition $d = d_0 < \cdots < d_{N_\tau} = e$ of I such that for each integer $i = 1, \cdots, N_\tau$, $(N_\tau > 2)$ the restriction of $\tau(\cdot)$ to the interval $J_i = [d_{i-1}, d_i)$, denoted by $\tau_i(\omega)$, is a C^2 function; as well as
ii) $\tau(J_i) = (d, e)$, that is, τ has N_τ monotonic onto maps τ_i;
iii) τ has a unique absolutely continuous invariant (or briefly ACI) measure, denoted by $f^*(\omega)d\omega$.

Several famous ergodic maps and their ACI measures are listed as follows.
(1) R-adic map $(N_\tau = R)$

$$\tau_R(\omega) = R\omega \bmod 1,\ R = 2, 3, 4, \cdots, \omega \in [0, 1], \quad f^*(\omega)d\omega = d\omega,$$

(2) logistic map $(N_\tau = 2)$ [1]

$$\mathrm{L}_2(\omega) = 4\omega(1-\omega), \omega \in [0, 1], \quad f^*(\omega)d\omega = \frac{d\omega}{\pi\sqrt{\omega(1-\omega)}},$$

(3) Chebyshev map of degree p $(N_\tau = p)$ [28–30]

$$T_p(\omega) = \cos(p\cos^{-1}\omega),\ p = 2, 3, 4, \cdots,\ \omega \in [-1, 1], \quad f^*(\omega)d\omega = \frac{d\omega}{\pi\sqrt{1-\omega^2}},$$

(4) Zigzag map with p branches $(N_\tau = p)$ [30]

$$N_p(\omega) = (-1)^{\lfloor p\omega \rfloor} p\omega,\ p = 2, 3, 4, \cdots,\ \omega \in [0, 1], \quad f^*(\omega)d\omega = d\omega,$$

where $\lfloor x \rfloor$ denotes the largest integer not greater than x and $N_2(\omega)$ is referred to as the tent map.

Now let us introduce the topological conjugation between two ergodic maps.

Definition 1 *(Topological Conjugation) [30,31] Two transformations $\overline{\tau} : \overline{I} \to \overline{I}$ and $\tau : I \to I$ on intervals $\overline{I}$ and I are called topological conjugate if there exists a homeomorphism h: $\overline{I} \overset{\text{onto}}{\longrightarrow} I$, such that*

$$\tau(\omega) = h \circ \overline{\tau} \circ h^{-1}(\omega). \tag{13}$$

Suppose $\tau(\cdot)$ and $\overline{\tau}(\cdot)$ have their ACI measures and denote them by $f^*(\omega)d\omega$ and $\overline{f}^*(\overline{\omega})d\overline{\omega}$, respectively. Then, under the *topological conjugation*, these ACI measures have the relation

$$f^*(\omega) = \left|\frac{dh^{-1}(\omega)}{d\omega}\right| \overline{f}^*(h^{-1}(\omega)). \tag{14}$$

Ulam and von Neumann [1] pointed out that the logistic map $L_2(\omega)$ and the tent map $N_2(\overline{\omega})$ are topological conjugate via $h^{-1}(\omega) = \frac{2}{\pi}\sin^{-1}\sqrt{\omega}$.
On the other hand, Grossmann and Thomae [30] observed that $T_p(\omega)$ and $N_p(\overline{\omega})$ are topological conjugate via $h(\overline{\omega}) = \cos\pi\overline{\omega}$. We introduce

Definition 2 *[31] The Perron-Frobenius operator P_τ acting on the function of bounded variation $H(\omega) \in L_1$ for $\tau(\omega)$ is defined as*

$$P_\tau H(\omega) = \frac{d}{d\omega}\int_{\tau^{-1}([d,\omega])} H(y)dy = \sum_{i=1}^{N_\tau} |g_i'(\omega)| H(g_i(\omega)), \tag{15}$$

where $g_i(\omega) = \tau_i^{-1}(\omega)$ is the $i-th$ preimage of ω.

Now let us consider a stationary real-valued sequences $\{H(X_n)\}_{n=0}^{\infty}$, where $X_n = \tau^n(\omega)$. It is known that the expectation of $H(\omega)$ defined by

$$\mathbf{E}[H(X_n)] = \int_I H(X_n) f^*(\omega) d\omega. \tag{16}$$

¿From the stationarity of process,we denote $\mathbf{E}[H(X_n)]$ by $\mathbf{E}[H(X)]$ or simply $\mathbf{E}[H]$.

Of major importance to the investigation of statistical properties of two sequences $\{G(\tau^n(\omega))\}_{n=0}^{\infty}$ and $\{H(\tau^n(\omega))\}_{n=0}^{\infty}$ is their 2nd-order cross-covariance function with delay $\ell = 0, 1, \cdots$ which is defined by

$$\rho(\ell; G, H) = \int_I (G(\omega) - \mathbf{E}[G])(H(\tau^\ell(\omega)) - \mathbf{E}[H]) f^*(\omega) d\omega. \tag{17}$$

This function denotes the auto-covariance function when $G = H$. The property of the Perron-Frobenius operator [31]

$$\int_I G(\omega) P_\tau\{H(\omega)\} d\omega = \int_I G(\tau(\omega)) H(\omega) d\omega, \quad G(\cdot) \in L^\infty \tag{18}$$

gives

$$\rho(\ell; G, H) = \int_I P_\tau^\ell\{G(\omega) f^*(\omega)\} H(\omega) d\omega - \mathbf{E}[G]\mathbf{E}[H], \tag{19}$$

where $G \circ \tau(\omega) = G(\tau(\omega))$ in (18) is referred to as the Koopman operator in a Hilbert space [31]. Next we introduce

Definition 3 *[12] The map is said to satisfy equidistributivity property (EDP) if the relation*

$$|g_i'(\omega)| f^*(g_i(\omega)) = \frac{f^*(\omega)}{N_\tau}, \quad 1 \le i \le N_\tau \tag{20}$$

holds.

Note that the EDP is proven to be invariant under the topological conjugation from a simple calculation. We get

Lemma 1 *[12] The maps $\tau(\cdot)$ with the EDP satisfy*

$$P_\tau\{(\Theta_t(\omega) - \mathbf{E}[\Theta_t])f^*(\omega)\} = \frac{s(\tau'(t))}{N_\tau}(\Theta_{\tau(t)}(\omega) - \mathbf{E}[\Theta_{\tau(t)}]) \times f^*(\omega), \tag{21}$$

where $s(\omega) = \begin{cases} -1 \text{ for } \omega < 0 \\ 1 \text{ for } \omega \geq 0 \end{cases}$, *referred to as the signum function.*

Lemma 1 and condition ii) of the PM onto map give

$$P_\tau\{\Theta_{d_i}(\omega)f^*(\omega)\} = \mathbf{E}[\Theta_{d_i}]f^*(\omega). \tag{22}$$

Furthermore we introduce [12,13]

Definition 4 *If for a class of maps with EDP, its associated function $H(\cdot)$ satisfies*

$$\frac{1}{N_\tau}\sum_{i=1}^{N_\tau} H(g_i(\omega)) = \mathbf{E}[H]. \tag{23}$$

then $H(\cdot)$ is said to satisfy the constant summation property (CSP).

Thus we get [12]

Lemma 2 *For a class of maps with EDP, the following three symmetric properties:*

S1 *the symmetric binary function $C_T(\omega)$, defined as*

$$t_r + t_{2M-r} = d + e, \; r = 0, 1, \cdots, M; \tag{24}$$

S2 *the symmetric ACI measure, defined as*

$$f^*(d + e - \omega) = f^*(\omega); \tag{25}$$

S3 *the symmetric map, defined as*

$$\tau(d + e - \omega) = \tau(\omega), \quad \omega \in I \tag{26}$$

give

$$P_\tau\{C_T(\omega)f^*(\omega)\} = \mathbf{E}[C_T]f^*(\omega) \tag{27}$$

which is a generalized version of (22).

Next, let $\boldsymbol{U} = U_0U_1 \cdots U_{m-1}$ be an arbitrary string of m binary digits where $U_n \in \{0, 1\}$ $(0 \leq n \leq m-1)$. Let $\boldsymbol{u}^{(r)} = u_0^{(r)}u_1^{(r)} \cdots u_{m-1}^{(r)}$ be the r-th string. Introducing a binary random variable $\Gamma_n(\omega; C_T, \boldsymbol{u}^{(r)}) = C_T(\omega)u_n^{(r)} + \overline{C}_T(\omega)\overline{u}_n^{(r)}$ for any binary function $C_T(\omega)$ with CSP, we can get [12]

Theorem: *The probability of the event* $\boldsymbol{u}^{(r)}$ *in an infinite binary sequence* $\{C_T(X_n)\}_{n=0}^{\infty}$ *is given by*

$$Prob(\boldsymbol{u}^{(r)}; C_T) = \int_I \left\{ \prod_{n=0}^{m-1} \Gamma_n(X_n; C_T, \boldsymbol{u}_m^{(r)}) \right\} f^*(\omega) d\omega$$
$$= \mathbf{E}[C_T]^s (1 - \mathbf{E}[C_T])^{m-s}, \tag{28}$$

where s *is the number of 1 in* $\{u_n^{(r)}\}_{n=0}^{m-1}$.

This implies $\{C_T(\tau^n(\omega))\}_{n=0}^{\infty}$ is a sequence of i.i.d. binary random variables with probability $\mathbf{E}[C_T]$. Figures 2 and 3 show examples of symmetric binary function $C_T(\omega)$ and sequence generator based on $C_T(\omega)$, respectively.

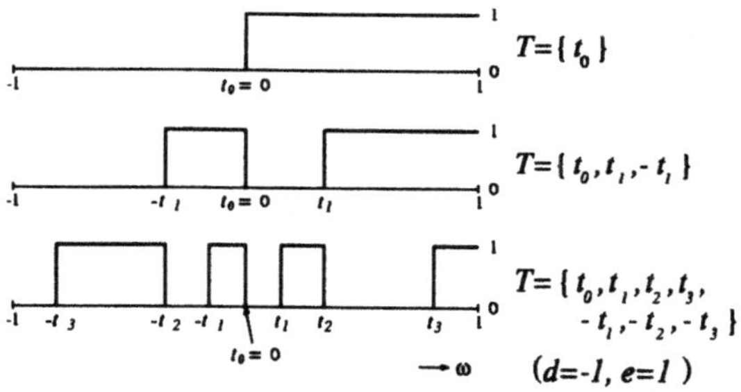

Fig. 2. *Independent* pulse functions defined by $C_T(\omega)$.

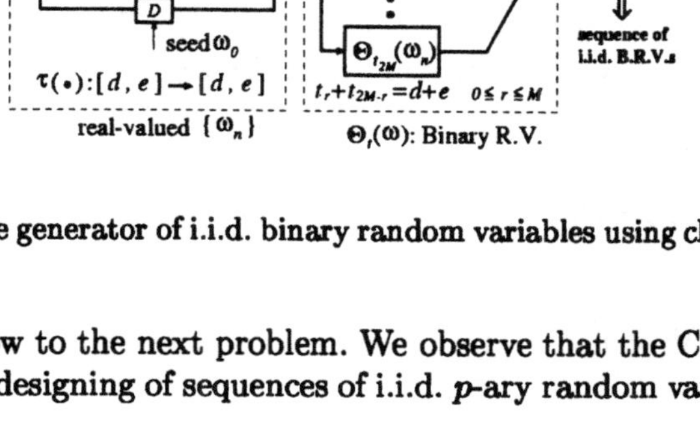

Fig. 3. Sequence generator of i.i.d. binary random variables using chaotic dynamics.

We turn now to the next problem. We observe that the CSP is of crucial importance in designing of sequences of i.i.d. p-ary random variables which is a generalized version of the binary case (see (27)) as follows [14,15]. To show this, we introduce several definitions.

Definition 5 *For piecewise monotonic onto maps $\tau(\cdot)$ with EDP, the partition $\{J_i\}_{i=1}^{N_\tau}$ in property* **i)** *is referred to as trivial partition.*

Remark 1 Each of N_τ indicators $I_{J_i}(\omega)$ for the *trivial partition* $\{J_i\}_{i=1}^{N_\tau}$ satisfies

$$P_\tau\{I_{J_i}(\omega)f^*(\omega)\} = \mathbf{E}[I_{J_i}]f^*(\omega),\ J_i = [d_{i-1}, d_i),\ i = 1, 2, \cdots, N_\tau, \quad (29)$$

that is, the random variable $I_{J_i}(\omega)$ satisfies the CSP, where $I_S(\omega)$ is an indicator with an interval $S \subset [d, e]$, defined by $I_S(\omega) = \begin{cases} 1 & \omega \in S \\ 0 & \omega \notin S \end{cases}$.

Definition 6 *Define a random variable taking a value in a set $\Gamma = \{\gamma_1, \cdots, \gamma_p\}$ by the linear combination of $I_{A_i}(\omega)$ given by $X(\omega) = \sum_{i=1}^{p} \gamma_i I_{A_i}(\omega)$ which is called an elementary random variable [5,6].*

The probability of a random variable X taking the value γ_i, denoted by $\text{Prob}\{X = \gamma_i\}$ is given by $\text{Prob}\{X = \gamma_i\} = \int_{A_i} f^*(\omega)d\omega = m(A_i)$.

Thus, the expectation $\mathbf{E}[X]$ is given by $\mathbf{E}[X] = \sum_{i=1}^{p} \gamma_i m(A_i)$.

Suppose that $p = N_\tau$ and $A_i = J_i (i = 1, 2, \cdots, N_\tau)$. Then, from (29), we can easily obtain

$$P_\tau\{X(\omega)f^*(\omega)\} = \mathbf{E}[X]f^*(\omega) \quad (30)$$

which is a generalized version of (27). Furthermore, it can be proven that $\{X(\tau^n(\omega))\}_{n=0}^{\infty}$ is a sequence of i.i.d. N_τ-ary random variables which is m-distributed.

Remark 2 For $p \le N_\tau$, nonlinear ergodic maps $\tau(\cdot)$ can generate a sequence of i.i.d. p-ary random variables by arbitrarily constructing new indicators such as $I_{J_i \cup J_j}(\omega)$. If $\text{Prob}(X = \gamma_i) = \frac{1}{p}$ for all $i = 1, 2, \cdots, p$, then $\{X(\tau^n(\omega))\}_{n=0}^{\infty}$ is a sequence of ***m*-*distributed*** ***p***-ary random variables.

It is an interesting issue from the theoretical and practical points of view to get a lot of sequences of i.i.d. p-ary random variables from a single chaotic real-valued orbit. However, the above trivial partition $\{J_i\}_{i=1}^{N_\tau}$ alone is incapable of producing a lot of elementary random variables from a single orbit. Thus we consider *nontrivial* partitions which can produce many kinds of elementary random variables.

Definition 7 *Let $\{\widetilde{J}_i\}_{i=1}^{N_\tau}$ be a nontrivial partition of J satisfying*

$$\widetilde{J}_i \bigcap \widetilde{J}_j = \phi,\ (i \ne j), \quad \bigcup_{i=1}^{N_\tau} \widetilde{J}_i = J. \quad (31)$$

If the above nontrivial partition $\{\widetilde{J}_i\}_{i=1}^{N_\tau}$ satisfies

$$P_\tau\{I_{\widetilde{J}_i}(\omega)f^*(\omega)\} = \mathbf{E}[I_{\widetilde{J}_i}]f^*(\omega),\ i = 1, 2, \cdots, N_\tau, \tag{32}$$

then it is obvious that an elementary random variable $X(\omega) = \sum_{i=1}^{N_\tau} \gamma_i I_{\widetilde{J}_i}(\omega)$ produces a sequence of i.i.d. N_τ-ary random variables. Hence we concentrate our attention on designing many kinds of nontrivial partitions satisfying (32). See for detail [14,15].

4 CSP and Statistical Independence

Such an important role of CSP is also observed in the discussions of statistical properties of a real-valued sequence as follows.

To evaluate independence of such a chaotic sequence, we must consider correlation coefficients between the N_1-th power sequence $\{X_n^{N_1}\}_{n=0}^\infty$ and the N_2-th power sequence $\{X_n^{N_2}\}_{n=0}^\infty$, defined by

$$\mathrm{Corr}(X_n^{N_1}, X_{n+\ell}^{N_2}) = \frac{\mathbf{E}[X_n^{N_1} X_{n+\ell}^{N_2}] - \mathbf{E}[X^{N_1}]\mathbf{E}[X^{N_2}]}{\mathbf{E}[X^{N_1+N_2}] - \mathbf{E}[X^{N_1}]\mathbf{E}[X^{N_2}]},$$
$$\ell = 1, 2, \cdots,\ N_1, N_2 \geq 0, \tag{33}$$

which is a nonlinear type of correlation of the $\{X_n\}_{n=0}^\infty$ sequence.

The notion of independence is of central importance in probability theory [4–6]. We know that the Kac's theorem [5] holds in the following.

Definition 8 *m stochastic real-valued random variables, $X_0, X_1, \cdots, X_{m-1}$ are independent if and only if for $\forall\, t_n \in \mathcal{R},\ 0 \leq n \leq m-1$*

$$\mathbf{E}[\{\prod_{n=0}^{m-1} \exp(jt_n X_n)\}] = \prod_{n=0}^{m-1} \mathbf{E}[\{\exp(jt_n X_n)\}], \tag{34}$$

where $\mathcal{R}$ denotes the set of real numbers.

Thus our main task is to evaluate the joint moment function

$$\mathbf{E}[X_n^{N_1} X_{n+\ell}^{N_2}] = \int_I \omega^{N_1}\{\tau^\ell(\omega)\}^{N_2} f^*(\omega)d\omega. \tag{35}$$

We get [26]

Lemma 3 *For a natural number N, N-th power sequence generated by the Chebyshev map of degree p, $T_p(\omega)$, satisfies CSP as follows:*

$$P_{T_p}\{\omega^N f^*(\omega)\} = \mathbf{E}[X^N]f^*(\omega),\ \textit{for } N < p, \tag{36}$$

$$P_{T_p}^\ell\{\omega^N f^*(\omega)\} = \mathbf{E}[X^N]f^*(\omega),\ \textit{for } {}^\forall N \textit{ and } \ell > \lfloor \log_p N \rfloor, \tag{37}$$

each of which yields respectively for a given N_1

$$\mathbf{E}[X_n^{N_1} X_{n+\ell}^{N_2}] = \mathbf{E}[X^{N_1}]\mathbf{E}[X^{N_2}],\ \textit{for } {}^\forall N_1 < p, {}^\forall N_2, {}^\forall \ell \in \mathcal{N}, \tag{38}$$

$$\mathbf{E}[X_n^{N_1} X_{n+\ell}^{N_2}] = \mathbf{E}[X^{N_1}]\mathbf{E}[X^{N_2}],\ \textit{for } {}^\forall \ell > \lfloor \log_p N_1 \rfloor,\ {}^\forall N_2 \in \mathcal{N}. \tag{39}$$

A real-valued sequence $\{X_n\}$ is referred to as being *p-th-order correlated* (respectively, $\lfloor \log_p N \rfloor$*-delay correlated*) [26] when (38) (respectively, (39)) holds. The explicit expression of preimages of $T_p(\omega)$, $g_i(\omega), (1 \leq i \leq p)$ give

Lemma 4 *[26] For a given integer n, $T_p(\omega)$ satisfies*

$$P_{T_p}\{T_n(\omega)f^*(\omega)\} = \begin{cases} T_{n/p}(\omega)f^*(\omega), & \text{for} \quad p|n \\ 0, & \text{for} \quad p\not|n, \end{cases} \tag{40}$$

where $p|n$ (respectively, $p\not|n$) means n is divisible (respectively, indivisible) by $p \neq 0$.

Using Lemmas 3 and 4, together with the famous relation [33]

$$e^{jt_0\omega} = \sum_{m=-\infty}^{\infty} j^m J_m(t_0) T_m(\omega), \tag{41}$$

we evaluate the characteristic function of $T_p(\omega)$ in a simpler form than the Geisel & Fairen's proof [32]

Corollary 1 *[26] The Chebyshev map of degree p, $T_p(\omega)$, satisfies*

$$\mathbf{E}[e^{jt_0X_n}e^{jt_1X_{n+\ell}}] = \mathbf{E}[e^{jt_0X}]\mathbf{E}[e^{jt_1X}] + \sum_{\substack{m=-\infty \\ m\neq 0}}^{\infty} j^{p^\ell m} J_{p^\ell m}(t_0)\, j^m J_m(t_1)$$
$$\text{for } \forall t_0, t_1 \in \Re, \quad \ell \in \mathcal{N}, \tag{42}$$

where $J_\nu(z)$ is the familiar Bessel function of degree ν defined by [33]

$$J_\nu(z) = \left(\frac{z}{2}\right)^\nu \sum_{r=0}^{\infty} \frac{(-z^2/4)^r}{r!(\nu+r)!}, \tag{43}$$

which implies that independence of a real-valued orbit generated by $T_p(\omega)$ is not guaranteed because the second term of the right-hand-side of (42), in general, always remains.

It is a natural question whether there are any other maps with EDP and the above correlated properties than the Chebyshev polynomial maps or not. In [27], we have given an affirmative answer to this question: there exist a class of maps in the form of rational functions. To review this, we begin by introducing the Katsura and Fukuda's result. Reference [34] gave a rational function version of the logistic map $L_2(\omega)$ with parameter k defined by

$$Q_2(\omega,k) = \frac{4\omega(1-\omega)(1-k^2\omega)}{(1-k^2\omega^2)^2}, \quad \omega \in [0,1] \tag{44}$$

with its ACI measure

$$f^*(\omega,k)d\omega = \frac{d\omega}{2K(k)\sqrt{\omega(1-\omega)(1-k^2\omega)}} \tag{45}$$

by means of a homeomorphism $h^{-1}(\omega,k)=\frac{1}{K(k)}\mathrm{sn}^{-1}(\sqrt{\omega},k)$, where $\mathrm{sn}(\omega,k)$ is the inverse function of the elliptic integral with modulus k ($|k|<1$) of the first kind in the Legendre-Jacobi normal form [33]

$$\omega=\int_0^{\mathrm{sn}(\omega,k)}\frac{dv}{\sqrt{(1-v^2)(1-k^2v^2)}},\tag{46}$$

and $K(k)$ is the complete elliptic integral of the first kind [33].

A homeomorphism $h^{-1}(\omega,k)=\frac{1}{2K(k)}\mathrm{cn}^{-1}(\omega,k)$ of $N_p(\overline{\omega})$ enables us to introduce the ***Jacobian elliptic Chebyshev rational maps*** with modulus k defined by

$$R_p(\omega,k)=\mathrm{cn}(p\,\mathrm{cn}^{-1}(\omega,k),k),\quad \omega\in[-1,1],\quad p=2,3,4,\cdots\tag{47}$$

with its ACI measure

$$f^*(\omega,k)d\omega=\frac{d\omega}{2K(k)\sqrt{(1-\omega^2)\{(1-k^2)+k^2\omega^2\}}},\tag{48}$$

where $cn(\omega,k)$ is the inverse function of the elliptic integral with modulus k of the first kind in the Legendre-Jacobi normal form [33], *i.e.*,

$$\omega=\int_{\mathrm{cn}(\omega,k)}^1\frac{dv}{\sqrt{(1-v^2)\{(1-k^2)+k^2v^2\}}}.\tag{49}$$

There are several classes of rational maps based on the inverse function of the elliptic integral as well as the Weierstrass elliptic function $\wp(\cdot)$. See for detail [27]. Furthermore, there is an attempt [35] to generalize the Katsura and Fukuda's result. Immediately we get

$$R_p(\omega,0)=T_p(\omega)\quad\text{with}\quad f^*(\omega,0)d\omega=\frac{d\omega}{\pi\sqrt{1-\omega^2}}\tag{50}$$

and show $R_p(\omega,k)$ satisfies the semi-group property

$$R_r(R_s(\omega,k),k)=R_{rs}(\omega,k).\tag{51}$$

Note that the Chebyshev polynomials are just one class of examples of this type.

Hence for any p ($p\geq 2$), the map $R_p(\omega,k)$ generates a sequence of i.i.d. p-ary random variables and satisfies the semi-group property like as $T_p(\omega)$. Furthermore, $R_{2^m}(\omega,k)$, $m=1,2,\cdots$ satisfies the linearly uncorrelated property of a real-valued trajectory $\{X_n\}_{n=0}^{\infty}$, $(X_{n+1}=R_{2^m}(X_n,k))$ [27]

$$\mathbf{E}[X_nX_{n+\ell}]=0,\ X_0=\omega,n=0,1,\cdots,\ \text{for }\ell>1;\tag{52}$$

even though, it is not easy to evaluate statistical properties of a chaotic real-valued sequence generated by $R_p(\omega,k)$.

5 Concluding Remarks

We have reviewed a simple sufficient condition for a random function to produce a sequence of i.i.d. $p \geq 2$-ary random variables using chaotic dynamics. Furthermore, we have made it clear that *constant summation property* results when the requirement of *equidistributivity property* is replaced by an independence requirement of real-valued random variables, as functions of a chaotic trajectory.

References

1. Ulam S. M., von Neumann J.(1947) On combination of stochastic and deterministic processes. Bull. Math. Soc. 53:1120
2. von Neumann J.(1951) Various techniques used in connection with random digit. In:Forsythe, G. E.(Summary written) National Bureau of Standards, Applied Math. Series, 12:36–38 or In: Reprinted in Collected Works of von Neumann,1963 5:768–770
3. Knuth D.(1981) The Art of Computer Programming 2, Seminumerical Algorithms, 2nd ed. Addison-Wesley, Reading, Massachusetts
4. Kac M.(1959) Statistical Independence in Probability Analysis and Number Theory. The Carus Mathematical Monographs, No.12
5. Loève M.(1977) Probability Theory I. Springer, Berlin Heidelberg
6. Billingsley P.(1995) Probability and Measure, Third Ed. John Wiley & Sons, New York
7. Simon M. K., Omura J. K., Scholtz R. A., Levitt B. K.(1994) Spread Spectrum Communications Handbook, Revised Ed. McGraw-Hill, New York
8. Massey J. L. (1988) An introduction to contemporary cryptology. Proc.IEEE 76:533–549
9. Sarwate D. V., Pursley M. B.(1980) Crosscorrelation properties of pseudorandom and related sequences. Proc. IEEE 68:593–619
10. Kohda T.,Tsuneda A.(1993) Pseudonoise Sequences by Chaotic Nonlinear Maps and Their Correlation Properties. IEICE Trans. E76-B: 855–862
11. Kohda T., Tsuneda A.(1994) Explicit Evaluations of Correlation Functions of Chebyshev Binary and Bit Sequences Based on Perron-Frobenius Operator. IEICE. Trans. E77-A: 1794–1800
12. Kohda T., Tsuneda A.(1997) Statistics of Chaotic binary Sequences. IEEE Trans. Information Theory 43:104–112
13. Kohda T.(1999) Sequences of I.I.D. Binary Random Variables Using Chaotic Dynamics. In: Ding C., Helleseth T., Niederreiter H. (Eds.) Sequences and their Applications, 1st International Conference of Sequences and Their Applications. Singapore, December 14–17,1998. Springer Berlin Heidelberg, 297–307
14. Kohda T., Tsuneda A.(1997) Design of sequences of p-ary random variables. 1997 IEEE International Symposium on Information Theory at Ulm, Germany, June 29–July 4, 1997. 17
15. Kohda T., Tsuneda A.(2000) Information Sources Using Chaotic Dynamics. In: Kennedy M.P., Rovatti R., Setti G.(Eds.) Chaotic Electronics in Telecommunications. CRC Press, Boca Raton,81–127

16. Broomhead,D.S., Huke,J.P., Muldoon,M.R.(1999) Codes for spread spectrum applications generated using chaotic dynamical systems. Dynamics and Stability of Systems, 14:95–105
17. Pecora L.M., Carroll T. L.(1990) Synchronization in chaotic systems. Physical Review Letters 64:821–824
18. Cuomo K.M., Oppenheim A. V.(1993) Circuit implementation of synchronized chaos with applications to communications. Physical Review Letters 71:65–68
19. Heidari-Bateni G., McGillem C.D.(1994) A chaotic direct-sequence spread-spectrum communication system.IEEE Trans. Communications 42:1524–527
20. Parlitz U., Ergezinger S.(1994) Robust communication based on chaotic spreading sequences. Physics Letters A 188:146–150
21. Mazzini G., Setti G., Rovatti R.(1997) Chaotic complex spreading sequences for asynchronous DS-CDMA part I : system modeling and results. IEEE Trans. Circuit Syst. CAS-44:937–947
22. Mazzini G., Rovatti R., Setti G.(1999) Interference Minimization by Autocrosscorrelation Shaping in Asynchronous DS-CDMA Systems: Chaos-based Spreading is Nearly Optimal. IEE Electronics Letters 35:1054-55
23. Götz M., Kelber K.,Schwarz W.(1997) Discrete-time chaotic encryption systems part I: statistical design approach. IEEE Trans. Circuit Syst. CAS-44:963–970
24. Umeno K., Kitayama K.(1999) Improvement of SNR with Chaotic Spreading Sequences for CDMA. 1999 IEEE Information Theory and Communication Workshop, South Africa, June 20–25,1999. 106
25. Chen C.C., Biglieri E.,Yao K.(2000) Design of Spread Spectrum Sequences Using Ergodic Theory. 2000 IEEE International Symposium on Information Theory at Sorrento, Italy, June 25–30, 2000. 379
26. Kohda T.,Tsuneda A., Lawrence A.J.(2000) Correlational properties of Chebyshev Chaotic Sequences. J. Time Series Analysis 21:181-191
27. Kohda T., Fujisaki H.(2001) Jacobian Elliptic Chebyshev Rational Maps. Physica D 148:242-254
28. Adler R.L., Rivlin T.J.(1964) Ergodic and mixing properties of Chebyshev polynomials. Proc. Amer.Math.Soc. 15:794-796
29. Rivlin T.J.(1990) Chebyshev Polynomials– From Approximation Theory to Algebra and Number Theory. John Wiley & Sons, New York
30. Grossmann S., Thomae S.(1977) Invariant distributions and stationary correlation functions of one-dimensional discrete processes. Z. Naturforsch. 32a:1353–1363
31. Lasota A., Mackey M.C.(1994) Chaos, Fractals, and Noise. Springer Berlin Heidelberg
32. Geisel T., Fairen V.(1984) Statistical Properties of Chaos in Chebyshev Maps. Physics Letters. 105A:263–266
33. Whittaker E.T., Watson G.N.(1935) A Course of Modern Analysis, Fourth Ed. Cambridge University Press, London
34. Katsura S., Fukuda W.(1985) Exactly Solvable Models Showing Chaotic Behaviour. Physica A 130:597-605
35. Umeno K.(1997) Method of Constructing Exactly Solvable Chaos. Phys. Rev. E 55:5280–5284

First–Order Optimal Approximation of Binary Sequences

Nicholas Kolokotronis[1], Panagiotis Rizomiliotis[1], and Nicholas Kalouptsidis[1]

National and Kapodistrian University of Athens, Department of Informatics and Telecommunications, Panepistimiopolis, 157 71 Athens, Greece

Summary. The determination of the minimum linear span sequence that differs from a given binary sequence, of period $N = 2^n - 1$, by at most one digit is discussed and three methods are presented: the sequential divisions method, the congruential equations method and the phase synchronization method. High level algorithm organizations are provided. Finally, guidelines on sequence characterization and design via the notion of robustness are given.

1 Introduction

Traditionally, binary sequences have been employed in numerous applications of communications and cryptography. Depending on the context these sequences are required to possess certain properties such as long period, equal number of zeros and ones, two–level autocorrelation and large linear span [7], [9]. Let $\boldsymbol{x}$ be a binary sequence of period $N = 2^n - 1$ and $L_{\boldsymbol{x}}$ be its linear span. $L_{\boldsymbol{x}}$ is defined as the order of the least order homogeneous linear recursion satisfied by $\boldsymbol{x}$. The Berlekamp–Massey algorithm needs $2N$ sequence digits in order to determine $L_{\boldsymbol{x}}$ and the linear feedback shift register (LFSR) associated to the least order homogeneous linear recursion [12]. Therefore, the linear span is a critical index for assessing the strength of a sequence against linear cryptanalytic attacks (such as the Berlekamp–Massey algorithm). However, even a large linear span does not ensure that a sequence is cryptographically secure. Consider the periodic extension of the length N vector $(0\ldots01)$. This sequence has linear span N but can be approximated by the all–zero sequence with linear span 0.

In many practical applications small deviations from a given sequence can be tolerated if substantial gains in the linear span are achieved. In this context the approximate realization problem becomes relevant. Let $x = (x_0, \ldots, x_{N-1})$ contain the first N elements of sequence $\boldsymbol{x}$ and e_m be a binary vector of weight 1, the single one being the mth digit, where $m \in \mathbb{Z}_N$ and $\mathbb{Z}_N = \{0, \ldots, N-1\}$. Define $y_m = x + e_m$, where $+$ denotes modulo 2 addition. The approximate realization problem is equivalent to the following minimization problem

$$L_{\boldsymbol{x}}^{(1)} \triangleq \min_{m \in \mathbb{Z}_N} L_{\boldsymbol{y}_m} . \tag{1}$$

It is conceivable that the approximation problem is of interest when $L_x^{(1)}$ is less than the linear span L_x of the original sequence. For this reason necessary conditions are provided in the subsequent sections. The optimal position $m = m^{\text{opt}}$ is referred to as the ***optimal shift*** and the resulting sequence $\boldsymbol{y}_{m^{\text{opt}}}$ is the first–order ***optimal approximation*** of $\boldsymbol{x}$, denoted by $\boldsymbol{x}^{(1)}$. The LFSR with feedback polynomial $f^{(1)}(z)$ that generates $\boldsymbol{x}^{(1)}$ is termed as the first–order ***approximate linear synthesizer*** of $\boldsymbol{x}$. Closely related to the above definitions is the notion of the ***single–error linear span profile*** that depicts the degree to which the linear span changes with respect to the error position.

The solution of the approximate realization problem is the main subject of this paper. Three methods are developed in order to determine the optimal shift, namely the sequential divisions method, the congruential equations method and the phase synchronization method. The sequential divisions method relies on the repetitive application of the Euclidean algorithm by factors of the minimal polynomial of $\boldsymbol{x}$. The congruential equations method works in the frequency domain and determines the optimal lag through a set of linear congruential equations. The solvability of these equations is analyzed and closed form expressions are derived. The phase synchronization method is based on the trace representation and builds upon cyclic equivalence in order to identify m^{opt}.

The three issues of characterization, algorithm implementation and sequence design are tightly related with the approximate realization problem. Characterization is concerned with the description of sequences $\boldsymbol{x}$ whose first–order approximations possess a given linear span. Of particular importance are those sequences for which $L_x^{(1)} \geq L_x$. Such sequences are called ***robust*** because their first–order approximations do not modify their complexity performance. Studies concerned with the single–error linear span profile, explicitly or implicitly address the characterization issue [2], [3] and [10]. In contrast to the above works this paper deals with the more interesting case of binary sequences with period $N = 2^n - 1$. These sequences are produced by typical nonlinear generators [7], [9]. Furthermore, besides $L_x^{(1)}$ and m^{opt}, the first–order optimal approximation $\boldsymbol{x}^{(1)}$ is also determined.

Algorithm implementation is primarily concerned with the design of efficient algorithms in order to determine $L_x^{(1)}$, m^{opt}, $\boldsymbol{x}^{(1)}$ and $f^{(1)}(z)$. An algorithm following a decoding approach is given in [2] for determining the optimal shift but its complexity is high. Furthermore, it does not provide any insight to the design of binary sequences which are robust to such approximation attacks. High level algorithm organizations for the proposed schemes are presented in this paper.

Sequence design deals with the design of sequences which, among other properties, feature high linear spans. This subject is beyond the scope of this work. However, some brief guidelines are given.

2 Background

In this section the trace representation and the Fourier series of a periodic sequence are reviewed in order to illuminate key concepts like linear span, cyclic equivalence and the characteristic phase.

Consider the binary sequence $\boldsymbol{x} = \{x_i\}_{i\geq 0}$, of period $N = 2^n - 1$, generated by a LFSR with feedback polynomial

$$f(z) = f_0(z) \cdots f_{K-1}(z) \tag{2}$$

where $f_\kappa(z)$, $\kappa \in \mathbb{Z}_K$, is an irreducible polynomial of degree n_κ and n_κ divides n. Throughout the paper we assume that $f(z)$ is the minimal polynomial of $\boldsymbol{x}$. This implies that the linear span of $\boldsymbol{x}$ is given by

$$L_{\boldsymbol{x}} = n_0 + \cdots + n_{K-1}\,. \tag{3}$$

Let α be a primitive element of $\mathbb{F}_{2^n}$ and $\alpha^{s_\kappa} \in \mathbb{F}^*_{2^{n_\kappa}}$ be a root of $f_\kappa(z)$. The integer s_κ is assumed to be the coset leader of the cyclotomic coset $\{s_\kappa, 2s_\kappa, \ldots, 2^{n_\kappa-1}s_\kappa\}$ modulo N. Sequence $\boldsymbol{x}$ also admits the representation [13]

$$x_i = \sum_{\kappa=0}^{K-1} x_i^\kappa = \sum_{\kappa=0}^{K-1} \mathrm{tr}_1^{n_\kappa}(\gamma_\kappa \alpha^{t_\kappa i}), \qquad \gamma_\kappa \in \mathbb{F}^*_{2^{n_\kappa}} \tag{4}$$

where t_κ is the coset leader of the cyclotomic coset of $-s_\kappa$ and $\mathbb{F}^*_{2^{n_\kappa}}$ consists of the nonzero elements of $\mathbb{F}_{2^{n_\kappa}}$. The function $\mathrm{tr}_1^n(\alpha) = \alpha + \alpha^2 + \cdots + \alpha^{2^{n-1}}$ is the trace function and it maps elements of $\mathbb{F}_{2^n}$ onto the prime subfield $\mathbb{F}_2$.

Let $x(z) = x_0 + x_1 z + \cdots + x_{N-1}z^{N-1}$ be the polynomial associated to vector $\boldsymbol{x}$. The Discrete Fourier Transform (DFT) of $\boldsymbol{x}$ is given by [1], [8]

$$X_j = \sum_{i=0}^{N-1} x_i \alpha^{ij} = x(\alpha^j) \tag{5}$$

where the exponents of α are taken modulo N. The invertible transform that relates the trace representation and the Fourier representation is unraveled by substituting (4) in (5). Then

$$X_j = \sum_{\kappa=0}^{K-1}\sum_{b=0}^{n_\kappa-1} \gamma_\kappa^{2^b} \sum_{i=0}^{N-1} \alpha^{(2^b t_\kappa + j)i} = \begin{cases} \gamma_\kappa^{2^b}\,, & \text{if } \exists\ (t_\kappa, b) \text{ such that} \\ & 2^b t_\kappa + j \equiv 0 \pmod N \\ 0\,, & \text{otherwise} \end{cases} \tag{6}$$

since if $2^b t_\kappa + j \not\equiv 0 \pmod N$, we have $\sum_{i=0}^{N-1} \alpha^{(2^b t_\kappa + j)i} = 0$. Clearly, equation $2^b t_\kappa + j \equiv 0 \pmod N$ has a solution (t_κ, b), with $b \in [0, n_\kappa - 1]$, if j belongs to the cyclotomic coset of s_κ.

As the above analysis indicates, the linear span of $\boldsymbol{x}$ is determined from its DFT sequence and in particular it equals the number of its non–zero spectral lines. From (5) and (6) we have

$$X_{-2^b t_\kappa} = x(\alpha^{-2^b t_\kappa}) = \gamma_\kappa^{2^b}\,. \tag{7}$$

Two sequences $\boldsymbol{x}$ and $\boldsymbol{y}$, of the same period N, are said to be cyclically equivalent, denoted by $\boldsymbol{x} \sim \boldsymbol{y}$, if and only if vectors x and y are related by a right (or left) cyclic shift.

Lemma 1. *Let $g_\gamma(z) = \mathrm{tr}_1^n(\gamma z^t)$, with $\gamma \in \mathbb{F}_{2^n}^*$, $t \in \mathbb{Z}_N$ and let d be the greatest common divisor of t and N. Define the set*

$$S = \{\boldsymbol{x}_c : \boldsymbol{x}_c = \{g_{\alpha^c}(\alpha^i)\}_{i\geq 0},\ c \in [0, d-1]\}.$$

Then $\forall\ \boldsymbol{x} = \{g_\gamma(\alpha^i)\}_{i\geq 0}$, $\exists\ c \in [0, d-1] : \boldsymbol{x} \sim \boldsymbol{x}_c$. The sequences in S are cyclically distinct and are called class leaders. Moreover, only $\boldsymbol{x}_0$ is constant on cyclotomic cosets.

Proof. Let $q \in \mathbb{Z}_N$ be an integer such that $\gamma = \alpha^q$. The set of all sequences $\boldsymbol{x} = \{g_\gamma(\alpha^i)\}_{i\geq 0}$ is partitioned into d equivalence classes since $\mathrm{ord}(\alpha^t) = \mathrm{ord}(\alpha^d) = N/d$. Application of the Euclidean algorithm gives

$$q = ud + c, \qquad c \in [0, d-1].$$

Consequently

$$\gamma = \alpha^{c+ud} = \alpha^{c-t\nu} \tag{8}$$

where

$$\nu = -u\left(\frac{t}{d}\right)^{\varphi(\frac{N}{d})-1} \bmod \frac{N}{d}$$

and $\varphi(\cdot)$ is the Euler's totient function. It is easily seen from (8) that $\boldsymbol{x} \sim \boldsymbol{x}_c$ and that ν equals the number of right shifts needed to obtain $\boldsymbol{x}$ from $\boldsymbol{x}_c$. □

All nonzero sequences whose minimal polynomial is primitive are cyclically equivalent. In this case $d = 1$ and $\boldsymbol{x}_0$ is called the *characteristic phase* of $\boldsymbol{x}$ [5].

3 The Sequential Divisions Method

It is known that the formal power series representation $\sum_{i=0}^{\infty} x_i z^i$ of sequence $\boldsymbol{x}$ is written as [8], [13]

$$\sum_{i=0}^{\infty} x_i z^i = \frac{x(z)}{1+z^N} = \frac{r(z)}{f(z)}, \qquad \deg(r) < \deg(f)$$

where $\gcd(r, f) = 1$ and $f(z)$ is given by (2). The above equation implies that

$$x(z) = r(z)\frac{1+z^N}{f(z)} = r(z)h(z) \tag{9}$$

where polynomial $h(z) = (1+z^N)/f(z)$ is relatively prime to $f(z)$. Let $\boldsymbol{y}_m$ and $\boldsymbol{e}_m$ be the periodic extensions of vectors y_m and e_m respectively. Obviously $\boldsymbol{y}_m = \boldsymbol{x} + \boldsymbol{e}_m$ and

$$\sum_{i=0}^{\infty} y_{m,i} z^i = \frac{x(z) + z^m}{1 + z^N} = \frac{y_m(z)}{1 + z^N} . \tag{10}$$

The above equation implies that the linear span of the approximation sequence $\boldsymbol{y}_m$ equals $L_{\boldsymbol{y}_m} = N - \deg(\gcd(y_m(z), 1 + z^N))$. Consequently, the minimization problem (1) becomes

$$L_{\boldsymbol{x}}^{(1)} = N - \max_{m \in \mathbb{Z}_N} \deg(\gcd(y_m(z), 1 + z^N)) . \tag{11}$$

Lemma 2. *The greatest common divisor of* $y_m(z)$ *and* $1 + z^N$ *divides* $f(z)$.

Proof. Let $g(z)$ be an irreducible factor of $\gcd(y_m(z), 1 + z^N)$. Then $g(z) \mid y_m(z)$ and $g(z) \mid 1 + z^N$. Since $\gcd(h, f) = 1$, $g(z)$ is either a factor of $h(z)$ or a factor of $f(z)$. Let $g(z)$ be a factor of $h(z)$. Then (10) leads to $g(z) \mid z^m$, since by hypothesis we have that $g(z) \mid y_m(z)$. This is a contradiction and hence $g(z)$ is necessarily a factor of $f(z)$. □

As m varies, Lemma 2 implies that $y_m(z)$ either belongs to the linear cyclic code generated by some factor of $f(z)$, or is relatively prime to $f(z)$.

Corollary 1. *The integer* $m = m^{opt}$ *for which* $y_{m^{opt}}(z)$ *is divisible by the highest degree factor of* $f(z)$ *constitutes an optimal shift.*

The above Corollary implies that $\deg(\gcd(y_m(z), 1 + z^N)) \leq L_{\boldsymbol{x}}$, which if combined with (11) gives

$$N - L_{\boldsymbol{x}} \leq L_{\boldsymbol{x}}^{(1)} . \tag{12}$$

Clearly, approximation is useful if $L_{\boldsymbol{x}}^{(1)} < L_{\boldsymbol{x}}$. From (12) we get that $N - L_{\boldsymbol{x}} \leq L_{\boldsymbol{x}}^{(1)} < L_{\boldsymbol{x}}$, which leads to the following necessary but not sufficient condition

$$N < 2L_{\boldsymbol{x}} . \tag{13}$$

The optimal shift is computed by means of Corollary 1 and the algorithm given in Fig. 1. The main procedure applies successive divisions by some factor of $f(z)$. Let the set $G = \{g_0(z), \dots, g_{2^K-1}(z)\}$ contain all possible factors of $f(z)$ ordered in decreasing order of their degrees, i.e. for $i < j$ it is $\deg(g_i) \geq \deg(g_j)$. The condition $g_i(z) \mid y_m(z)$ is then tested for all m starting from $g_0(z) = f(z)$. The algorithm returns the optimal shift m^{opt} and the attainable linear span $L_{\boldsymbol{x}}^{(1)}$.

The complexity of the sequential divisions algorithm is mainly attributed to the computation of the remainder of the division of $y_m(z)$ by $g_i(z)$ for all $m \in \mathbb{Z}_N$. Considerable computational reductions can be achieved by the method described next.

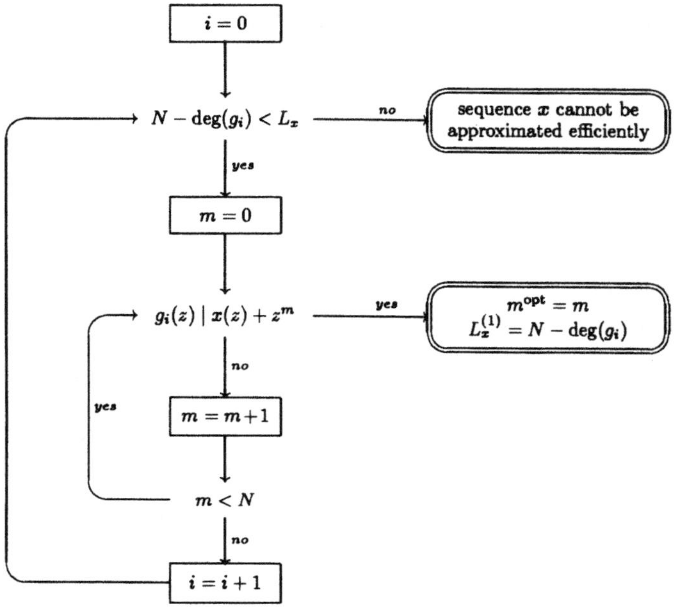

Fig. 1. The sequential divisions algorithm. It outputs the position m where the error should be made as well as the linear span of the approximation sequence

Definition 1. Let $\mathbf{F}$ be the generator matrix, in non–systematic form [1], of the linear cyclic code $\mathcal{F}$ and let $n \geq 0$. The matrix obtained by adding (resp. deleting the last) n rows and columns is called extended (resp. shortened) generator matrix and is denoted by $\mathbf{F}^{(n)}$ (resp. $\mathbf{F}^{(-n)}$).

Let $\mathbf{F}$ and $\widehat{\mathbf{F}}$ be the generator matrices of the $(N, N-k)$ linear cyclic codes generated by $f(z)$ and $z^k f(z^{-1})$ respectively. Then, $\mathbf{F}$ and $\widehat{\mathbf{F}}$ are related by

$$\mathbf{F} = \mathbf{J}_{N-k}\,\widehat{\mathbf{F}}\,\mathbf{J}_N \tag{14}$$

where $\mathbf{J}_n$ is the $n \times n$ order reversing matrix

$$\mathbf{J}_n = \begin{bmatrix} & & 1 \\ & \cdot^{\cdot^{\cdot}} & \\ 1 & & \end{bmatrix}.$$

Suppose that $f(z)$ is factored as $f(z) = g_i(z)g_j(z)$, where $g_i, g_j \in G$ and $\gcd(g_i, g_j) = 1$. Let $\mathbf{G}_i$ and $\mathbf{G}_j$ be the generator matrices of the $(N, N-k_i)$ and $(N, N-k_j)$ linear cyclic codes generated by $g_i(z)$ and $g_j(z)$ respectively. Clearly, $k = k_i + k_j$ and $\mathbf{F}$ is factored as

$$\mathbf{F} = \mathbf{G}_j^{(-k_i)}\,\mathbf{G}_i = \mathbf{G}_i^{(-k_j)}\,\mathbf{G}_j\,. \tag{15}$$

Theorem 1. *Let the minimal polynomial $f(z)$ of sequence x be factored as above and let $m \in \mathbb{Z}_N$ be an error position such that*

$$g_i(z) \mid y_m(z) . \tag{16}$$

Then

$$e_m\, b_j^{\mathrm{T}} = x\, b_j^{\mathrm{T}} \tag{17}$$

where b_j is the $1 \times N$ vector $b_j = x^{(k_j)}\, \mathbf{J}_{N+k_j}\, (\mathbf{G}_j^{(k_j)})^{\mathrm{T}}$ and $x^{(k_j)}$ is the periodic extension of x by k_j digits.

Proof. It follows from (9) that $x(z)$ belongs to the linear cyclic code $\mathcal{H}$ generated by $h(z)$. Let $\mathbf{H}$ be its generator matrix. Then $x = r\,\mathbf{H}$, where the vector r is associated to polynomial $r(z)$ in (9). The generator matrix $\widehat{\mathbf{F}}$ of the dual code $\mathcal{H}^{\perp}$ is generated by $z^{L_x} f(z^{-1})$. Thus $\widehat{\mathbf{F}}\, x^{\mathrm{T}} = 0$, where x^{T} is the transpose of vector x. Then from (14) and (15) the last equation gives

$$\mathbf{G}_i^{(-k_j)}\, \mathbf{G}_j\, \mathbf{J}_N\, x^{\mathrm{T}} = 0 . \tag{18}$$

Since the linear code is cyclic the periodic extension of the codeword x satisfies the same parity check equation with x except it applies on a cyclic shift of x. Therefore (18) leads to

$$\mathbf{G}_i\, \mathbf{G}_j^{(k_j)}\, \mathbf{J}_{N+k_j}\, (x^{(k_j)})^{\mathrm{T}} = 0 . \tag{19}$$

By assumption y_m belongs to the linear cyclic code generated by $\mathbf{G}_i$. Thus, it holds $x + e_m = u\,\mathbf{G}_i$. By substitution in (19) we finally obtain

$$(x + e_m)\, \mathbf{G}_j^{(k_j)}\, \mathbf{J}_{N+k_j}\, (x^{(k_j)})^{\mathrm{T}} = 0$$

which concludes the proof. □

The use of (17) simplifies the sequential divisions algorithm. The right hand–side of (17) is a binary constant c_j, whereas the left hand–side depends on the error position. Let $b_{j,m}$ denote the mth element of vector b_j. Then (17) is rewritten as

$$b_{j,m} = c_j . \tag{20}$$

For example, at the first iteration of the sequential divisions algorithm, where $g_i(z) = f(z)$ and $g_j(z) = 1$, relations (16) and (20) become

$$f(z) \mid y_m(z) \tag{21}$$

and

$$x_{N-1-m} = x_{(N-1)/2}$$

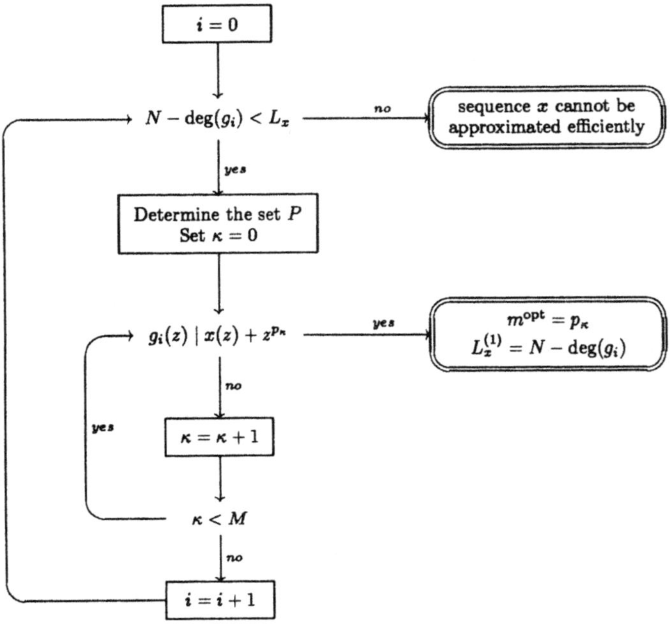

Fig. 2. The enhanced version of the sequential divisions algorithm. For each $g_i \in G$ the set $P = \{p_0, \dots, p_{M-1}\}$ of possible error positions is determined

respectively. The scheme resulting from Theorem 1 is summarized in Fig. 2. However, the reduction in the complexity occurs at the cost of identifying the set $P = \{p_0, \dots, p_{M-1}\}$ of sequence positions which satisfy (20). It is clear that about half the positions will survive this test.

The number of the positions can be further reduced. Indeed, consider the sequence $\boldsymbol{x}^* \sim \boldsymbol{x}$ obtained from $\boldsymbol{x}$ by performing τ right shifts. Let $c_j^* \in \mathbb{F}_2$ be the new constant corresponding to $\boldsymbol{x}^*$ and let $P^* = \{p_0^* - \tau, \dots, p_{R-1}^* - \tau\}$ be the new set of sequence positions which satisfy (20). If τ is chosen so that $c_j^* \neq c_j$, then the optimal shift belongs to $P \cap P^*$.

4 The Congruential Equations Method

The minimization problem (1) can be alternatively solved by means of a system of congruential equations, which is presented next.

Lemma 3. *Let r_κ be the integer satisfying $\alpha^{r_\kappa} = x(\alpha^{s_\kappa})$, where α^{s_κ} is a root of $f(z)$. Then m^{opt} satisfies a subset of the equations*

$$s_\kappa m \equiv r_\kappa \pmod{N}, \qquad \kappa \in \mathbb{Z}_K. \tag{22}$$

Proof. Evaluation of (21) at the roots of $f(z)$ gives the K congruential equations $\alpha^{r_\kappa} + \alpha^{s_\kappa m} \equiv 0 \pmod{\alpha^N + 1}$ which lead directly to (22). Clearly,

$L_x^{(1)}$ satisfies (12) with equality if and only if the system (22) has a solution $m = m^{\text{opt}}$. If not, then $N - L_x < L_x^{(1)}$ and m^{opt} satisfies only a subset of the equations in (22). □

By Definition 1, the subset of congruential equations satisfied by m^{opt} maximizes the sum of the degrees of the minimal polynomials of α^{s_κ} for which $s_\kappa m^{\text{opt}} \equiv r_\kappa \pmod N$.

The solutions of each congruential equation will be conveniently referred to as *local solutions*, while the solution of the system is called *global solution*. Existence and uniqueness of the global solution is next analyzed. To simplify the exposition we treat in detail the case $K = 2$.

Suppose that $f(z) = f_0(z) f_1(z)$ where $f_0(z)$ and $f_1(z)$ are primitive, then each equation of (22) has a local solution if and only if $d_\kappa = \gcd(s_\kappa, N)$ divides r_κ. Since $d_\kappa = 1$ every congruential equation admits the unique local solution

$$m_\kappa = r_\kappa s_\kappa^{\varphi(N)-1} \bmod N, \qquad \kappa = 0, 1\,.$$

Then, a global solution $m^{\text{opt}} = m_0 = m_1$ exists if and only if

$$r_0 s_1 \equiv r_1 s_0 \pmod N\,. \tag{23}$$

Suppose that $f_0(z)$ and $f_1(z)$ are irreducible but not necessarily primitive. The congruential equation corresponding to $f_\kappa(z)$ has d_κ local solutions given by

$$m_{\kappa,i_\kappa} = \frac{r_\kappa}{d_\kappa}\Big(\frac{s_\kappa}{d_\kappa}\Big)^{\varphi(\frac{N}{d_\kappa})-1} + \frac{N}{d_\kappa} i_\kappa \bmod N, \qquad \kappa = 0, 1$$

where $i_\kappa \in [0, d_\kappa - 1]$. A global solution exists, i.e. $m^{\text{opt}} = m_{0,i_0} = m_{1,i_1}$ for some pair (i_0, i_1), if and only if

$$\frac{N}{d_0} i_0 - \frac{N}{d_1} i_1 \equiv \Big(\frac{s_0}{d_0}\Big)^{\varphi(\frac{N}{d_0})-1} \Big(\frac{s_1}{d_1}\Big)^{\varphi(\frac{N}{d_1})-1} \Big(\frac{r_1}{d_1}\frac{s_0}{d_0} - \frac{r_0}{d_0}\frac{s_1}{d_1}\Big) \pmod N\,. \tag{24}$$

This diophantine equation has a solution if and only if

$$\gcd\Big(\frac{N}{d_0}, \frac{N}{d_1}, N\Big) \mid \Big(\frac{s_0}{d_0}\Big)^{\varphi(\frac{N}{d_0})-1} \Big(\frac{s_1}{d_1}\Big)^{\varphi(\frac{N}{d_1})-1} \Big(\frac{r_1}{d_1}\frac{s_0}{d_0} - \frac{r_0}{d_0}\frac{s_1}{d_1}\Big)\,. \tag{25}$$

Since $\gcd(N/d_\kappa, (s_\kappa/d_\kappa)^{\varphi(N/d_\kappa)-1}) = 1$ we subsequently examine whether N/d_κ has any factors in common with $(s_{1-\kappa}/d_{1-\kappa})^{\varphi(N/d_{1-\kappa})-1}$, for $\kappa = 0, 1$. But the integer $s_{1-\kappa}/d_{1-\kappa}$ contains no factors in common with N, which implies that $\gcd(N/d_\kappa, (s_{1-\kappa}/d_{1-\kappa})^{\varphi(N/d_{1-\kappa})-1}) = 1$. From

$$\gcd\Big(\frac{N}{d_0}, \frac{N}{d_1}, N\Big) = \gcd\Big(\frac{N}{d_0}, \frac{N}{d_1}\Big) = \frac{N}{\operatorname{lcm}(d_0, d_1)}$$

and the above analysis, condition (25) becomes

$$\frac{N}{\operatorname{lcm}(d_0,d_1)} \mid \frac{r_1}{d_1}\frac{s_0}{d_0} - \frac{r_0}{d_0}\frac{s_1}{d_1} \Leftrightarrow N \mid \frac{r_1 s_0 - r_0 s_1}{\gcd(d_0,d_1)}\,.$$

Thus, we have proved the following Theorem.

Theorem 2. *The system of congruential equations*

$$\begin{aligned} s_0 m &\equiv r_0 \pmod{N} \\ s_1 m &\equiv r_1 \pmod{N} \end{aligned}$$

has a global solution if and only if $d_\kappa = \gcd(s_\kappa, N) \mid r_\kappa$ *for* $\kappa = 0, 1$ *and*

$$\det\left(\begin{bmatrix} s_0 & r_0 \\ s_1 & r_1 \end{bmatrix}\right) / \gcd(d_0, d_1) \equiv 0 \pmod{N}\,. \tag{26}$$

Condition (26) is extremely useful when large fields are involved because in this case the computation of the Euler's totient function becomes intractable. The above condition can be extended to the general case ($K > 2$), as seen below.

Corollary 2. *The system* (22) *of congruential equations has a solution if and only if* $d_\kappa = \gcd(s_\kappa, N) \mid r_\kappa$ *for all* $\kappa \in \mathbb{Z}_K$ *and there exists an integer* $m \in \mathbb{Z}_N$ *such that*

$$r - m \cdot s = 0 \bmod N \tag{27}$$

where $r = (r_0, \ldots, r_{K-1})$ *and* $s = (s_0, \ldots, s_{K-1})$. *Then* $m = m^{opt}$.

The approach presented in this section is closely related to the sequential divisions scheme. If vectors r and s are linearly dependent, then by Lemma 3 and (27) all roots of $f(z)$ are also roots of $y_{m^{\text{opt}}}(z)$, which implies that $f(z) \mid y_{m^{\text{opt}}}(z)$. This corresponds to the first step of the sequential divisions algorithm. If r and s are linearly independent, then m^{opt} satisfies a subset of the equations in (22). Thus, there exists some $g_i(z) \in G$ such that $g_i(z) \mid y_{m^{\text{opt}}}(z)$, where polynomial $g_i(z)$ is the largest degree factor of $f(z)$ with this property.

5 The Phase Synchronization Method

Let I be the set of coset leaders of the cyclotomic cosets modulo N, and suppose that $|I| = V$. Furthermore, assume that $t_0, \ldots, t_{K-1}$ are the first K elements of I and that $\gamma_\kappa = \alpha^{q_\kappa}$, where γ_κ are defined in (4) and $\kappa \in \mathbb{Z}_K$. Then (4) takes the form

$$x_i = \sum_{\kappa=0}^{K-1} \operatorname{tr}_1^{n_\kappa}(\alpha^{q_\kappa + t_\kappa i})\,. \tag{28}$$

It is easily seen that the trace representation of the error sequence e_m is given by

$$e_{m,i} = \sum_{\kappa=0}^{V-1} \mathrm{tr}_1^{n_\kappa}(\alpha^{t_\kappa(i-m)}) .$$

Adding the error sequence to x we get

$$y_{m,i} = \sum_{\kappa=0}^{K-1} \mathrm{tr}_1^{n_\kappa}((\alpha^{q_\kappa} + \alpha^{-mt_\kappa})\alpha^{t_\kappa i}) + \sum_{\kappa=K}^{V-1} \mathrm{tr}_1^{n_\kappa}(\alpha^{t_\kappa(i-m)}) . \tag{29}$$

It is clear from the above formula that the approximate realization problem becomes equivalent to

$$L_x^{(1)} = N - \max_{m\in\mathbb{Z}_N} \sum_{\kappa=0}^{K-1} w_{m,\kappa} n_\kappa \tag{30}$$

where $w_{m,\kappa} = 1$ if $q_\kappa = -mt_\kappa \bmod N$ and $w_{m,\kappa} = 0$ otherwise.

The maximum reduction of the linear span occurs if $q_\kappa = -m^{\mathrm{opt}} t_\kappa \bmod N$ for all $\kappa \in \mathbb{Z}_K$. According to Lemma 1, all component sequences of x must satisfy $x^\kappa \sim x_0^\kappa$ and be obtained from x_0^κ by applying m^{opt} right shifts. In this case, $L_x^{(1)}$ satisfies (12) with equality, and by (29) we get

$$x_i^{(1)} = \sum_{\kappa=K}^{V-1} \mathrm{tr}_1^{n_\kappa}(\alpha^{t_\kappa(i-m^{\mathrm{opt}})}) .$$

In this case, the component sequences are said to be *synchronized* . Hence, we have proved the following.

Theorem 3. *Let us consider the binary sequence* $x = \{\sum_{\kappa=0}^{K-1} g_{\alpha^{q_\kappa}}(\alpha^i)\}_{i\geq 0}$, *where* $g_{\alpha^{q_\kappa}}(z) = \mathrm{tr}_1^{n_\kappa}(\alpha^{q_\kappa} z^{t_\kappa})$. *Let* $m \in \mathbb{Z}_N$ *be an integer such that*

$$q + m \cdot t = 0 \bmod N \tag{31}$$

where $q = (q_0, \dots, q_{K-1})$ *and* $t = (t_0, \dots, t_{K-1})$. *Then the optimal shift is given by* $m = m^{opt}$.

Relation (27) can be directly derived from (31) and vice–versa. Indeed, for all $\kappa \in \mathbb{Z}_K$ there exists an integer b_κ such that $s_\kappa + 2^{b_\kappa} t_\kappa \equiv 0 \pmod N$, where $b_\kappa \in [0, n_\kappa - 1]$. Let $b = (2^{b_0}, \dots, 2^{b_{K-1}})$. Then

$$- s = b \odot t \bmod N \tag{32}$$

where $\odot$ denotes the Schur–Hadamard product. From (7) and Lemma 3 we get that $\alpha^{r_\kappa} = \gamma_\kappa^{2^{b_\kappa}}$, which, in turn, implies that

$$r = b \odot q \bmod N . \tag{33}$$

```
for m = 0,1,...,N-1
    L[m] = N
    for i = 0,1,...,K-1
        c[i] = q[i] + m · t[i] mod N
        if c[i] > 0
           w[i] = 0
        else
           w[i] = 1
        endif
        L[m] = L[m] - w[i] · n[i]
    endfor
endfor
```

Fig. 3. An algorithm for calculating the single–error linear span profile. The vector $L = (L[0], \dots, L[N-1])$ depicts how the linear span of x changes with respect to the error position m

Combining (32) and (33), relation (27) leads to

$$\begin{aligned} r - m \cdot s = 0 \bmod N &\Leftrightarrow b \odot \{q + m \cdot t\} = 0 \bmod N \\ &\Leftrightarrow q + m \cdot t = 0 \bmod N \end{aligned}$$

since $\gcd(2^{b_\kappa}, N) = 1$ for all $\kappa \in \mathbb{Z}_K$. Hence, the system of congruential equations (31) can be solved by applying the method of section 4, instead of (27).

As a result of (30), a simple algorithm for computing the single–error linear span profile of sequence x is given in Fig. 3. Vectors q and t are defined in Theorem 3, while $n = (n_0, \dots, n_{K-1})$. It's algorithmic complexity is $O(N)$, but it's time complexity is very low, since it does not employ the Berlekamp–Massey algorithm [2].

Example 1. Consider the binary sequence $x = \{g(\alpha^i)\}_{i \geq 0}$, of period 63 and linear span $L_x = 54$, where

$$\begin{aligned} g(z) &= \mathrm{tr}_1^6(\alpha^{28} z) + \mathrm{tr}_1^6(\alpha^{21} z^3) + \mathrm{tr}_1^6(\alpha^{14} z^5) + \mathrm{tr}_1^6(\alpha^{7} z^7) + \mathrm{tr}_1^6(\alpha^{56} z^{11}) \\ &\quad + \mathrm{tr}_1^6(\alpha^{49} z^{13}) + \mathrm{tr}_1^6(\alpha^{42} z^{15}) + \mathrm{tr}_1^6(\alpha^{14} z^{23}) + \mathrm{tr}_1^6(\alpha^{49} z^{31}) . \end{aligned} \tag{34}$$

The vectors $q = (28, 21, 14, 7, 56, 49, 42, 14, 49)$ and $t = (1, 3, 5, 7, 11, 13, 15, 23, 31)$ are linearly dependent, and thus $L_x^{(1)} = 63 - 54 = 9$. Figure 4 indicates that this lower bound is achieved for $m^{\mathrm{opt}} = 35$.

6 Sequence Characterization and Design Issues

Sequence characterization and design issues are of great importance in applications like cryptography [4], [10] and [14]. When designing sequences for

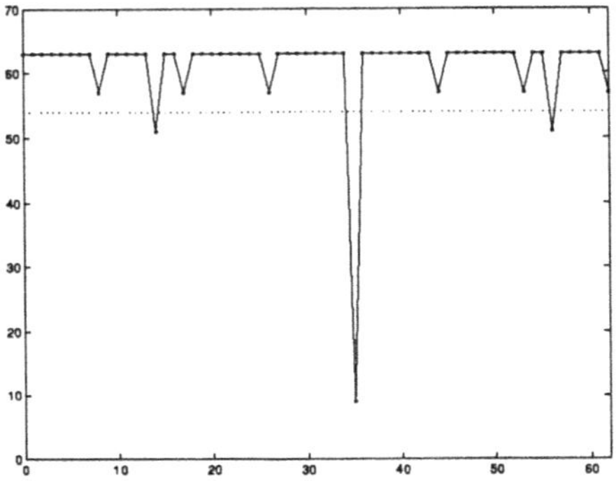

Fig. 4. The single–error linear span profile of sequence $\boldsymbol{x}$. The dotted line illustrates its current linear span $L_x = 54$

key–stream generators our goal is not only to ensure that the key–stream is characterized by large linear span, but also that small changes in the digits of the sequence do not cause any reductions in its linear span. Otherwise, an approximation attack could be proved beneficial. In this respect the following Definition becomes relevant.

Definition 2. The binary sequence $\boldsymbol{x}$, of period $N = 2^n - 1$, is said to be robust to single–error approximations if $L_x^{(1)} \geq L_x$.

Theorem 4. *Let the binary sequence $\boldsymbol{x}$ has linear span satisfying $L_x \leq N - n$. The following statements are equivalent:*

1. *Sequence $\boldsymbol{x}$ is robust and $L_x^{(1)} \geq N - n$.*
2. *For any pair $\boldsymbol{x}^{\kappa_0}$, $\boldsymbol{x}^{\kappa_1}$ of component sequences of $\boldsymbol{x}$ such that $\boldsymbol{x}^{\kappa_0} \sim \boldsymbol{x}_0^{\kappa_0}$ and $\boldsymbol{x}^{\kappa_1} \sim \boldsymbol{x}_0^{\kappa_1}$, it holds*

   $$(\nu_{\kappa_0} - \nu_{\kappa_1}) \cdot \mathrm{lcm}(d_{\kappa_0}, d_{\kappa_1}) \not\equiv 0 \pmod{N}$$

   where $\boldsymbol{x}^{\kappa_0}$ (resp. $\boldsymbol{x}^{\kappa_1}$) is obtained from $\boldsymbol{x}_0^{\kappa_0}$ (resp. $\boldsymbol{x}_0^{\kappa_1}$) by applying ν_{κ_0} (resp. ν_{κ_1}) right shifts.
3. *There exists no reducible factor $g_i(z) \in G$ of $f(z)$ such that $g_i(z) \mid y_m(z)$ for any $m \in \mathbb{Z}_N$.*

According to Theorem 4 robustness is also secured if maximal length sequences are avoided as component sequences in favor of sequences $\boldsymbol{x}^{\kappa} \not\sim \boldsymbol{x}_0^{\kappa}$, which ensures that no local solutions exist for the congruential system (22). This remark leads to the following Corollary.

Corollary 3. *Consider the sequence $\boldsymbol{x}$ whose all component sequences $\boldsymbol{x}^{\kappa}$ are not cyclically equivalent to $\boldsymbol{x}_0^{\kappa}$. Then $L_x^{(1)} = N$ and sequence $\boldsymbol{x}$ is robust.*

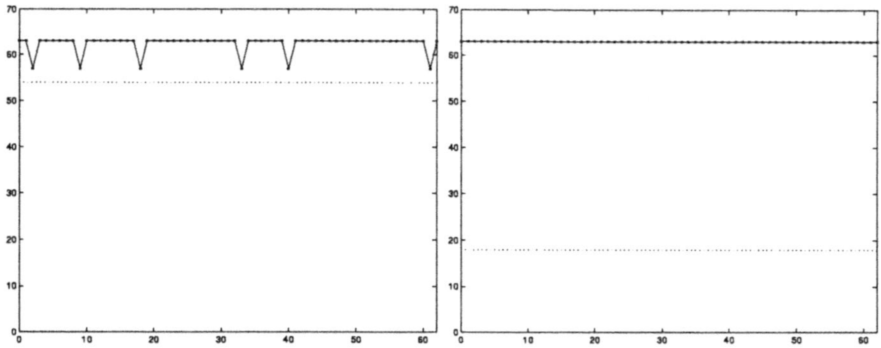

Fig. 5. The single–error linear span profile of the sequences in Examples 2 (at the left) and 3 (at the right), with linear span $L_x = 54$ and $L_x = 18$ respectively

The preceding Theorem enables us to enhance the complexity features of a sequence by suitable modification of the phase shifts of its component sequences. This is illustrated next.

Example 2. Consider the binary sequence of Example 1, generated by the polynomial $g(z)$ in (34). Guided by Theorem 4 $g(z)$ is modified to $g'(z)$. Then the following expression results

$$\begin{aligned} g'(z) = {} & \mathrm{tr}_1^6(\alpha^{23}z) + \mathrm{tr}_1^6(\alpha^{34}z^3) + \mathrm{tr}_1^6(\alpha^{10}z^5) + \mathrm{tr}_1^6(\alpha^{45}z^7) + \mathrm{tr}_1^6(\alpha^{41}z^{11}) \\ & + \mathrm{tr}_1^6(\alpha^{18}z^{13}) + \mathrm{tr}_1^6(\alpha^{62}z^{15}) + \mathrm{tr}_1^6(\alpha^{45}z^{23}) + \mathrm{tr}_1^6(\alpha^{48}z^{31}) . \end{aligned} \tag{35}$$

The minimum linear span obtained by changing one digit of x is $L_x^{(1)} = 63 - 6 = 57$, which is shown at the left–hand side of Fig. 5.

Example 3. The sequence x generated by the second, fourth and seventh term of (35)

$$g''(z) = \mathrm{tr}_1^6(\alpha^{34}z^3) + \mathrm{tr}_1^6(\alpha^{45}z^7) + \mathrm{tr}_1^6(\alpha^{62}z^{15})$$

has a flat single–error linear span profile, namely $L_x^{(1)} = N$. This is shown at the right–hand side of Fig. 5.

7 Conclusions

The determination of the sequence with the minimum linear span that differs from a given binary sequence, of period $N = 2^n - 1$, by at most one digit is discussed and three methods are presented. It is seen that sequence $x^{(1)}$ approximates x efficiently, i.e. $L_x^{(1)} < L_x$, if a sufficient number of its component sequences are synchronized. Finally, criteria for securing robust approximation are provided.

Extension of the above results to more general sequences with elements in arbitrary fields as well as to sequences generated by nonlinear automata is currently explored.

References

1. Blahut, R. E. (1984) Theory and Practice of Error Control Codes. Addison–Wesley Inc.
2. Cusick, T. W., Ding, C. et al. (1998) Stream Ciphers and Number Theory. North–Holland Mathematical Library, Elsevier Science
3. Ding, C., Xiao, G. et al. (1991) The Stability Theory of Stream Ciphers. Lecture Notes in Computer Science, Springer–Verlag
4. Games, R. A., Chan, A. H. (1983) A fast algorithm for determining the complexity of a binary sequence with period 2^n. IEEE Transactions on Information Theory **29**(1), 144–146
5. Golomb, S. W. (1967) Shift Register Sequences. Holden–Day Inc., San Francisco
6. Golomb, S. W. (1980) On the classification of balanced binary sequences of period $2^n - 1$. IEEE Transactions on Information Theory **26**(6), 730–732
7. Groth, E. J. (1971) Generation of binary sequences with controllable complexity. IEEE Transactions on Information Theory **17**(3), 288–296
8. Kalouptsidis, N. (1996) Signal Processing Systems. Telecommunications and Signal Processing Series, John Wiley & Sons Inc.
9. Key, E. L. (1976) An analysis of the structure and complexity of nonlinear binary sequence generators. IEEE Transactions on Information Theory **22**(6), 732–736
10. Kurosawa, K., Sato, F. et al. (2000) A relationship between linear complexity and k-error linear complexity. IEEE Transactions on Information Theory **46**(2), 694–698
11. Lidl, R., Niederreiter, H. (1996) Finite Fields. Encyclopedia of Mathematics and its Applications **20**, Cambridge University Press
12. Massey, J. L. (1969) Shift register synthesis and BCH decoding. IEEE Transactions on Information Theory **15**(1), 122–127
13. Rueppel, R. A. (1986) Analysis and Design of Stream Ciphers. Communications and Control Engineering Series, Springer–Verlag
14. Stamp, M., Martin, C. F. (1993) An algorithm for the k-error linear complexity of binary sequences with period 2^n. IEEE Transactions on Information Theory **39**(4), 1398–1401

On the Uniformity of Distribution of Congruential Generators over Elliptic Curves

Edwin El Mahassni and Igor Shparlinski

Department of Computing, Macquarie University
NSW 2109, Australia
{eelmaha,igor}@ics.mq.edu.au

Summary. We show that the elliptic curve analogue of the linear congruential generator produces uniformly distributed sequences. The proof is based on a recent estimate of D. Kohel and I. E. Shparlinski of character sums over points of elliptic curves.

1 Introduction

Let p be a prime and let $m \geq 1$ be an integer. We denote by $\mathbb{F}_q$ the field of $q = p^m$ and by $\mathbb{F}_p$ the field of $q = p$ elements which we also identify with the set $\{0, 1, \ldots, p-1\}$.

Let w_0 and g be given elements of $\mathbb{F}_q$. We recall that the ***linear congruential generator of pseudorandom numbers*** is the sequence $w_1, w_2, \ldots$ of elements of $\mathbb{F}_q$ defined by the recurrence relation

$$w_n = g w_{n-1}, \qquad n = 1, 2, \ldots, \tag{1}$$

with the ***initial value*** w_0.

Let $\mathbf{E}$ be an elliptic curve over $\mathbb{F}_q$, given by an affine *Weierstrass equation* of the form

$$y^2 + (a_1 x + a_3) y = x^3 + a_2 x^2 + a_4 x + a_6,$$

see [1,22]. It is known, see [1,22], that the set $\mathbf{E}(\mathbb{F}_q)$ of $\mathbb{F}_q$-rational points of $\mathbf{E}$ form an Abelian group under an appropriate composition rule (which we denote by $\oplus$) and with the point at infinity $\mathcal{O}$ as the neutral element. We also recall that

$$|N - q - 1| \leq 2q^{1/2},$$

where $N = |\mathbf{E}(\mathbb{F}_q)|$ is the number of $\mathbb{F}_q$-rational points, including the point at infinity $\mathcal{O}$.

Let $G \in \mathbf{E}$ be a point of order t, that is, t is the size of the cyclic group generated by G. As above, for a given *initial value* $W_0 \in \mathbf{E}(\mathbb{F}_q)$ one can define the *elliptic curve congruential generator of pseudorandom numbers* as the sequence $W_1, W_2, \ldots$ of points of $\mathbf{E}(\mathbb{F}_q)$ satisfying the recurrence relation

$$W_n = G \oplus W_{n-1}, \qquad n = 1, 2, \ldots, \tag{2}$$

with the *initial value* W_0, see [8,10]. Obviously this sequence is purely periodic with period t.

It is know that the linear congruential generator produces cryptographically weak sequences, see [6,12,13,16], thus partially motivated by this fact an elliptic curve congruential generator has been introduced in [10] and then in [8] where some attractive properties of this generator and similar generators have been established. On the other hand, one of the advantages of the linear congruential generator (1) has been a variety of results about the distribution of its elements [14,18,19], Here we use some recent bounds of exponential sums from [15] to establish similar results for the elliptic curve congruential generator (2).

In the case $q = p$ we establish the uniformity of distribution of the components of the points $W_n = (u_n, v_n)$, $n = 1, \ldots, t$, generated by (1) which corresponds to the generator proposed in [10]. We also consider the other special partial case $p = 2$, which corresponds to the version of this generator proposed in [8].

We remark that properties of elliptic curve analogues of some other pseudorandom sequences have been studied in [20,21].

2 Preparations

Let $W_n = (u_n, v_n) \in \mathbb{F}_q^2$ be a sequence generated by (1). Our main tool is bounds of exponential sums of the form

$$S(r,s) = \sum_{n=1}^{t} \exp\left(2\pi i \mathrm{Tr}(ru_n + sv_n)/p\right), \qquad (r,s) \in \mathbb{F}_q^2,$$

where

$$\mathrm{Tr}(z) = z + z^p + \ldots + z^{p^{m-1}}$$

is the trace of $z \in \mathbb{F}_q$ in $\mathbb{F}_p$.

The following result is a partial case of Corollary 1 of [15].

Lemma 1. *Let $(r,s) \in \mathbb{F}_q^2$ and let $f(P) = rx + sy$ where x and y are the components of the point $Q = P \oplus W_0$, $Q = (x,y) \in \mathbf{E}(\mathbb{F}_q)$, be a nonconstant* *function on* $\mathbf{E}(\mathbb{F}_q)$*, then the bound*

$$|S(r,s)| = O\left(q^{1/2}\right)$$

holds.

For a set $\mathcal{M} \subseteq \mathbb{F}_p^2$ we define the *discrepancy* $D(\mathcal{M})$ modulo p as

$$D(\mathcal{M}) = \sup_{\mathcal{B}\subseteq[0,1]^2} \left| \frac{N(\mathcal{B})}{\#\mathcal{M}} - |\mathcal{B}| \right|,$$

where $N(\mathcal{B})$ is the number of pairs

$$\left(\frac{m_1}{p}, \frac{m_2}{p}\right), \qquad (m_1, m_2) \in \mathcal{M},$$

which hit the box $\mathcal{B} = [\alpha_1, \beta_1] \times [\alpha_2, \beta_2] \subseteq [0,1]^2$ of size $|\mathcal{B}| = (\beta_1 - \alpha_1)(\beta_2 - \alpha_2)$.

We make use of a very special case of the *Koksma–Szüsz inequality*, see Theorem 1.21 of [4] (see also Lemma 3.9 of [18] and Chapter 3 of [19].

Lemma 2. *For any set* $\mathcal{M} \subseteq \mathbb{F}_p^2$*, the bound*

$$D(\mathcal{M}) \ll \frac{1}{p} + \frac{1}{\#\mathcal{M}} \sum_{\substack{r,s=0\\ r,s>0}}^{p-1} \frac{1}{\bar{r}\,\bar{s}} \left| \sum_{(m_1,m_2)\in\mathcal{M}} \mathbf{e}\left(rm_1 + sm_2\right) \right|$$

holds, where $\bar{a} = \max\{|a|, 1\}$.

3 Main Results

Let $m = 1$, that is, $q = p$. We denote by $\Delta = D(\mathcal{W})$ the discrepancy of the set

$$\mathcal{W} = \left\{W_n = (u_n, v_n) \in \mathbb{F}_p^2 \;:\; n = 1, 2, \ldots, t\right\},$$

where the sequence $(W_n)_{n=1}^{t}$ is given by (2).

Theorem 1. *The bound* $\Delta = O(t^{-1}p^{1/2}\log^2 p)$ *holds.*

Proof. We show that if $(r,s) \in \mathbb{F}_p^2 \backslash \{(0,0)\}$ then the function $f(P) = rx + sy$ where $P \oplus W_0 = (x,y) \in \mathbf{E}(\mathbb{F}_p)$, is nonconstant over $\mathbf{E}(\mathbb{F}_p)$. We remark that when P runs through the whole set $\mathbf{E}(\mathbb{F}_p)$ the point $Q = P \oplus W_0$ runs through the whole set $\mathbf{E}(\mathbb{F}_p)$ as well. So it is enough to show that $f(Q) = rx + sy$, $Q = (x, y) \in \mathbf{E}(\mathbb{F}_p)$, is nonconstant over $\mathbf{E}(\mathbb{F}_p)$. Assume the opposite, that is, that $rx + sy = c$ for some $c \in \mathbb{F}_p$ and all points $Q = (x, y) \in \mathbf{E}(\mathbb{F}_p)$. Then the system of equations

$$\begin{aligned} x^3 + Ax + B &= y^2; \\ rx + sy &= c; \end{aligned}$$

has $N - 1 \geq p - 2p^{1/2}$ solutions.

On the other hand, if $s \neq 0$ then it implies $x^3 + Ax + B = s^{-2}(c - rx)^2$ which has at most 3 solution in x (and for each of them y is defined uniquely by the second equation). If $s = 0$ then $r \neq 0$ and x is defined uniquely from the second equation, and for this value there are at most 2 values of y defined by the second equation. Thus the above system of equations has at most 3 solutions for any $(r, s) \in \mathbb{F}_p^2 \backslash \{(0, 0)\}$. Assuming that p is large enough thus $N - 1 \geq p - 2p^{1/2} > 3$ we obtain that Lemma 1 applies. Therefore, from Lemma 2 we derive

$$\Delta = O\left(\frac{1}{p} + \frac{1}{t} \sum_{\substack{r,s=0 \\ r,s>0}}^{p-1} \frac{1}{\overline{r}\,\overline{s}} |S(r, s)|\right) = O\left(\frac{1}{p} + \frac{p^{1/2} \log^2 p}{t}\right),$$

and the desired result follows.

Now let $p = 2$, that is $q = 2^m$. For $\mu = 0, 1$ we denote by N_μ the number of solution of the equation

$$\mathrm{Tr}\,(v_n) = \mu, \qquad n = 1, \ldots, t.$$

Theorem 2. *The bound*

$$\max_{\mu=0,1} |N_\mu - t/2| = O(q^{1/2})$$

holds.

Proof. For any pair $(r, s) \in \mathbb{F}_{2^m}^2$, we have

$$N_0 + N_1 = t \qquad \text{and} \qquad N_0 - N_1 = S(0, 1).$$

As before we apply Lemma 1 and the desired result follows.

4 Random Padding and Pollard-type Algorithms on Elliptic Curves

In several encryption applications, it may be necessary to pad messages to some standard format containing redundancy. Electronic voting is such an example: the vote is typically a very short binary message $\mu \in \{0,1\}^k$, where k is small in comparison to the security parameters and key length of the corresponding system. In the extreme case we can have $k = 1$ (a "yes/no" election). Encrypting μ directly is obviously insecure if a deterministic public key system is used; an adversary can try all 2^k possibilities for μ. One can partly solve this adding randomness; encrypt a long random string ρ subject to the condition that the k least significant bits of ρ coincide with μ. However, there may be reasons why a completely random pad is not allowed. Again, we may take the voting scenario as an example of this, where some parts of the padding typically are fixed to avoid malicious users generating random "garbage" votes. Moreover, the results of [2] indicate that even probabilistic encryption of "short" plain texts may not be secure (at least this holds for the El Gamal cryptosystem over $\mathbb{F}_p^*$).

If one uses say RSA, the above is easily implemented, and the security of the vote encoding follows from results on the bit-security of the RSA cryptosystem, see [5,9,11].

Now let us consider the case of elliptic curve cryptography. It is now no longer directly obvious how to find a "sufficiently random" point P in the group $\mathbf{H} \subset \mathbf{E}$ generated G subject to such a condition on the bit-patterns on the components of this point. The most obvious way is to compute rG for a randomly selected r until one of the components of $P = rG$ (say the x-component for elliptic curves over $\mathbb{F}_p$, $p > 2$, or the y-component for elliptic curves over $\mathbb{F}_{2^m}$) has the required bit-pattern. Our method can be used to show that if the order t of G satisfies $t \geq 2^k p^{1/2+\varepsilon}$ then on average 2^k points must be generated before this happens. After that the corresponding point P can be encrypted using, for example, the El Gamal cryptosystem on elliptic curves. The method of [3] can be used to prove the bit security of the El Gamal cryptosystem on elliptic curves and thus the security of the whole scheme. More details about the above mentioned application to electronic voting can be found in [17].

We remark that Theorem 2 cannot extended to the sequence (u_n) because for $q = 2^m$ Lemma 1 does not apply to the function $f(P) = rx$. In fact it is known that the x-components of points on an elliptic curve over $\mathbb{F}_{2^m}$ has some redundancy and in many case $\mathrm{Tr}(u_n)$ is constant, see Chapter 4 of [1]. This fact has also been note in [23] and its implication have been discussed for Pollard like algorithms for the discrete logarithm problem on elliptic curves [7,24]. The paper [23] proposes several ways to overcome this problem. The results of this work show that the trace of the y-component is not biases. In fact the same method can be used to show a more general statement. Let us fixed a basis of $\mathbb{F}_{2^m}$ over $\mathbb{F}_2$ and consider strings of $k < (0.5 - \varepsilon)m$ coordinates of y-component at some fixed positions. Using Lemma 1 one can easily show that these strings are uniformly distributed when the corresponding point $P = (x, y)$ runs through a sufficiently large subgroup of $\mathbf{E}$. Thus the y-component seems to be more appropriate for such algorithms.

5 Remarks

We observe that the implicit constant in the above estimate can easily be evaluated.

We note that when $t < p^{1/2} \log p$ the result of Theorem 1 is trivial. On the other hand, using the bounds of Theorem 3.4 and Theorem 5.5 of [14] for the linear congruential generator (1), one can obtain nontrivial results for sequences of period $t \geq p^{1/3+\varepsilon}$ and $t \geq p^{\varepsilon}$ for all and almost all primes p, respectively. Obtaining similar improvements of Theorem 1 is a challenging problem.

On the other hand, for the fields of characteristic $p = 2$ Theorem 2 provides the result (which is trivial for $t < q^{1/2}$) of about the same strength which those known for the linear congruential generator (1), see [18,19].

For a special type of elliptic curves a more accurate result than that of Theorem 2 has been obtained in [8] but that technique cannot be extended to other curves.

Studying the multidimensional distribution of k-tuples

$$(W_n, \ldots, W_{n+k-1}), \qquad n = 1, 2, \ldots,$$

is an important question as well.

Finally, it is also of interest to study other properties of the sequence (2), in particular to estimate its linear complexity; for some special curves over a field of characteristic two it has been done in [8].

Acknowledgement. The authors thank Mats Näslund for his interest and fruitful discussions, in particular, for the idea to apply our results to padding on elliptic curves. The aforementioned application to electronic voting is due to him as well, and has in fact been implemented as described in [17]. We are also grateful to Nigel Smart for pointing out possible applications of our results to Pollard-type algorithms on elliptic curves.

References

1. I. Blake, G. Seroussi, and N. Smart, *Elliptic Curves in Cryptography*, London Mathematical Society, Lecture Notes Series, **265**, Cambridge University Press, 1999.
2. D. Boneh, A. Joux and P. Q. Nguyen, 'Why textbook ElGamal and RSA encryption are insecure', *Lect. Notes in Comp. Sci.*, Springer-Verlag, Berlin, **1976** (2000), 30–43.
3. D. Boneh and I. E. Shparlinski, 'On the unpredictability of bits of the elliptic curve Diffie–Hellman scheme', *Proc. Crypto'2001*, Springer-Verlag, Berlin, 2001 (to appear).
4. M. Drmota and R. Tichy, *Sequences, discrepancies and applications*, Springer-Verlag, Berlin, 1997.
5. R. Fischlin and C. P. Schnorr, 'Stronger security proofs for RSA and Rabin bits', *J. Cryptology* **13** (2000), 221–244.
6. A. M. Frieze, J. Håstad, R. Kannan, J. C. Lagarias and A. Shamir, 'Reconstructing truncated integer variables satisfying linear congruence', *SIAM J. Comp.*, **17** (1988), 262–280.
7. R. Gallant, R. Lambert and S. Vanstone, 'Improving the parallelized Pollard lambda search on anomalous binary curves', *Math. Comp.*, **69** (2000), 1699–1705.
8. G. Gong, T. A. Berwson and D. A. Stinson, 'Elliptic curve pseudorandom sequence generators', *Lect. Notes in Comp. Sci.*, Springer-Verlag, Berlin, **1758** (2000), 34–49.
9. M. I. González Vasco and M. Näslund, 'A survey of hard core functions', *Proc. Workshop on Cryptography and Computational Number Theory*, Singapore 1999, Birkhäuser, 2001, 227–256.
10. S. Hallgren, 'Linear congruential generators over elliptic curves', *Preprint CS-94-143*, Dept. of Comp. Sci., Cornegie Mellon Univ., 1994, 1–10.
11. J. Håstad and M. Näslund, 'The security of individual RSA bits', *Proc 39th IEEE Symp. on Foundations of Comp. Sci.*, 1998, 510–519.
12. A. Joux and J. Stern, 'Lattice reduction: A toolbox for the cryptanalyst', *J. Cryptology* **11** (1998), 161–185.
13. H. Krawczyk, 'How to predict congruential generators', *J. Algorithms*, **13** (1992), 527–545.
14. S. V. Konyagin and I. Shparlinski, *Character sums with exponential functions and their applications*, Cambridge Univ. Press, Cambridge, 1999.
15. D. R. Kohel and I. E. Shparlinski, 'Exponential sums and group generators for elliptic curves over finite fields', *Lect. Notes in Comp. Sci.*, Springer-Verlag, Berlin, **1838** (2000), 395–404.
16. J. C. Lagarias, 'Pseudorandom number generators in cryptography and number theory', *Proc. Symp. in Appl. Math.*, Amer. Math. Soc., Providence, RI, **42** (1990), 115–143.
17. F. Lindholm, *Evaluation and Implementation of Secure Electronic Voting*, MSc Thesis, TRITA-NA-E0107, Stockholm University, 2001.
18. H. Niederreiter, 'Quasi-Monte Carlo methods and pseudo-random numbers', *Bull. Amer. Math. Soc.*, **84** (1978), 957–1041.
19. H. Niederreiter, *Random number generation and quasi–Monte Carlo methods*, SIAM, Philadelphia, 1992.

20. I. E. Shparlinski, 'On the Naor–Reingold pseudo-random number function from elliptic curves', *Appl. Algebra in Engin., Commun. and Computing*, **11** (2000), 27–34.
21. I. E. Shparlinski and J. H. Silverman, 'On the linear complexity of the Naor–Reingold pseudo-random function from elliptic curves', *Designs, Codes and Cryprography* (to appear).
22. J. H. Silverman, *The arithmetic of elliptic curves*, Springer-Verlag, Berlin, 1995.
23. N. Smart, 'A note on the x-coordinate of points on an elliptic curve in characteristic two', *Inform. Proce. Letters* (to appear).
24. M. Wiener and R. Zuccherato, 'Faster attacks on elliptic curve cryptosystems', *Lect. Notes in Comp. Sci.*, Springer-Verlag, Berlin, **1556** (1999), 190–200.

Further Constructions of Resilient Boolean Functions with Very High Nonlinearity

Subhamoy Maitra[1] and Enes Pasalic[2]

[1] Indian Statistical Institute, 203, B.T. Road, Calcutta 700 035, India.
e-mail: subho@isical.ac.in
[2] Lund University, P.O. Box 118, 221 00 Lund, Sweden.
e-mail: enes@it.lth.se

Summary. One well known method of generating key stream sequences for stream ciphers is to combine the outputs of several Linear Feedback Shift Registers (LFSR) using a combining Boolean function. Here we concentrate on the design of good combining Boolean functions. We provide resilient Boolean functions with currently best known nonlinearity. These functions were not known earlier and the problems of their existence were posed as open questions in literature. Some of the functions we construct here achieve the provable upper bound on nonlinearity for resilient Boolean functions. Our technique interlinks mathematical results with classical computer search.

Keywords: Stream Ciphers, Boolean functions, Nonlinearity, Correlation Immunity, Resiliency.

1 Introduction

Construction of correlation immune and resilient (balanced correlation immune) Boolean functions has been an interesting research area from mid eighties [19,20,8,2,18]. These functions have immediate applications in stream cipher systems. Very recently Sarkar and Maitra [17] have provided weight divisibility results on resilient Boolean functions which in turn present nontrivial upper bounds on the nonlinearity of such functions. Similar kinds of results related to weight divisibility and upper bound on nonlinearity of resilient and correlation immune Boolean functions have also been presented independently by Tarannikov [21] and Zheng and Zhang [22]. Currently the weight divisibility results have been settled by Carlet [4] and Carlet and Sarkar [5] for resilient and correlation immune Boolean functions involving the algebraic degree too. These weight divisibility results have direct consequences towards the upper bound on nonlinearity of these functions and on the other hand, construction of such functions achieving this bound on nonlinearity strengthens the tightness of the upper bound results.

In a more practical direction, these functions have immediate applications in stream cipher cryptosystems. A standard model of stream cipher [19,20,6] combines the outputs of several independent Linear Feedback Shift Register

(LFSR) sequences using a nonlinear Boolean function to produce the key stream. This key stream is bitwise XORed with the message bit stream to produce the cipher. The decryption machinery is identical to the encryption machinery. Getting the kind of Boolean function which we propose here provides the best possible trade-off among the parameters important to resist the known cryptanalytic techniques (see the papers [20,12,9,3] and the references in these papers).

It is now well accepted that for a Boolean function to be used in stream cipher systems, it must satisfy the properties balancedness, high nonlinearity, high algebraic degree and high order of correlation immunity (see Section 2 for definitions). All of the above mentioned parameters are important in resisting different kinds of attacks. Also it is not possible to get the best possible values for each of these parameters separately and there are certain trade-offs involved among the above parameters. Siegenthaler showed [19] that for an n-variable balanced function, of degree d and order of correlation immunity m ($1 \leq m \leq n-2$), we have $m + d \leq n - 1$. Currently, the exact nature of trade-off among order of correlation immunity, nonlinearity and algebraic degree has also been investigated [17,21,22,4,5]. Earlier, a series of papers (see [18,7,13,16] and the references in these papers) have approached the construction problem by fixing the number of variables and the order of correlation immunity and then trying to design balanced Boolean functions with as high nonlinearity as possible. However, the most recent papers [17,21,22,4,5,14,10] concentrate on construction of functions achieving the upper bound on nonlinearity. It should be noted that the current results in construction of resilient functions achieving upper bounds on nonlinearity concentrate on high order of resiliency. In fact for n-variable, m-resilient functions, the value of m is in the range $m > \frac{n}{2} - 2$ and all these functions have three valued Walsh spectra. However, the functions we consider here are of low order of resiliency and the Walsh spectra values are not three valued.

In this paper, for the first time we construct an 8-variable, 1-resilient Boolean function with nonlinearity 116. Earlier all the 8-variable resilient functions of different orders (except order 1) with maximum possible algebraic degree and maximum possible nonlinearity (equal to the upper bound) were known. We here close the issue by proving the case for order 1 also. Our construction uses 5-variable Boolean functions with nonlinearity 11 as input and then concatenates two such functions to construct 6-variable unbalanced 1st order correlation immune function with nonlinearity 26. From such a function we provide a construction of 8-variable unbalanced 1st order correlation immune function with nonlinearity 116. We then exploit a randomized search technique using bit manipulation and linear transformation to get 8-variable, 1-resilient function with nonlinearity 116. Construction of this function was posed as an open question in [17].

Next a method is provided to construct highly nonlinear 1-resilient functions. For 10 variables, this gives nonlinearity 488 functions with algebraic degree both 6 and 8. Construction of such functions have also been left as open questions in [17]. Moreover, we provide a general construction method for 1-resilient, degree $n-2$ functions on n variables, $n \geq 12$ even, which gives better nonlinearity than what presented in [16].

We conclude with n-variable, m-resilient functions for $1 < m \leq \frac{n}{2} - 2$. An existing recursive construction [21,14] is analysed in detail which generates resilient functions on higher number of variables from resilient functions on lower number of variables. The efficiency of this construction depends on proper choice of an initial function. We identify functions with good properties (the 1-resilient functions we present in this paper and the functions from [16]) as initial input to this recursive construction. As example, we use the $(8,1,6,116)$ functions in *desired* form to construct the $(11,3,7,976)$ functions. Also using the functions of [16] as initial ones, we provide a method to construct $(n, m, n-m-1, 2^{n-1}-2^{\frac{2n+m-3}{4}})$ functions when $2n-3m \equiv 3 \bmod 4$.

2 Preliminaries

By Ω_n we mean the set of all Boolean functions of n-variables. A Boolean function f of n variables maps the elements of $\{0,1\}^n$ to $\{0,1\}$. One representation of n-variable Boolean function is by a binary string of length 2^n. Consider the following enumeration of the elements of $\{0,1\}^n : \sigma_0, \ldots, \sigma_{2^n-1}$, where σ_i is the n-bit binary representation of i. Note that the enumeration of $\{0,1\}^{n+1}$ can be described recursively as follows.

$$\begin{array}{ll} \{0,1\}^1 & : 0, 1; \\ \{0,1\}^{n+1} & : 0\sigma_0, \ldots, 0\sigma_{2^n-1}, 1\sigma_0, \ldots, 1\sigma_{2^n-1}. \end{array}$$

Since a Boolean function f is a map from $\{0,1\}^n$ to $\{0,1\}$, it is completely specified by specifying the values of f for the elements σ_i. Thus f is completely specified by the string

$$f(\sigma_0), \ldots, f(\sigma_{2^n-1}).$$

Conversely, we may consider any binary string of length 2^n to uniquely define a Boolean function with respect to the enumeration of $\{0,1\}^n$ given above.

Definition 1. The addition operator over $GF(2)$ is denoted by $\oplus$. An n-variable Boolean function $f(X_n, \ldots, X_1)$ can be considered to be a multivariate polynomial over $GF(2)$. This polynomial can be expressed as a sum of products representation of all distinct k-th order products $(0 \leq k \leq n)$ of the variables. More precisely, $f(X_n, \ldots, X_1)$ can be written as $a_0 \oplus (\bigoplus_{i=1}^{i=n} a_i X_i) \oplus (\bigoplus_{1 \leq i \neq j \leq n} a_{ij} X_i X_j) \oplus \ldots \oplus a_{12\ldots n} X_1 X_2 \ldots X_n$ where the coefficients $a_0, a_{ij}, \ldots, a_{12\ldots n} \in \{0,1\}$. This representation of f is called the algebraic normal form (ANF) of f. The number of variables in the highest order product term with nonzero coefficient is called the algebraic degree, or simply degree of f.

For sake of notational convenience by f we will denote both the ANF and the string representation of f. The binary string representation of Boolean functions is going to be used in our search algorithms. The most frequently used operation will be the concatenation of two Boolean functions. Hence we briefly explain this. Let f_0, f_1 be two n-variable functions, represented by bit strings of length 2^n. By $f = f_0 f_1$ we will denote the $(n+1)$-variable function f, defined by

$$f(0\sigma_i) = f_0(\sigma_i)$$
$$f(1\sigma_i) = f_1(\sigma_i)$$

In view of the above, we prefer to write $f(X_{n+1}, X_n, \ldots, X_1)$ instead of the more usual notation $f(X_1, \ldots, X_n, X_{n+1})$. Algebraically the concatenation operation corresponds to the following operation.

$$f(X_{n+1}, X_n, \ldots, X_1) = (1 \oplus X_{n+1}) f_0(X_n, \ldots, X_1) \oplus X_{n+1} f_1(X_n, \ldots, X_1).$$

We interpret a Boolean function as a binary string of length 2^n. To save space we also represent this binary string of length 2^n as a string of hexadecimal digits of length $\frac{2^n}{4} = 2^{n-2}$.

For binary strings S_1, S_2 of same length λ, we denote by $\#(S_1 = S_2)$ (respectively $\#(S_1 \neq S_2)$), the number of places where S_1 and S_2 are equal (respectively unequal). The Hamming distance between S_1, S_2 is denoted by $d(S_1, S_2)$, i.e. $d(S_1, S_2) = \#(S_1 \neq S_2)$. Another important measure of distance between S_1, S_2 is $wd(S_1, S_2)$, defined as, $wd(S_1, S_2) = \#(S_1 = S_2) - \#(S_1 \neq S_2)$. This is also referred as Walsh distance between two strings. The quantity $wd(,)$ is the correlation between S_1 and S_2 and it is not a metric in the usual sense. Note that, $wd(S_1, S_2) = \lambda - 2\,d(S_1, S_2)$. Also the Hamming weight or simply the weight of a binary string S is the number of ones in S. This is denoted by $wt(S)$. An n-variable function f is said to be balanced if its output column in the truth table contains equal number of 0's and 1's, i.e., $wt(f) = 2^{n-1}$.

Definition 2. Functions of degree at most one are called affine functions. An affine function with constant term equal to zero is called a linear function. The set of all n-variable affine (respectively linear) functions is denoted by $A(n)$ (respectively $L(n)$). The nonlinearity of an n variable function f is $nl(f) = min_{g \in A(n)}(d(f, g))$, i.e. the distance from the set of all n-variable affine functions.

Definition 3. Let $X = (X_n, \ldots, X_1)$ and $\omega = (\omega_n, \ldots, \omega_1)$ both belong to $\{0,1\}^n$ and $X \cdot \omega = X_n\omega_n \oplus \ldots \oplus X_1\omega_1$. Let $f(X)$ be a Boolean function on n variables. Then the Walsh transform of $f(X)$ is a real valued function over $\{0,1\}^n$ that can be defined as $W_f(\omega) = \sum_{X \in \{0,1\}^n} (-1)^{f(X) \oplus X \cdot \omega}$. For a Boolean function f, we define $NZ(f) = \{\omega \mid W_f(\omega) \neq 0\}$, where W_f is the Walsh transform of f.

Definition 4. [8] A function $f(X_n, \ldots, X_1)$ is m-th order correlation immune (CI) iff its Walsh transform satisfies $W_f(\omega) = 0$, for $1 \leq wt(\omega) \leq m$. The function f is balanced iff $W_f(0) = 0$. Balanced m-th order correlation immune functions are called m-resilient functions. Thus, a function $f(X_n, \ldots, X_1)$ is m-resilient iff its Walsh transform satisfies $W_f(\omega) = 0$, for $0 \leq wt(\omega) \leq m$.

The relationship between Walsh transform and $wd(,)$ is given by $W_f(\omega) = wd(f, \bigoplus_{i=1}^{i=n} \omega_i X_i)$.

By an (n, m, d, x) function we denote an n-variable, m-resilient function with degree d and nonlinearity x. By $(n, 0, d, x)$ function we mean a balanced n-variable function with degree d and nonlinearity x. By $[n, m, d, x]$ function we denote an n-variable unbalanced correlation immune function of order m, nonlinearity x and degree d. In the above notation a component is replaced by a '$-$', if it is not specified, e.g., $(n, m, -, x)$, if the degree is not specified.

Next we present a standard technique of constructing correlation immune functions [13]. The method is as follows. *We will refer this method as LT-method throughout this paper.*

Given a function $f \in \Omega_n$, we define $S_f = \{\omega \in \{0,1\}^n \mid W_f(\omega) = 0\}$, where W_f is the Walsh transform of f. If there exists n linearly independent vectors in S_f, then one can construct a nonsingular $n \times n$ matrix B_f whose rows are linearly independent vectors from S_f. Let, $C_f = B_f^{-1}$. Now if we construct a function $f'(X) = f(C_f X)$, then both f', f *have the same nonlinearity and algebraic degree.* Moreover, $W_{f'}(\omega) = 0$ for $wt(\omega) = 1$, where $W_{f'}$ is the Walsh Transform of f'. This ensures that f' is 1st order correlation immune. See [13] for proof of this result. Also if f is balanced then f' is balanced and hence 1-resilient.

3 Construction of $(8, 1, 6, 116)$ Functions

In this section we present the construction method of $(8, 1, 6, 116)$ function. For this purpose we first consider how to construct $[6, 1, 5, 26]$ function with weight 30. Next we provide the construction of $[8, 1, 5, 116]$ function having weight 124. Using these $[8, 1, 5, 116]$ functions, we present the construction of $(8, 1, 6, 116)$ functions. Note that we use similar kind of idea for the constructions of $[6, 1, 5, 26]$ function (see [14]) and $[8, 1, 5, 116]$ function (see [10]) that we have developed earlier. However, we choose the weight of the functions with more care in this initiative.

3.1 Construction of $[6, 1, 5, 26]$ & $[8, 1, 5, 116]$ functions

Consider a 6-variable function h with algebraic degree 5 and nonlinearity 26. Note that by proper permutation of input variables, the function h can always be written as $(1 \oplus X_6)h_1 \oplus X_6 h_2$, where h_1, h_2 are 5-variable functions with algebraic degree 5. Concentrate on a specific degree 5 term in the ANF of h. Clearly there is one variable X_i which is not present in that degree 5 term. Let us consider some permutation of the input variables such that X_i goes to X_6. Then by fixing $X_6 = 0$ we get the function h_1 and putting $X_6 = 1$ we get the function h_2. It is clear to see that the algebraic degree of h_1, h_2 must be 5. It is also important to note that the nonlinearity of both h_1, h_2 must be 11. The functions h_1, h_2 are of degree 5 and hence of odd weight. This gives that the nonlinearity of h_1, h_2 are odd. Now if the nonlinearity of either h_1 or h_2 is less than 11, then the function h cannot have nonlinearity 26.

We initially concentrate on [1, Table 1], where there are just 4 out of 49 equivalence classes with nonlinearity 11. These functions are of weight 11 also. Consider such a function h'_1. Now, there are linear functions l of 5-variables, such that $d(h'_1, l) = 15$ and hence $wt(h'_1 \oplus l) = 15$. In this manner we can select four 5-variable functions from four different representative classes [1] which are of nonlinearity 11 and weight 15. *The choice of weight 15 function is from the motivation that they are closed to balancedness, and its importance will be clearer in the next subsections.* Let the representative functions are as follows which we present as output column of the truth tables in hexadecimal format, $h_1 =$ 565656A8, $h_2 =$ 6633550E, $h_3 =$ 635F7240, $h_4 =$ 724E6351.

Thus, concatenating each representative of the i-th class, $i = 1, \ldots, 4$, with all nonsingular affine transformations applied to a representant of the j-th class, $j = i, \ldots, 4$, we will get a lot of 6-variable functions h. We will choose those with weight 30 and nonlinearity 26. Now, if the function h satisfies $wd(h, l) = 0$ for any set of six independent 6-variable linear functions l, then we can use the LT-method as in Section 2 to get $[6, 1, 5, 26]$ functions with weight 30.

At this point let us mention the following well known [15,10] technical result with proof.

Proposition 1. *Let h be an n-variable function, where n is even and $d \geq 2$. Consider the function $g(X_{n+2}, \ldots, X_1) = X_{n+2}X_{n+1} \oplus h(X_n, \ldots, X_1)$, i.e., the truth table of g is of the form $hhhh^c$. Then g has the nonlinearity $2^n + 2x$. Moreover, if h is an $(n, 1, d, x)$ function, then g is an $(n+2, 1, d, 2^n + 2x)$ function.*

Proof. Note that for any affine function $\lambda \in A(n+2)$, we can write λ in any one of the forms $llll, ll^cll^c, lll^cl^c, ll^cl^cl$, where $l \in A(n)$. Now consider $\lambda = llll$. Then, $d(g, \lambda) = d(hhhh^c, llll) = d(h, l) + d(h, l) + d(h, l) + d(h^c, l) = 2d(h, l) + d(h, l) + d(h^c, l) = 2x + 2^n$. The result is similar for λ of other forms also. This gives the nonlinearity result.

Since, h is 1-resilient, h^c is also 1 resilient. Thus, it is easy to see that $g = hhhh^c$ is also 1-resilient.

Since, g is of the form $g(X_{n+2}, \ldots, X_1) = X_{n+2}X_{n+1} \oplus h(X_n, \ldots, X_1)$, the algebraic degree of g is the same as the algebraic degree of h. □

Now we concentrate on $[8,1,5,116]$ functions. We mention the construction method provided in [10]. Let h be an $[n,1,d,x]$ function, where n is even. Consider the function $g(X_{n+2},\ldots,X_1) = X_{n+2}X_{n+1} \oplus h(X_n,\ldots,X_1)$, i.e. the truth table of g is of the form $hhhh^c$. Now consider the following matrix C_g in terms of LT-method as given in Section 2.

$$C_g = \begin{bmatrix} 1\,0\,0\,.\,.\,0\,0\,1 \\ 1\,0\,0\,.\,.\,0\,1\,0 \\ 0\,0\,0\,.\,.\,1\,0\,0 \\ \cdot\;\cdot\;\cdot\;\cdot\;\cdot\;\cdot\;\cdot\;\cdot \\ \cdot\;\cdot\;\cdot\;\cdot\;\cdot\;\cdot\;\cdot\;\cdot \\ 0\,0\,1\,.\,.\,0\,0\,0 \\ 0\,1\,0\,.\,.\,0\,0\,0 \\ 1\,0\,0\,.\,.\,0\,0\,0 \end{bmatrix}$$

Consider $X = (X_{n+2},\ldots,X_1)$ and we interpret it as a column vector here. Hence, the function $g'(X) = g(C_gX)$ is an $[n+2,1,d,2^n+2x]$ function [13]. Also if the weight of h is y, then weight of g is $2^n + 2y$, and g' also has the same weight.

Hence, if we start from a $[6,1,5,26]$ function h with weight 30, it is possible to construct an $[8,1,5,116]$ function g' with weight 124. To get $(8,1,6,116)$ function we will modify the $[8,1,5,116]$ functions. We will attempt this by the following two methods.

3.2 Attempt for (8, 1, 6, 116) functions : Method 1

Given a function f of n variables, we can very well represent it as a bit string of length 2^n (the output column of the truth table). In that case, by $f[i]$ we mean the i-th bit in the array, where $0 \le i \le 2^n - 1$.

Theorem 1. *Let f be an $[8,1,5,116]$ function with weight 124. Then it is possible to construct a function g, which is either an $(8,1,6,112)$ function or an $(8,1,6,116)$ function.*

Proof. We take the initial idea of this proof from palindromically related functions in [11]. Let $n = 8$. Suppose there exists i_0, i_1 with $i_0 + i_1 = 2^n - 1$, $0 \le i_0 \le 2^{n-2} - 1$ and j_0, j_1 with $j_0 + j_1 = 2^n - 1$, $2^{n-2} \le j_0 \le 2^{n-1} - 1$, such that, $f[i_0] = f[i_1] = f[j_0] = f[j_1] = 0$, $g[i_0] = g[i_1] = g[j_0] = g[j_1] = 1$, and $g[k] = f[k]$ if $k \ne i_0, i_1, j_0, j_1$. Note that the function f is not balanced and hence cannot be of the form $f_1 f_1^c$, where $f_1, f_2 \in \Omega_7$. Thus the function f must satisfy the condition $f[i_0] = f[i_1] = f[j_0] = f[j_1] = 0$, and we can construct g by changing the bits from 0 to 1.

Using the result of palindromically related functions [11], it can be seen that if f is correlation immune, then g is also correlation immune. Also it is clear that $wt(g) = wt(f) + 4 = 128$. Hence, g is balanced. So g is 1-resilient.

Since f is of algebraic degree 5, if we write $f = f_1 f_2 f_3 f_4$, the concatenation of four 6-variable functions, then $wt(f_i)$ is even for $1 \leq i \leq 4$. Now, the way we have changed the bits, if we write $g = g_1 g_2 g_3 g_4$, the concatenation of four 6-variable functions, then $wt(g_i)$ is odd for $1 \leq i \leq 4$. Thus, algebraic degree of g must be at least 6. Also using Siegenthaler's inequality [19] on algebraic degree of resilient functions, we know that the maximum algebraic degree of g is 6. Thus g is of algebraic degree 6.

As we have changed 4 bit positions, $nl(f) - 4 \leq nl(g) \leq nl(f) + 4$. We know that the nonlinearity of 1-resilient functions must be divisible by 4 [17]. Hence, $nl(g)$ can be 112 or 116 or 120. However, it can not be 120, since bent functions are not balanced. Thus the result. □

Now we can generate a large database of $[8, 1, 5, 116]$ functions f and then can modify the functions to get g as given in Theorem 1. This will generate $(8, 1, 6, x)$ functions with $x = 112$ or 116.

3.3 Attempt for (8, 1, 6, 116) functions : Method 2

Let f be an $[8, 1, 5, 116]$ function with weight 124. Select any four bits i_1, i_2, i_3, i_4 of f such that $f[i_1] = f[i_2] = f[i_3] = f[i_4] = 0$. Then construct a function g, such that, $g[i_1] = g[i_2] = g[i_3] = g[i_4] = 1$ and g is same as f for all other bits. Note that g is balanced. Now check the nonlinearity of f. If it stays at 116, then we try to find out a set of 8 independent nonzero linear functions of 8-variables and if such a set exists, we use LT-method used in Section 2 to get an $(8, 1, -, 116)$ function. We know [4,5] that the maximum possible nonlinearity of an n-variable (n even), m-resilient ($m \leq \frac{n}{2} - 2$), degree d function is $2^{n-1} - 2^{\frac{n}{2}-1} - 2^{m+1+\lfloor \frac{n-m-2}{d} \rfloor}$. Thus, the degree of an $(8, 1, -, 116)$ function must be 6, since if the degree is less than 6, then the nonlinearity will be at most 112. Hence any $(8, 1, -, 116)$ function must be an $(8, 1, 6, 116)$ function.

3.4 An (8, 1, 6, 116) function

In our experiment, all the functions we got using Method 1 are $(8, 1, 6, 112)$ functions. Though Method 1 is not successful for this purpose, we will later show (see Remark 1 in Section 4) that this method can be used successfully indeed in some other situations. The $(8, 1, 6, 116)$ functions are found from the Method 2 only. The following is one example.

Let us start with the $[6, 1, 5, 26]$ function $h =$ `3A1A6C17E8D1582D`, which has weight 30. From this we get the $[8, 1, 5, 116]$ function of weight 124, `1E111E11EEEEE11EEEE1111EE111E1E1877787888878887778877778888887777`. Then we use Method 2 on this $[8, 1, 5, 116]$ function to get one $(8, 1, 6, 116)$ function, `6F4FC635EE280B7135159C4BB472512BCA8A932DD2E4A84D90D0C977CABEF217`.

This construction completely solves the maximum nonlinearity issue of resilient functions on 8 variables. The following table shows the maximum nonlinearity corresponding to each order of resiliency and note that it is possible to construct such functions. Also it is very clear from [21,4,5] that all these functions possess the maximum possible algebraic degree $(7-m)$ where m is the order of resiliency. Each of these functions achieves the nonlinearity equal to the upper bound.

order of resiliency	1	2	3	4	5	6
maximum achievable nonlinearity	116	112	112	96	64	0

Table 1. Nonlinearity results for 8-variable Boolean functions.

4 1-resilient Functions on Even Number of Variables

At this point we already have $(8, 1, 6, 2^{8-1} - 2^{\frac{8}{2}} + 2^{\frac{8}{2}-2})$ functions. Thus, using such a function as the initial one we get the following result.

Theorem 2. *For even $n \geq 8$, it is possible to construct $(n, 1, 6, 2^{n-1} - 2^{\frac{n}{2}} + 2^{\frac{n}{2}-2})$ functions.*

Proof. Given an $(m, 1, 6, 2^{m-1} - 2^{\frac{m}{2}} + 2^{\frac{m}{2}-2})$ function we can use Proposition 1 to get $(m+2, 1, 6, 2^m + 2(2^{m-1} - 2^{\frac{m}{2}} + 2^{\frac{m}{2}-2}) = 2^{m+1} - 2^{\frac{m+2}{2}} + 2^{\frac{m+2}{2}-2})$ function, which is the inductive step. The initial condition is proved using $(8, 1, 6, 2^{8-1} - 2^{\frac{8}{2}} + 2^{\frac{8}{2}-2} = 116)$ function. □

Putting $n = 10$, from Theorem 2 we get $(10, 1, 6, 488)$ function, which has also been placed as an open problem in [17,14]. Also this function has an important significance in terms of weight divisibility results. From the weight divisibility results [4,5], it is known that the maximum possible nonlinearity of an n-variable, m-resilient, degree d function is $2^{n-1} - 2^{\frac{n}{2}-1} - 2^{m+1+\lfloor\frac{n-m-2}{d}\rfloor}$. For the $(10, 1, 6, -)$ function, $n = 10, m = 1, d = 6$ and hence the maximum possible nonlinearity is $2^9 - 2^4 - 2^3 = 488$. Thus, the nonlinearity 488 is the maximum possible nonlinearity of any $(10, 1, 6, -)$ function. Hence, this result in turn shows the tightness of the upper bound on nonlinearity obtained from weight divisibility results [4,5].

Note that the construction of Theorem 2 provides functions with algebraic degree 6. We modify the construction in Theorem 2 to construct 1-resilient n-variable $(n \geq 10)$ functions with maximum possible algebraic degree $(n - 2)$. The function $(8, 1, 6, 116)$ is already optimized with respect to algebraic degree.

Theorem 3. *For even $n \geq 10$, it is possible to construct $(n, 1, n-2, 2^{n-1} - 2^{\frac{n}{2}} + 2^{\frac{n}{2}-2} - 4)$ functions.*

Proof. We will consider an $(n, 1, 6, 2^{n-1} - 2^{\frac{n}{2}} + 2^{\frac{n}{2}-2})$ function f'. Note that this function f' is of the form $hhhh^c$, where h is an $(n-2)$ variable function (see proofs of Proposition 1 and Theorem 2). First note that we can not directly use the function f' in this construction. However, we will use a function f which has the same parameter as f'. We will also construct f from f' itself. We will come to that at the end of this proof.

Now, the proof of the theorem goes in the same line as in Theorem 1. We need an $(n, 1, 6, 2^{n-1} - 2^{\frac{n}{2}} + 2^{\frac{n}{2}-2})$ function f with the following condition. Consider i_0, i_1 with $i_0 + i_1 = 2^n - 1$, $0 \le i_0 \le 2^{n-2} - 1$ and j_0, j_1 with $j_0 + j_1 = 2^n - 1$, $2^{n-2} \le j_0 \le 2^{n-1} - 1$, such that $f[i_0] = f[i_1] = a$, and $f[j_0] = f[j_1] = b$, with $a + b = 1$. Note that a, b can take the values 0 or 1. We call i_0, i_1 and j_0, j_1 two pairs of palindromic positions since considering f as a binary string of length 2^n (output column of the truth table), we find those i_0, i_1 (similarly j_0, j_1) at same distances from top and bottom. Then we can construct a function g as follows, $g[i_0] = g[i_1] = 1 - a$, $g[j_0] = g[j_1] = 1 - b$, and $g[k] = f[k]$ if $k \neq i_0, i_1, j_0, j_1$.

Using the result of palindromically related functions [11], it can be seen that if f is correlation immune, then g is also correlation immune. Also it is clear that $wt(g) = wt(f) = 2^{n-1}$. Hence, g is balanced. So g is 1-resilient.

Since f is of algebraic degree 6, if we write $f = f_1 f_2 f_3 f_4$, the concatenation of four $(n-2)$-variable functions, then $wt(f_i)$ is even for $1 \le i \le 4$. Now, the way we have changed the bits, if we write $g = g_1 g_2 g_3 g_4$, the concatenation of four $(n-2)$-variable functions, then $wt(g_i)$ is odd for $1 \le i \le 4$. Thus, algebraic degree of g must be at least $n - 2$. Also using Siegenthaler's inequality [19] on algebraic degree of resilient functions, we know that the maximum algebraic degree of g is $n - 2$. Thus g is of algebraic degree $n - 2$.

Since we have changed 4 bit positions, $nl(f) - 4 \le nl(g) \le nl(f) + 4$. We know that the nonlinearity of 1-resilient functions must be divisible by 4 [17]. Hence, $nl(g)$ can be $2^{n-1} - 2^{\frac{n}{2}} + 2^{\frac{n}{2}-2} - 4$, *or* $2^{n-1} - 2^{\frac{n}{2}} + 2^{\frac{n}{2}-2}$, *or* $2^{n-1} - 2^{\frac{n}{2}} + 2^{\frac{n}{2}-2} + 4$. We here conservatively consider the least possible value.

Now we come back to construction of f from f'. Note that f' is of the form $hhhh^c$, where h is an $(n-2)$ variable function. Thus we can not get i_0, i_1 (with $i_0 + i_1 = 2^n - 1, 0 \le i_0 \le 2^{n-2} - 1$) such that $f'[i_0] = f'[i_1] = a$. Let $h = h_1 h_2$. Thus, $f' = h_1 h_2 h_1 h_2 h_1 h_2 h_1^c h_2^c$. We consider $f = h_1 h_1 h_1 h_1^c h_2 h_2 h_2 h_2^c$. Note that f is basically the same function as f' with permutation of input variables. Hence, f is an $(n, 1, 6, 2^{n-1} - 2^{\frac{n}{2}} + 2^{\frac{n}{2}-2})$ function. Now note that, if $h_1 h_2$ has two pairs of palindromic positions, then the function f will also have two pairs of palindromic positions. We start initially from the $(8, 1, 6, 116)$ function, which has two such pairs of palindromic positions. Hence the existence of such an f is guaranteed for all even $n \ge 10$. □

Remark 1. Using Theorem 3, it is possible to get a $(10, 1, 8, 484)$ *function. Same quality result is available in [16, Theorem 9]. Now it is important to note that, in the proof of Theorem 3 we mentioned that this is the least possible value of nonlinearity in our construction. In fact we exhausted all possibilities by changing two pairs of palindromic positions as given in Theorem 3 for an* $(10, 1, 6, 488)$ *function and found the nonlinearity 488. Thus, we could construct a* $(10, 1, 8, 488)$ *function which is completely described as follows,*

6F4FC675EE280B7135159C4BB472512B6F4FC635EE280B7135159C4BB472512B
6F4FC635EE280B7135159C4BB472512B90B0398A11D7F48ECAEA63B44B8DAED4
CA8A932DD2E4A84D90D0C977C8BEF217CA8A932DD2E4A84D90D0C977CABEF217
CA8A932DD2E4A84D90D0C977CABEF21735756CD22D1B57B26F2F368837410DE8.

This is a better result than [16, Theorem 9]. Similarly there is a possibility of getting $(n, 1, n-2, 2^{n-1} - 2^{\frac{n}{2}} + 2^{\frac{n}{2}-2})$ *or* $(n, 1, n-2, 2^{n-1} - 2^{\frac{n}{2}} + 2^{\frac{n}{2}-2} + 4)$ *functions, if one can select proper palindromic positions.*

Note that in [16], 1-resilient functions with nonlinearity $2^{n-1} - 2^{\frac{n}{2}} + 2^{\frac{n}{2}-2}$ could be achieved for $n \geq 12$, however, the maximum possible algebraic degree for that construction was $\frac{n}{2} + 2$ [16, Theorem 7], which is much less than the optimized degree $(n-2)$. Here we concentrate on functions with optimized degree only.

For $n \geq 12$, the currently best known nonlinearity achieved by 1-resilient functions [16, Theorem 8] with maximum possible algebraic degree $(n-2)$ is $2^{n-1} - 2^{\frac{n}{2}} + y$, where y is the maximum possible nonlinearity of an $(\frac{n}{2} - 1)$-variable 1-resilient function with algebraic degree $(\frac{n}{2} - 3)$. We estimate y as $2^{\frac{n}{2}-2} - 2^{\frac{n}{4}-2} - 4$ [17], the upper bound of nonlinearity for an $(\frac{n}{2} - 1)$-variable function which is 1-resilient. So, the currently best known nonlinearity achieved by 1-resilient functions [16] with maximum possible algebraic degree $(n-2)$ is $2^{n-1} - 2^{\frac{n}{2}} + 2^{\frac{n}{2}-2} - 2^{\frac{n}{4}-2} - 4$.

We here achieve the nonlinearity $2^{n-1} - 2^{\frac{n}{2}} + 2^{\frac{n}{2}-2} - 4$ for 1-resilient functions with maximum possible algebraic degree $(n-2)$.

Hence, considering the functions with maximum possible algebraic degree, we find that for $n \geq 12$ *the nonlinearity achieved in this paper for* 1*-resilient Boolean function (algebraic degree* $n-2$*) is* $2^{n-1} - 2^{\frac{n}{2}} + 2^{\frac{n}{2}-2} - 4$*, which is strictly greater than the nonlinearity* $2^{n-1} - 2^{\frac{n}{2}} + 2^{\frac{n}{2}-2} - 2^{\frac{n}{4}-2} - 4$ *achieved in [16] for* 1*-resilient Boolean function (algebraic degree* $n-2$*). The constructions* $(8, 1, 6, 116)$ *and* $(10, 1, 8, 488)$ *provides better nonlinearity than [16] for the cases* $n = 8, 10$ *and also these are placed as important open questions in [17].*

5 Functions with Order of Resiliency $m > 1$

Here we analyse an existing construction technique [21,14] for generating resilient functions on higher number of variables from functions on lower number of variables. This construction was first proposed in [21] and then simplified in [14]. The construction provides $(n+3)$-variable, $(m+2)$-resilient function from n-variable, m-resilient initial function. In [21], the requirement was two initial functions with some specific properties and this was later modified in [14] where the requirement was only one initial function. This initial function [14] has to be in *desired* form. We show that the conditions mentioned in the *desired* form [14] need to be interpreted in different manners for different cases. Also we will use this recursive construction to provide resilient functions of higher order which were not known earlier. Let us now start with the definition and basic construction method [14].

Definition 5. An $(n, m, d, -)$ function f is in *desired* form if it is of the form $f = (1 + X_n)f_1 + X_n f_2$, where f_1, f_2 are $(n-1, m, d-1, -)$ functions.

Construction 1. [14] *Let f be an (n, m, d, x) function in* **desired** *form, where f_1, f_2 are both $(n-1, m, d-1, -)$ functions. Let $F = ff^cf^cf$, or written in ANF, $F = X_{n+2}+X_{n+1}+f$. Let $G = ghh^cg^c$ where $g = f_1f_1^c$ and $h = f_2f_2^c$. In ANF the function G is given by, $G = (1 + X_{n+2} + X_{n+1})f_1 + (X_{n+2} + X_{n+1})f_2 + X_{n+2} + X_n$. In the language of [21], the function G above is said to depend quasilinearly on the pair of variables (X_{n+2}, X_{n+1}). We construct a function H in $n+3$ variables in the following way, $H = (1+X_{n+3})F + X_{n+3}G$. Then the function H constructed from f is an $(n+3, m+2, d+1, 2^{n+1}+4x)$ function in the* **desired** *form.*

For the functions F, G, we have $NZ(F) \cap NZ(G) = \emptyset$, i.e., the set of nonzero Walsh coefficients of the constituent functions of H are disjoint. This has been clearly explained in [14]. However, this is not very clear what is the relationship among $NZ(f_1), NZ(f_2)$, where f_1, f_2 are the constituent functions of the initial function f. We will look into this problem below.

Proposition 2. *Let $m > \frac{n}{2} - 2$ and f be an $(n, m, n-m-1, 2^{n-1} - 2^{m+1})$ function in* desired *form. Also $f = (1 \oplus X_n)f_1 \oplus X_n f_2$, where f_1, f_2 are $(n-1, m, n-m-2, -)$ functions. Then $NZ(f_1) \cap NZ(f_2) = \emptyset$.*

Proof. Note that, $nl(f_1) \geq nl(f) - 2^{n-2} = 2^{n-2} - 2^{m+1}$. Given the upper bound of nonlinearity of a resilient function [17], $nl(f_1) \leq 2^{n-2} - 2^{m+1}$. Thus, $nl(f_1) = 2^{n-2} - 2^{m+1}$. The nonlinearity for f_2 is same. Also from [17] we know that the Walsh spectra of f, f_1, f_2 are three valued $0, \pm 2^{m+2}$.

Now consider, $NZ(f_1) \cap NZ(f_2) \neq \emptyset$. Then there exists an $(n-1)$-variable linear function l (may be degenerate) such that $wd(f_1, l) \neq 0$ and also $wd(f_2, l) \neq 0$. If both $wd(f_1, l), wd(f_2, l)$ are of same sign, then the Walsh spectra of f will contain the value $+2^{m+3}$ or -2^{m+3}, which is a contradiction. If $wd(f_1, l), wd(f_2, l)$ are of different sign, then without loss of generality, consider, $wd(f_1, l) = 2^{m+2}$ and $wd(f_2, l) = -2^{m+2}$. Now, $wd(f, X_n \oplus l) = wd(f_1, l) - wd(f_2, l) = 2^{m+3}$, which is again not possible. □

The above proposition guarantees that if we get an n-variable, m-resilient $(m > \frac{n}{2} - 2)$ Boolean function f with maximum possible nonlinearity in *desired* from, then the constituents functions f_1, f_2 (which are $(n-1)$-variable, m-resilient functions with maximum possible nonlinearity) will satisfy the condition $NZ(f_1) \cap NZ(f_2) = \emptyset$.

However, Proposition 2 does not always hold for the functions where the Walsh spectra is not three valued. This happens in the following two cases. (1) Consider the range $m > \frac{n}{2} - 2$, where the functions do not possess the maximum possible nonlinearity. Then the Walsh spectra is not three valued.

(2) Consider the range $m \leq \frac{n}{2} - 2$, and the functions may or may not possess the maximum possible nonlinearity. Here the Walsh spectra is not three valued irrespective of the value of nonlinearity, whether it is maximum possible or not.

There is also a fundamental difference in the resulting functions out of this construction for different order of resiliency. (1) For $m > \frac{n}{2} - 2$, if one can start with a *desired* function with maximum possible nonlinearity, then the construction when repeatedly used will generate an infinite sequence of functions which are of maximum possible nonlinearity. (2) For $m \leq \frac{n}{2} - 2$, even if one can start with a *desired* function with maximum possible nonlinearity, then the construction when repeatedly used will generate an infinite sequence of functions which are not guaranteed to be of maximum possible nonlinearity.

We also like to clarify the issue on nonlinearity of f_1, f_2, the constituent functions of f. Given the (n, m, d, x) function f, we get, both $nl(f_1), nl(f_2) \geq x - 2^{n-2}$. Now we have the following cases. (1) If the function f satisfy the maximum possible nonlinearity and $m > \frac{n}{2} - 2$, then $nl(f_1), nl(f_2) = x - 2^{n-2}$. (2) For $m > \frac{n}{2} - 2$, if the function f does not possess maximum possible nonlinearity, then it is possible that either of f_1, f_2 can have nonlinearity strictly greater than $x - 2^{n-2}$. (3) For $m \leq \frac{n}{2} - 2$, it is possible that either of f_1, f_2 can have nonlinearity strictly greater than $x - 2^{n-2}$ irrespective of whether it achieves the maximum possible nonlinearity or not.

Depending on the discussion made above, it is important to clarify the concept of *desired* functions once again. *Given an n-variable, m-resilient function f of the form $f = (1 + X_n)f_1 + X_n f_2$, the only requirement is that the constituent $(n-1)$-variable functions f_1, f_2 should also be m-resilient. It is not required to overload any restriction on Walsh spectra of f_1, f_2. This is automatically taken care of.*

5.1 Use of *desired* Form

To use the above construction on the $(8, 1, 6, 116)$ functions we found here, we need a function in *desired form*. One way is to get an $(8, 1, 6, 116)$ function as a concatenation of two $(7, 1, 5, 52)$ functions f_1, f_2. However, this is not automatically guaranteed. As example, the $(8, 1, 6, 116)$ function presented in Subsection 3.4 is a concatenation of two $(7, 0, 5, 52)$ functions, not $(7, 1, 5, 52)$ functions.

We searched the database of $(8, 1, 6, 116)$ functions and generate functions with permutations of input variables. Note that these functions are also $(8, 1, 6, 116)$ functions. We check whether these new functions are in *desired* form. The following $(8, 1, 6, 116)$ function is in *desired* form with nonintersecting Walsh spectra of its constituent $(7, 1, 5, 52)$ functions. The function is, `A0BE6E1D8191B45FFA8289477E76D2056C72A6D1B2A3876C364E658B4DC5E136`.
This is the way we found $(8, 1, 6, 116)$ functions in desired form. Thus using Construction 1, we get an $(11, 3, 7, 976)$ function. Note that the previously best known nonlinearity for an $(11, 3, 7, -)$ function is 960 [16, Section 6].

It is of interest to concentrate on the following $(8, 1, 6, 116)$ function f. Here, $NZ(f_1) \cap NZ(f_2) \neq \emptyset$, f_1 is a $(7, 1, 5, 52)$ function and f_2 is $(7, 1, 5, 56)$. This $(8, 1, 6, 116)$ function is in *desired* form with intersecting Walsh spectra of its constituent functions. The function is,
CF6730989E329A3A4C795C59190CE6F3AC592DB1F90C871B309FD8E165CA724B.

We list the functions in the series provided by repeated use of the Construction 1. These functions are $(8, 1, 6, 116)$, $(11, 3, 7, 976)$, $(14, 5, 8, 8000)$, $(17, 7, 9, 64768), \ldots$ etc.

Next we concentrate on the functions provided in [16, Theorem 10b] and the Construction 1 [21,14] given above to get highly nonlinear resilient functions with odd order. This is the first time where Construction 1 is used for the functions with $m \leq \frac{n}{2} - 2$ in generalized manner.

Theorem 4. *It is possible to construct* $(n, m, n-m-1, 2^{n-1} - 2^{\frac{2n+m-3}{4}})$ *functions when* $2n - 3m \equiv 3 \bmod 4$.

Proof. We start with a p-variable (p odd) function f provided in [16, Theorem 10b] with order of resiliency 1 and nonlinearity $2^{p-1} - 2^{\frac{p-1}{2}}$. It is clear that the functions provided in [16, Theorem 10b] are in *desired* form as each of f_1, f_2 are also 1-resilient functions. Also the algebraic degree of the function f is $p - 2$.

Thus given some odd $m > 1$, we will calculate the value $t = \frac{m-1}{2}$. As the Construction 1 increases the order of resiliency by 2 in each step and the number of variables by 3 in each step we have to start with a function on $p = n - 3t$ variables and order of resiliency 1. We will then recursively use the Construction 1 by t times.

Now we will show that in this series of construction, we will always get $(n_i, m_i, n_i - m_i - 1, 2^{n_i-1} - 2^{\frac{2n_i+m_i-3}{4}})$ functions. For $i = 0$, we take $n_0 = p, m_0 = 1$. This satisfies the nonlinearity $2^{n_i-1} - 2^{\frac{2n_i+m_i-3}{4}} = 2^{p-1} - 2^{\frac{p-1}{2}}$. This is the base case of the induction. Now consider that this is true for all i less than or equal to j. Thus we can construct $(n_j, m_j, n_j - m_j - 1, 2^{n_j-1} - 2^{\frac{2n_j+m_j-3}{4}})$ functions. Now we apply Construction 1 on this. So we will get a function $(n_j + 3, m_j + 2, n_j - m_j, 2^{n_j+1} + 4(2^{n_j-1} - 2^{\frac{2n_j+m_j-3}{4}}))$. Note that $n_{j+1} = n_j + 3$ and $m_{j+1} = m_j + 2$. Now, $2^{n_j+1} + 4(2^{n_j-1} - 2^{\frac{2n_j+m_j-3}{4}}) = 2^{n_j+2} - 2^{\frac{2n_j+m_j-3}{4}+2} = 2^{n_j+2} - 2^{\frac{2(n_j+3)+(m_j+2)-3}{4}} = 2^{n_{j+1}-1} - 2^{\frac{2n_{j+1}+m_{j+1}-3}{4}}$. This proves the inductive step.

Now we have, $t = \frac{m-1}{2}$ and $p = n - 3t$. Note that p is odd. Thus, $n - \frac{3m-3}{2}$ is odd, which gives, $2n - 3m + 3 \equiv 2 \bmod 4$, i.e. $2n - 3m \equiv 3 \bmod 4$. □

As an immediate corollary we get the following result putting $m = 3$.

Corollary 1. *For even* $n \geq 8$, *it is possible to construct* $(n, 3, n-4, 2^{n-1} - 2^{\frac{n}{2}})$ *functions which are in* **desired** *form.*

Note that in [16, Theorem 10d], $(n, 2, n-3, 2^{n-1} - 2^{\frac{n}{2}})$ functions have been reported. Our result provides same nonlinearity with order of resiliency 3. In [17], construction of $(8,3,4,112)$ and $(10,3,6,480)$ functions have been given. These are special cases of Corollary 1. For $n = 12$, we get $(12,3,8,1984)$ function. This is better than $(12,3,8,1968)$ function given in [16, Section 6].

It is important to see that we can go on repeating Construction 2 and we will get a sequence of *desired* functions. These functions will initially start with the functions with $m \leq \frac{n}{2} - 2$ and then in the later part of the sequence they land into $m > \frac{n}{2} - 2$. As example, consider the sequence, $(10,3,6,480)$, $(13,5,7,3968)$, $(16,7,8,32256)$, $(19,9,9,260096)\ldots$, where though the nonlinearities are not achieving the upper bound, they provide a very interesting construction of suboptimal functions with very high nonlinearity, which were not known earlier. Also these functions don't have 3-valued Walsh spectra.

6 Conclusion

In this initiative we present different construction ideas for resilient Boolean functions with very high nonlinearity. We here mainly concentrate on low order of resiliency. As specific examples on small number on input variables, we are able to construct $(8,1,6,116)$, $(10,1,6,488)$, $(10,1,8,488)$, $(11,3,7,976)$, and $(12,3,8,1984)$ functions which were not known earlier. We also provide some generalized construction methods for functions on higher number of variables. Our results either supersede or solve important open questions presented in literature. Moreover, these Boolean functions have immediate application in design of stream cipher systems.

Acknowledgement : The authors are grateful to the anonymous reviewers for their comments that have improved both the technical quality and presentation of this paper.

References

1. E. R. Berlekamp and L. R. Welch. Weight distributions of the cosets of the (32, 6) Reed-Muller code. *IEEE Transactions on Information Theory*, IT-18(1):203–207, January 1972.
2. P. Camion, C. Carlet, P. Charpin, and N. Sendrier. On correlation immune functions. In *Advances in Cryptology - CRYPTO'91*, pages 86–100. Springer-Verlag, 1992.
3. A. Canteaut and M. Trabbia. Improved fast correlation attacks using parity-check equations of weight 4 and 5. In *Advances in Cryptology - EUROCRYPT 2000*, number 1807 in Lecture Notes in Computer Science, pages 573–588. Springer Verlag, 2000.
4. C. Carlet. On the coset weight divisibility and nonlinearity of resilient and correlation immune functions. In *Sequences and Their Applications*, SETA 2001.

5. C. Carlet and P. Sarkar. Spectral domain analysis of correlation immune and resilient Boolean functions. Accepted in *Finite Fields and Its Applications*, 2001.
6. C. Ding, G. Xiao, and W. Shan. *The Stability Theory of Stream Ciphers*. Number 561 in Lecture Notes in Computer Science. Springer-Verlag, 1991.
7. E. Filiol and C. Fontaine. Highly nonlinear balanced Boolean functions with a good correlation-immunity. In *Advances in Cryptology - EUROCRYPT'98*. Springer-Verlag, 1998.
8. X. Guo-Zhen and J. Massey. A spectral characterization of correlation immune combining functions. *IEEE Transactions on Information Theory*, 34(3):569–571, May 1988.
9. T. Johansson and F. Jonsson. Fast correlation attacks through reconstruction of linear polynomials. In *Advances in Cryptology - CRYPTO 2000*, number 1880 in Lecture Notes in Computer Science, pages 300–315. Springer Verlag, 2000.
10. S. Maitra. Correlation immune Boolean functions with very high nonlinearity. *Cryptology ePrint Archive, eprint.iacr.org, No. 2000/054*, October 27, 2000.
11. S. Maitra and P. Sarkar. Hamming weights of correlation immune Boolean functions. *Information Processing Letters*, 71(3-4):149–153, 1999.
12. W. Meier and O. Staffelbach. Fast correlation attack on stream ciphers. In *Advances in Cryptology - EUROCRYPT'88*, volume 330, pages 301–314. Springer-Verlag, May 1988.
13. E. Pasalic and T. Johansson. Further results on the relation between nonlinearity and resiliency of Boolean functions. In *IMA Conference on Cryptography and Coding*, number 1746 in Lecture Notes in Computer Science, pages 35–45. Springer-Verlag, 1999.
14. E. Pasalic, S. Maitra, T. Johansson and P. Sarkar. New constructions of resilient and correlation immune Boolean functions achieving upper bounds on nonlinearity. In *Workshop on Coding and Cryptography*, Paris, January 2001.
15. O. S. Rothaus. On bent functions. *Journal of Combinatorial Theory, Series A*, 20:300–305, 1976.
16. P. Sarkar and S. Maitra. Construction of nonlinear Boolean functions with important cryptographic properties. In *Advances in Cryptology - EUROCRYPT 2000*, number 1807 in Lecture Notes in Computer Science, pages 485–506. Springer Verlag, 2000.
17. P. Sarkar and S. Maitra. Nonlinearity bounds and constructions of resilient Boolean functions. In *Advances in Cryptology - CRYPTO 2000*, number 1880 in Lecture Notes in Computer Science, pages 515–532. Springer Verlag, 2000.
18. J. Seberry, X. M. Zhang, and Y. Zheng. On constructions and nonlinearity of correlation immune Boolean functions. In *Advances in Cryptology - EUROCRYPT'93*, pages 181–199. Springer-Verlag, 1994.
19. T. Siegenthaler. Correlation-immunity of nonlinear combining functions for cryptographic applications. *IEEE Transactions on Information Theory*, IT-30(5):776–780, September 1984.
20. T. Siegenthaler. Decrypting a class of stream ciphers using ciphertext only. *IEEE Transactions on Computers*, C-34(1):81–85, January 1985.
21. Y. V. Tarannikov. On resilient Boolean functions with maximum possible nonlinearity. *Proceedings of INDOCRYPT*, LNCS volume 1977, 19-30, 2000.
22. Y. Zheng and X. M. Zhang. Improving upper bound on nonlinearity of high order correlation immune functions. In *SAC 2000*, Lecture Notes in Computer Science. Springer Verlag, 2000.

On Certain 3-Weight Cyclic Codes Having Symmetric Weights and a Conjecture of Helleseth

Gary McGuire

Department of Mathematics, National University of Ireland, Maynooth, Co. Kildare, Ireland

Summary. When the binary cyclic code of length $2^m - 1$ with check polynomial $m_1(x)m_t(x)$ has exactly three nonzero weights, we discuss the symmetry of the two outer weights about the middle weight. This question is connected to a conjecture of Helleseth, that the crosscorrelation function of two m-sequences of length $2^m - 1$ must take at least four values when m is a power of 2. We present partial results on this problem, and some equivalent conditions to the symmetry of the weights.

1 Introduction

Helleseth [7] conjectured that two binary m-sequences of length $n = 2^m - 1$ cannot have a 3-valued crosscorrelation function when m is a power of 2. This conjecture is still open. Helleseth made a more general conjecture in [7] concerning p-ary sequences, but we consider only $p = 2$ in this article. In Calderbank-McGuire-Poonen-Rubinstein [3] it is proved that the crosscorrelation function cannot take the values $-1, -1 + A, -1 - A$. This seems a reasonable assumption to make on the values, because in all known examples of a 3-valued crosscorrelation function (when m is not a power of 2 of course), the values have this form.

One way to complete the proof of the conjecture is to prove that if a crosscorrelation function is 3-valued, then the values must have the form $-1, -1 + A, -1 - A$, for some A.

If we assume that -1 is a crosscorrelation value, and that there are three values, then the values will have the form $-1, -1 + A, -1 - B$, for some positive A and B.

In this article we will consider the following question: if the crosscorrelation function of two binary m-sequences takes the three values $-1, -1 + A, -1 - B$, for some positive A and B, must $A = B$?

We will present some equivalent conditions to the condition $A = B$. We will study the crosscorrelation function by studying the corresponding binary cyclic codes of length $n = 2^m - 1$. Most of the time we will study these codes for arbitrary m, and section 6 considers the case that m is a power of 2.

2 The MacWilliams Identities

We assume the reader is familiar with the basic properties of the crosscorrelation function, which can be found in [7]. We take the approach via cyclic codes to the crosscorrelation function. Let ω be a primitive element in the finite field $GF(2^m)$. Let $m_i(x)$ denote the minimal polynomial of ω^i over $GF(2)$. Let $C^\perp$ be the binary cyclic code of length $n = 2^m - 1$ generated by the product $m_1(x)m_t(x)$, where $t \geq 3$ is an odd integer. Knowledge of the weight distribution of C is equivalent to knowledge of the crosscorrelation spectrum of the two m-sequences with characteristic polynomials $m_1(x)$ and $m_t(x)$, see [7]. If t is not relatively prime to $n = 2^m - 1$ then the sequence defined by $m_t(x)$ will not be an m-sequence.

We **assume throughout this article** that C has 3 weights $w_1 = 2^{m-1} - a, w_2 = 2^{m-1}$ and $w_3 = 2^{m-1} + b$, where a and b are positive integers.

The **question we wish to consider in this article** is: must $a = b$?

The question of whether $A = B$ from the introduction is equivalent to this question of whether $a = b$.

Let A_i be the number of codewords of weight i in C, and let B_i be the number of codewords of weight i in $C^\perp$. Since $C^\perp$ is a subcode of the Hamming code we have $B_1 = B_2 = 0$.

The first thing we do is see what the MacWilliams identities (or Pless power moment identities) can tell us. They yield

$$\sum_{i=0}^{2^m-1} A_i = 2^{2m}$$

$$\sum_{i=0}^{2^m-1} iA_i = 2^{2m-1}(2^m - 1)$$

$$\sum_{i=0}^{2^m-1} i^2 A_i = 2^{2m-2}2^m(2^m - 1).$$

Since C has three weights w_1, w_2, w_3, these equations become

$$\begin{pmatrix} 1 & 1 & 1 \\ w_1 & w_2 & w_3 \\ w_1^2 & w_2^2 & w_3^2 \end{pmatrix} \begin{pmatrix} A_{w_1} \\ A_{w_2} \\ A_{w_3} \end{pmatrix} = \begin{pmatrix} 2^{2m} - 1 \\ (2^m - 1)2^{2m-1} \\ (2^m - 1)2^{3m-2} \end{pmatrix}.$$

The system can clearly be solved for the A_i in terms of m and a and b. We get

$$A_{w_1} = \frac{2^{m-1}(2^m - 1)(2^{m-1} + b)}{a(a+b)} \tag{1a}$$

$$A_{w_2} = \frac{(2^m - 1)(2^{m-1}(a-b) - 2^{2m-2} + ab + 2^m ab)}{ab} \tag{1b}$$

$$A_{w_3} = \frac{2^{m-1}(2^m - 1)(2^{m-1} - a)}{b(a+b)}. \tag{1c}$$

We can then use the next equation from the MacWilliams identities to give us

$$B_3 = \left(\frac{2^m-1}{3}\right)\left(2^{m-1} - 1 - \frac{w_1 w_3}{2^{m-1}}\right) \tag{2}$$

which can be written

$$B_3 = \left(\frac{2^m-1}{3}\right)\left(a - b - 1 + \frac{ab}{2^{m-1}}\right). \tag{3}$$

It follows from equation (3) that 2^{m-1} divides ab.

We are interested in possible values of a and b. We record the following result in the case of symmetric weights. See also Corollary 5.

Theorem 1. *If $a = b$ then a is a power of 2.*

Proof: Putting $a = b$ in (1b) gives

$$\begin{aligned} A_{w_2} &= \frac{(2^m-1)(-2^{2m-2} + a^2(1+2^m))}{a^2} \\ &= (2^m-1)\left(-\frac{2^{2m-2}}{a^2} + (2^m+1)\right). \end{aligned}$$

Since $w_2 = 2^{m-1}$ is relatively prime to $n = 2^m - 1$ we know that $2^m - 1$ divides A_{w_2}, so we conclude that a^2 divides 2^{2m-2}, which gives the result. □

3 Codewords of Weight 3

A codeword of weight 3 in $C^\perp$ is a cyclic shift of a codeword of weight 3 with a 1 in the first coordinate. Since the parity check matrix of $C^\perp$ is

$$\begin{pmatrix} 1 & \omega & \omega^2 & \dots & \omega^{n-1} \\ 1 & \omega^t & \omega^{2t} & \dots & \omega^{(n-1)t} \end{pmatrix}, \tag{4}$$

a codeword of weight 3 means that there exist $x, y \in GF(2^m)$ with $x \neq y$ and $x, y \neq 0, 1$ such that

$$1 + x + y = 0, \quad 1 + x^t + y^t = 0.$$

In other words, we have a solution to $x^t + (x+1)^t = 1$ in $GF(2^m)$ apart from 0 and 1.

If we define

$$h_t(x) := x^t + (x+1)^t + 1$$

then the number of distinct roots of $h_t(x)$ in $GF(2^m)$ will be related to B_3, the number of codewords of weight 3. We will make this precise shortly.

Observe that the polynomial $h_t(x)$ is invariant under the transformations

$$f(x) \mapsto f(x+1), \qquad f(x) \mapsto x^{deg(f)} f\left(\frac{1}{x}\right).$$

If $\alpha \neq 0, 1$ is a root of $h_t(x)$, this implies that $\alpha + 1$ and $1/\alpha$ are also roots of $h_t(x)$. The transformations $\alpha \mapsto \alpha + 1$ and $\alpha \mapsto 1/\alpha$ generate the group $SL_2(2)$, and so this group of order 6 acts on the set of roots of $h_t(x)$. The images of a root α under $SL_2(2)$ are

$$\alpha, \alpha+1, \frac{1}{\alpha}, \frac{\alpha+1}{\alpha}, \frac{\alpha}{\alpha+1}, \frac{1}{\alpha+1},$$

and it is easy to check that these images are all distinct iff $\alpha \notin GF(4)$. Note that α and its images have the same order. This type of thing has been studied before, in [6], [2], [5], for example. Some of our results may be found there.

If m is even and 3 does not divide t then every element of $GF(4)$ is a root of $h_t(x)$ in $GF(2^m)$. The elements of $GF(4)$ correspond to a codeword with $n/3$ distinct cyclic shifts. All other codewords of weight 3 have n distinct cyclic shifts.

The cyclic group of order n acts on the codewords of weight 3. Let T be the number of orbits of length n. If m is even and 3 does not divide t there is one other orbit of length $n/3$, as we said in the previous paragraph. So

$$B_3 = \begin{cases} \frac{n}{3} + nT, & \text{if } m \text{ is even and 3 does not divide } t; \\ nT, & \text{if } m \text{ is odd or 3 divides } t. \end{cases} \tag{5}$$

Let S denote the set of distinct roots of $h_t(x)$ that are not in $GF(4)$.

Theorem 2. *T is equal to the number of orbits of $SL_2(2)$ on S.*

Proof: Intuitively this is clear. It is not hard to see a $1-1$ correspondence between the orbits of C_n of length n and the orbits of $SL_2(2)$ of length 6, as follows.

Recall that ω is a primitive element of $GF(2^m)$. A codeword of weight 3 in an orbit of length n means that there are i, j such that

$$(\omega^i)^t + (\omega^j)^t + (\omega^i + \omega^j)^t = 0$$

where $j - i \notin \{0, n/3, 2n/3\}$. We can now see that cyclic shifts of the codeword correspond to multiplication of this equation by powers of ω, so that C_n is generated by $x \mapsto \omega x$. Dividing this equation by $(\omega^i)^t$ moves one of the 1's to the first position, and the other two 1's are in the positions corresponding to ω^{j-i} and $1 + \omega^{j-i}$. This codeword is a cyclic shift of the original codeword. Similarly for dividing by $(\omega^j)^t$ and $(\omega^i + \omega^j)^t$.

It is now clear that when a codeword in that orbit has a 1 in the first position, the 6 possible positions for the other 1's are the images under $SL_2(2)$ of ω^{i-j}, which is a root of $h_t(x)$.

In the same way, a root of $h_t(x)$ not in $GF(4)$ determines 6 such positions and hence a single orbit of length n. This completes the proof of theorem 2. □

Let $N(m,t)$ denote the number of distinct roots of $h_t(x)$ in $GF(2^m)$.

Corollary 1.

$$N(m,t) = \begin{cases} 4+6T, & \textit{if } m \textit{ is even and } 3 \textit{ does not divide } t; \\ 2+6T, & \textit{if } m \textit{ is odd or } 3 \textit{ divides } t. \end{cases}$$

Corollary 2.

$$B_3 = \left(\frac{2^m-1}{3}\right)\left(\frac{1}{2}N(m,t)-1\right).$$

In particular, $d(C^\perp) \geq 4$ *if and only if* $N(m,t) = 2$.

Corollary 2 follows from Corollary 1 and equation (5).

Corollary 3. *If* m *is even and* 3 *does not divide* t *then* $d(C^\perp) = 3$.

Corollary 4. *The following hold.*

1. $N(m,t) = 2(a-b) + \frac{ab}{2^{m-2}}$
2. $a = \frac{2^{m-1}\left(\frac{1}{2}N(m,t)+b\right)}{2^{m-1}+b}$
3. $b = \frac{2^{m-1}\left(a-\frac{1}{2}N(m,t)\right)}{2^{m-1}-a}$

Proof: To prove 1 use the expressions for B_3 in Corollary 2 and equation 3. Parts 2 and 3 follow by solving for a and b. □

Theorem 3. *The following are equivalent.*

1. $a = b$
2. $N(m,t) = \frac{a^2}{2^{m-2}}$
3. $N(m,t) = \frac{b^2}{2^{m-2}}$

Proof: If $a = b$ then Corollary 4 part 1 immediately gives parts 2 and 3. If 2. holds, then substitution into Corollary 4 part 2 gives $a = b$. If 3. holds, then substitution into Corollary 4 part 3 gives $b = a$. □

Corollary 5. *Suppose* $a = b$.

1. *If* m *is even then* $a = 2^s$ *where* $s \geq m/2$.
2. *If* m *is odd then* $a = 2^s$ *where* $s \geq (m-1)/2$.

Proof: Assume m is even and $a = b$. By Theorem 3 part 2, $N(m,t)$ is a square. By Corollary 1, $N(m,t)$ is at least 2. Therefore $N(m,t) = (\frac{a}{2^{m/2-1}})^2 \geq 4$. By Theorem 1 a is a power of 2, giving the result.

Now assume m is odd and $a = b$. By Theorem 3 part 2, we know that $a^2/2^{m-2}$ is a positive integer. Since m is odd we may conclude that $a^2/2^{m-1}$ is a positive integer, and therefore that $2^{(m-1)/2}$ divides a. By Theorem 1 a is a power of 2, giving the result. □

We remark that when $m = 6$ and $t = 5$ or 13, C is a 3-weight code with weights 24, 32 and 40. In this case $a = b = 2^{m/2}$, so this is an example with $s = m/2$.

When m is odd the dual code of the 2-error-correcting BCH code has weights $2^{m-1}, 2^{m-1} \pm 2^{(m-1)/2}$, so this is an example with $s = (m-1)/2$.

We will use the following theorem in later sections.

Theorem 4. *Let C_i denote the cyclotomic coset modulo* 15 *containing i.*

1. *If $t \in C_1$ then $h_t(x)$ has three irreducible binary factors of degree* 4.
2. *If $t \in C_7$ then $h_t(x)$ has no irreducible binary factors of degree* 4.

Proof: If $t \in C_1$ let $w \in \{1,2,4,8\}$ satisfy $w \equiv t \pmod{15}$. Then $x^t = x^w$ for all $x \in GF(16)$. If $\alpha \in GF(16)$ then

$$h_t(\alpha) = \alpha^t + (\alpha+1)^t + 1 = \alpha^w + (\alpha+1)^w + 1 = (\alpha + (\alpha+1) + 1)^w = 0.$$

Therefore if $t \in C_1$ the splitting field of $h_t(x)$ contains $GF(16)$, which implies that $h_t(x)$ has three irreducible factors of degree 4.

If $t \in C_7$ let $w \in \{7,14,13,11\}$ satisfy $w \equiv t \pmod{15}$. Then for each $\alpha \in GF(16) - GF(4)$ we check that $\alpha^w + (\alpha+1)^w + 1 \neq 0$. □

4 $h_t(x)$

If we could predict the factorisation of $h_t(x)$ into irreducible factors over $GF(2)$, given t, then we could calculate $N(m,t)$ and consequently B_3 using the following theorem.

Theorem 5. *If $h_t(x)$ has an irreducible factor of degree k, then the roots of this factor are in $GF(2^m)$ if and only if k divides m.*

The proof is clear. However, a description of the factorisation of $h_t(x)$ for arbitrary t seems difficult. The polynomial $h_t(x)$ has been discussed many times before, for example in [2], [5].

Table 1 gives the factorization of $h_t(x)$ into binary irreducibles for $t \leq 65$. The notation $(4)^2$ means that an irreducible of degree 4 occurs as a factor of multiplicity 2. Theorem 4 can be verified for the values in the table.

For some values of t we can explain the patterns in the factorisation. We shall sketch a few results, but presumably they are not new.

Table 1. Factorisation of $h_t(x)$over $GF(2)$

t	$h_t(x) = x^t + (x+1)^t + 1$
5	$(1)(1)(2)$
7	$(1)(1)(2)^2$
9	$(1)(1)(3)(3)$
11	$(1)(1)(2)(3)(3)$
13	$(1)(1)(2)^5$
15	$(1)(1)(3)^2(3)^2$
17	$(1)(1)(2)(4)(4)(4)$
19	$(1)(1)(2)^2(4)(4)(4)$
21	$(1)(1)(6)(6)(6)$
23	$(1)(1)(2)(3)(3)(4)(4)(4)$
25	$(1)(1)(2)^8(3)(3)$
27	$(1)(1)(8)(8)(8)$
29	$(1)(1)(2)(3)^4(3)^4$
31	$(1)(1)(2)^2(4)^2(4)^2(4)^2$
33	$(1)(1)(5)(5)(5)(5)(5)(5)$
35	$(1)(1)(2)(5)(5)(5)(5)(5)(5)$
37	$(1)(1)(2)^5(3)(3)(9)(9)$
39	$(1)(1)(3)(3)(5)(5)(5)(5)(5)(5)$
41	$(1)(1)(2)(12)(12)(12)$
43	$(1)(1)(2)^2(3)^2(3)^2(8)(8)(8)$
45	$(1)(1)(14)(14)(14)$
47	$(1)(1)(2)(4)(4)(4)(5)(5)(5)(5)(5)(5)$
49	$(1)(1)(2)^{17}(4)(4)(4)$
51	$(1)(1)(3)(3)(14)(14)(14)$
53	$(1)(1)(2)(3)(3)(4)(4)(4)(10)(10)(10)$
55	$(1)(1)(2)^2(9)(9)(15)(15)$
57	$(1)(1)(3)^9(3)^9$
59	$(1)(1)(2)(27)(27)$
61	$(1)(1)(2)^5(4)^4(4)^4(4)^4$
63	$(1)(1)(5)^2(5)^2(5)^2(5)^2(5)^2(5)^2$
65	$(1)(1)(2)(3)(3)(6)(6)(6)(6)(6)(6)(6)(6)(6)$

Theorem 6. *If* $t = 2^i + 1$ *then* $h_t(x) = x^{2^i} + x$.

The proof of theorem 6 is straightforward.

Corollary 6. *If* $t = 2^i + 1$ *then* $N(m,t) = 2^e$ *where* $e = gcd(i,m)$.

The decimation $t = 2^{2i} - 2^i + 1$ is known as the Kasami decimation.

Theorem 7. *If* $t = 2^{2i} - 2^i + 1$ *then*

$$h_t(x) = x(x+1) \prod_{\alpha \in GF(2^i), \alpha \neq 0,1} (x-\alpha)^{2^i+1}.$$

Proof:

$$\begin{aligned}
h_t(x) &= (x+1)^t + x^t + 1 \\
&= (x+1)(x+1)^{2^i(2^i-1)} + x^t + 1 \\
&= (x+1)(x^{2^i}+1)^{2^i-1} + x^t + 1 \\
&= (x+1)(x^{2^i(2^i-1)} + x^{2^i(2^i-2)} + \cdots + x^{2^i} + 1) + x^t + 1 \\
&= x(x^{2^i(2^i-2)} + x^{2^i(2^i-3)} + \cdots + x^{2^i} + 1) + \\
&\qquad (x^{2^i(2^i-1)} + x^{2^i(2^i-2)} + \cdots + x^{2^i}) \\
&= x(x^{2^i-2} + x^{2^i-3} + \cdots + x + 1)^{2^i} + \\
&\qquad (x^{2^i-1} + x^{2^i-2} + \cdots + x)^{2^i} \\
&= x\left(\frac{x^{2^i-1}-1}{x-1}\right)^{2^i} + x^{2^i}(x^{2^i-2} + x^{2^i-3} + \cdots + x + 1)^{2^i} \\
&= (x + x^{2^i})\left(\frac{x^{2^i-1}-1}{x-1}\right)^{2^i} \\
&= x(x^{2^i-1}-1) \prod_{\alpha \in GF(2^i), \alpha \neq 0,1} (x-\alpha)^{2^i} \\
&= x(x-1) \prod_{\alpha \in GF(2^i), \alpha \neq 0,1} (x-\alpha)^{2^i+1}.
\end{aligned}$$

□

Corollary 7. *If* $t = 2^{2i} - 2^i + 1$ *then* $N(m,t) = 2^e$ *where* $e = gcd(i,m)$.

A more general result than Theorem 7 corresponding to codewords of weight 4 can be found in [8].

Remark. When $t = 2^i + 1$ or $t = 2^{2i} - 2^i + 1$, and m/e is odd, it is known by results of Kasami, Gold, and Welch that C is a 3-weight code with weights $2^{m-1} - 2^{(m+e-2)/2}$, 2^{m-1}, and $2^{m-1} + 2^{(m+e-2)/2}$. Also, it is shown in [8] that when $t = 2^i + 1$ or $t = 2^{2i} - 2^i + 1$, the code $C^\perp$ has minimum distance 5 if and only if $e = gcd(m, i) = 1$. Therefore, if we choose m and i so that $e = 1$ and $m/e = m$ is odd, we obtain a code $C^{\perp}$ with minimum distance 5 whose dual code is a 3-weight code. By a theorem of Goethals and van Tilborg, we may conclude that $C^{\perp}$ is uniformly packed. The uniformly packed codes with $t = 2^i + 1$ are not equivalent to the codes with $t = 2^{2i} - 2^i + 1$, because (for example) they have different automorphism groups. The automorphism group of a code with $t = 2^i + 1$ is easily seen to contain the affine group, whereas the group of a code with $t = 2^{2i} - 2^i + 1$ does not.

Theorem 8. *If $t = 2^i - 1$ then the distinct roots of $h_t(x)$ are precisely the elements of $GF(2^{i-1})$.*

Proof:

$$\begin{aligned}(x+1)h_t(x) &= (x+1)^{t+1} + x^{t+1} + x^t + x + 1\\ &= (x+1)^{2^i} + x^{t+1} + x^t + x + 1\\ &= x^{2^i-1} + x\\ &= x(x^{2^{i-1}-1} + 1)^2.\end{aligned}$$

□

The reader can verify this theorem for the relevant values of t using Table 1.

Corollary 8. *If $t = 2^i - 1$ then $N(m,t) = 2^d$ where $d = gcd(i-1, m)$.*

Theorem 9. *If $t = 2^i + 2^j - 1$ then the distinct roots of $h_t(x)$ are precisely the elements of $GF(2^i) \cup GF(2^j)$.*

Proof:

$$\begin{aligned}(x+1)h_t(x) &= (x+1)^{t+1} + x^{t+1} + x^t + x + 1\\ &= (x+1)^{2^i}(x+1)^{2^j} + x^{t+1} + x^t + x + 1\\ &= x^{t+1} + x^{2^i} + x^{2^j} + 1 + x^{t+1} + x^t + x + 1\\ &= x(x^{2^i-1} + x^{2^j-1} + x^{2^i+2^j-2} + 1)\\ &= x(x^{2^i-1} + 1)(x^{2^j-1} + 1).\end{aligned}$$

□

The reader can verify this theorem for the relevant values of t using Table 1. These values are 5, 7, 9, 11, 15, 17, 19, 23, 31, 35, 39, 47.

Corollary 9. *If $t = 2^i + 2^j - 1$ then $N(m,t) = 2^{d_1} + 2^{d_2} - 2^{d_3}$ where $d_1 = gcd(i, m)$, $d_2 = gcd(j, m)$, and $d_3 = gcd(i, j, m)$.*

5 Krawtchouk Analysis

Again $C^{\perp}$ is the binary cyclic code of length $n = 2^m - 1$ generated by $m_1(x)m_t(x)$ and we assume that C has 3 weights $w_1 = 2^{m-1} - a, w_2 = 2^{m-1}$ and $w_3 = 2^{m-1} + b$. Some of the results in this section are due to A. R. Calderbank.

Consider the annihilator polynomial of C

$$F(x) = 2^{2m}\left(1 - \frac{x}{w_1}\right)\left(1 - \frac{x}{w_2}\right)\left(1 - \frac{x}{w_3}\right) \tag{6}$$

whose roots are the weights. We express $F(x)$ in its Krawtchouk expansion

$$F(x) = \alpha_0 P_0(x) + \alpha_1 P_1(x) + \alpha_2 P_2(x) + \alpha_3 P_3(x) \tag{7}$$

where $P_k(x) = \sum_{j=0}^{k}(-1)^j \binom{x}{j}\binom{n-x}{k-j}$ is the k-th binary Krawtchouk polynomial. We assume the covering radius of $C^{\perp}$ is 3 (the covering radius is at most 3 since C has 3 weights).

For example, when $t = 13$ and $m = 6$, we have $\alpha_0 = \alpha_1 = -11/10$ and $\alpha_2 = \alpha_3 = 1/10$.

Theorem 10. *With notation as above, the following are equivalent.*

1. $a = b$.
2. $\alpha_2 = \alpha_3$.
3. $\alpha_0 = \alpha_1$.
4. $B_4 = \binom{n-3}{4} B_3$.
5. $N(m,t) = \frac{a^2}{2^{m-2}}$.
6. $N(m,t) = \frac{b^2}{2^{m-2}}$.

Proof: By Theorem 3 parts (1), (5) and (6) are equivalent. We next show that (1) and (2) are equivalent.

Substituting for each $P_k(x)$ in equation 7 and equating coefficients of x in equations 6 and 7 gives

$$\frac{4}{3}\alpha_3 = \frac{2^{m+1}}{w_1 w_3} \tag{8a}$$

$$\alpha_2 + n\alpha_3 = \frac{2^m}{w_1 w_3}(w_1 + w_3 + 2^{m-1}) \tag{8b}$$

$$2\alpha_1 + 2n\alpha_2 + (n^2 - n + \frac{2}{3})\alpha_3 = \frac{2^{m+1}}{w_1 w_3}(w_1 w_3 + 2^{m-1}(w_1 + w_3)) \tag{8c}$$

$$\alpha_0 + n\alpha_1 + \binom{n}{2}\alpha_2 + \binom{n}{3}\alpha_3 = 2^{2m} \tag{8d}$$

and these can be used to express each α_i in terms of m, a, and b.

Dividing equation 8b by equation 8a and rearranging gives

$$\frac{\alpha_2}{\alpha_3} = \frac{2}{3}(w_1 + w_3 + 2^{m-1}) - n.$$

It is easily seen from this equation that

$$\alpha_2 = \alpha_3 \Longleftrightarrow w_1 + w_3 = n + 1.$$

Since $w_1 + w_3 = n + 1 \Longleftrightarrow a = b$, this proves that (1) is equivalent to (2).

Next, putting $x = 2^{m-1}$ in equation 7 and using the fact that $F(2^{m-1}) = 0$ gives the equation

$$\alpha_1 - \alpha_0 = \binom{n-1}{2}(\alpha_3 - \alpha_2) \tag{9}$$

after substituting for the $P_k(x)$ and simplifying. It is clear from equation 9 that (2) and (3) are equivalent.

To complete the proof we will show that (1) and (4) are equivalent. This can be proved using an expression for B_4 found from the MacWilliams identities, but we shall give another proof using Delsarte theory

The distribution matrix B for $C^\perp$ is the 2^n-by-$(n+1)$ matrix with entries $B_i(x)$, where $B_i(x)$ is the number of codewords of $C^\perp$ at distance i from x, for any $x \in GF(2)^n$, $0 \le i \le n$. Many rows of B are identical; for example, $B_i(x) = B_i$ if $x \in C^\perp$.

A result from Delsarte theory [4] states that

$$\sum_{i=0}^{3} \alpha_i B_i(x) = 1 \tag{10}$$

for all $x \in GF(2)^n$.

If we use equation (10) when $x \in C^\perp$ we obtain

$$\alpha_0 + \alpha_3 B_3 = 1. \tag{11}$$

Using this equation and equations (8a), (8b), (8c), (8d), yields an expression for B_3 in terms of m, a, b. The same expression as equation (3) is found.

For a minimum weight 1 coset $C^\perp + x$, we have $B_2(x) = 3B_3/n$ because the code is cyclic. Similarly $B_3(x) = 4B_4/n$. Thus the first four positions of row x, for any weight 1 coset $C^\perp + x$, are

$$(0, 1, \frac{3B_3}{n}, \frac{4B_4}{n}).$$

The fact that $B_2(x)$ is constant for all x of weight 1 means that the codewords of weight 3 form a 1-design.

Equation (10) will now tell us that

$$\alpha_1 + \frac{3B_3}{n}\alpha_2 + \frac{4B_4}{n}\alpha_3 = 1. \tag{12}$$

From the Krawtchouk expansion of $xF(x)$, we get

$$B_4 = \left(\frac{n\alpha_3 - 3\alpha_2}{4\alpha_3}\right)B_3 + \frac{n(\alpha_0 - \alpha_1)}{4\alpha_3}. \tag{13}$$

Completion of proof of Theorem 10: If $a = b$ then $\alpha_1 = \alpha_0$ and $\alpha_2 = \alpha_3$, and putting these into equation (13) gives (4).

Conversely, suppose $B_4 = (\frac{n-3}{4})B_3$ is true. Putting this into equation (12) gives

$$\alpha_1 + \frac{3B_3}{n}(\alpha_2 - \alpha_3) + B_3\alpha_3 = 1.$$

Now use equation (11) to get

$$\alpha_0 - \alpha_1 = (\frac{3B_3}{n})(\alpha_2 - \alpha_3).$$

Comparing this to equation (9) we conclude that either both sides are 0, in which case we are done, or else $B_3 = n(n-1)/6$. But the latter option is impossible, because from equation (2) we have

$$B_3 = \binom{n}{3}\left(\frac{n-1}{2} - \frac{w_1w_3}{2^{m-1}}\right),$$

and it follows that $B_3 = n(n-1)/6$ is impossible as it would imply $w_1w_3 = 0$.

Remark. The values of $B_i(x)$ for a coset $C^\perp + x$ of minimum weight 2 are difficult to calculate and depend on x, in general. If it were possible to find $B_i(x)$ for weight 2 cosets, it might be possible to prove that $\alpha_2 = \alpha_3$.

In the case of the Kasami decimation $t = 2^{2i} - 2^i + 1$, it was shown by Kasami that the C is a 3-weight code when m/e is odd, where $e = gcd(m, i)$. In this case the weights are $2^{m-1} - 2^{(m+e-2)/2}$, 2^{m-1}, and $2^{m-1} + 2^{(m+e-2)/2}$, so here $a = b$. If we assume $e = 1$ and m odd, then $d(C^\perp) \geq 4$ by Corollary 7 and Corollary 2. Then, by conditions 1 and 4 of Theorem 10 we may conclude that $B_4 = 0$ since $B_3 = 0$. Hence $d(C^\perp) = 5$ when m is odd and $gcd(m, i) = 1$.

As we remarked earlier, in [8] a more general result is shown, namely that $d(C^\perp) = 5$ if and only if $gcd(m, i) = 1$.

6 When m is a power of 2

We **assume for the rest of this section** that m is a power of 2, and as usual that C is a 3-weight code, with weights w_1, w_2, w_3 as before. We also assume that t is relatively prime to $2^m - 1$, so in particular 3 does not divide t. By Corollary 1, $N(m, t) = 4 + 6T$.

We can now prove one special case of Helleseth's conjecture. By the Welch bound, the lowest crosscorrelation values occur when the minimum distance of $C^\perp$ is 5, so the case that $C^\perp$ has distance 5 is of particular interest.

Theorem 11. *If m is a power of 2, and t is relatively prime to $n = 2^m - 1$, and the minimum distance of $C^{\perp}$ is 4 or 5, then C cannot be a 3-weight code.*

Proof: This follows immediately from Corollary 3. □

The only irreducible factors of $h_t(x)$ that contribute to codewords of weight 3 are ones whose degree is a power of 2. Let N_i be the number of irreducible factors of $h_t(x)$ of degree 2^i. Then by Theorem 5 (or by the Frobenius automorphism)

$$N(m,t) = 4 + \sum_{i=2}^{\infty} 2^i N_i. \tag{14}$$

Recall that equation (3) implies that $2^{m-1}|ab$. We can improve this.

Lemma 1. $2^m|ab$

Proof. By Corollary 1 we have $N(m,t) = 4 + 6T$, and combining this with equation (14) shows that T is even. The two expressions 1 and 3 for B_3 yield

$$3T = a - b - 2 + \frac{ab}{2^{m-1}}. \tag{15}$$

We know that a and b are even since all weights are even, and T is even, so $ab/2^{m-1}$ must be even. □

Corollary 10. *Both a and b are divisible by 4.*

Proof: If not then by Lemma 1 one of a, b is divisible by 2^{m-1}, which is ridiculous. □

Helleseth's conjecture, in the case that t is congruent modulo $2^m - 1$ to any element of the cyclotomic coset containing -1, follows from the last corollary. This is because, in that case, there are two roots of $m_1(x)m_t(x)$ that have product 1. By McEliece's theorem (see [3]) not all weights are divisible by 4, contradicting Corollary 10. Indeed, Lachaud and Wolfmann have shown that all possible crosscorrelation values occur when the decimation is -1.

Let ν_2 denote the 2-adic valuation on the integers, i.e., $\nu_2(n) = k$ where 2^k divides n and 2^{k+1} does not divide n. The previous Lemma can be stated as $\nu_2(ab) \geq m$. Note that $\nu_2(a) = \nu_2(w_1)$ and $\nu_2(b) = \nu_2(w_3)$ and $2 \leq \nu_2(a), \nu_2(b) < m - 1$.

Let C_i denote the cyclotomic coset modulo 15 containing i. Recall that t is either in C_1 or C_7, since $gcd(t, 15) = 1$.

Lemma 2. *Let C_i denote the cyclotomic coset modulo 15 containing i, and assume m is a power of 2.*

1. If $t \in C_1$ *then* $\nu_2(ab) > m$.
2. If $t \in C_7$ *then* $\nu_2(ab) = m$.

Proof: (1) By Theorem 4 we have $N_2 = 3$. Therefore by equation (14)

$$N(m,t) = 4 + 6T = 16 + 8N_3 + 16N_4 + 32N_5 + \cdots$$

which gives

$$3T = 6 + 4N_3 + 8N_4 + 16N_5 + \cdots.$$

With $T = 2R$ we get

$$3(R-1) = 2N_3 + 4N_4 + 8N_5 + \cdots$$

which implies that $R-1$ is even, so R is odd and $T \equiv 2 \pmod 4$. Now taking equation (15) modulo 4 gives

$$\frac{ab}{2^{m-1}} \equiv 0 \pmod 4$$

proving part (1).

(2) By Theorem 4 we know $N_2 = 0$. Therefore by equation (14)

$$N(m,t) = 4 + 6T = 4 + 8N_3 + 16N_4 + 32N_5 + \cdots$$

which gives

$$3T = 4N_3 + 8N_4 + 16N_5 + \cdots.$$

With $T = 2R$ we get

$$3R = 2N_3 + 4N_4 + 8N_5 + \cdots$$

which implies that R is even, so $T \equiv 0 \pmod 4$. Now taking equation (15) modulo 4 gives

$$\frac{ab}{2^{m-1}} \equiv 2 \pmod 4$$

proving part (2). □

We note that in case (1), if $t = 287$ then $N_3 = 30$ and $T \equiv 2 \pmod 8$. We do not have an example where $T \equiv 6 \pmod 8$. In case (2), if $t = 37$ then $N_3 = 0$ and $T \equiv 0 \pmod 8$, while if $t = 43$ then $N_3 = 3$ and $T \equiv 4 \pmod 8$.

Corollary 11. *If* $t \in C_1$ *then* $8|a$ *and* $8|b$.

Proof: If not, say $\nu_2(a) = 2$ and then by part (1) of the previous Lemma we get $\nu_2(b) \geq m - 1$ which is impossible. □

We also have the following theorem from Calderbank-McGuire-Poonen-Rubinstein [3].

Theorem 12. *If m is a power of 2, then either $\nu_2(a) < m/2$ or $\nu_2(b) < m/2$. In other words, $2^{m/2}$ does not divide both a and b.*

As an application of our results we can give a "computer-free" proof of the result we want in the case $m = 8$. Other partial results are available but we omit the details.

We assume C is a 3-weight code with all the same notation as earlier. If $t \in C_7$ then the result is proved in [1]. So we assume $t \in C_1$.

By Theorem 12 one of a, b is divisible by 8 and not by 16. By Corollary 11 both a and b are divisible by 8. By Lemma 2 we have $2^9 | ab$. Since $w_2 = 2^7$ we cannot have 2^7 dividing a or b. It follows that $\nu_2(a) = 3$ and $\nu_2(b) = 6$, or else $\nu_2(b) = 3$ and $\nu_2(a) = 6$.

Suppose $\nu_2(a) = 6$ (a similar argument works if $\nu_2(b) = 6$). Then $a = 64$ since the existence of any odd prime divisors of a would imply $a > 2^7$. We also have $b = 8k$ where k is odd. Calculation of A_{w_1} or A_{w_3} from equation (1a) gives a nonintegral answer for one of these, for each possible value of k. This contradiction proves that C cannot be a 3-weight code.

Note. Blokhuis and Calderbank [1] have proven Helleseth's conjecture in the case $t \in C_7$.

Acknowledgements. The author thanks Robert Calderbank for introducing him to this problem and sharing his ideas.

References

1. A. Blokhuis, A. R. Calderbank, personal communication.
2. A. Canteaut, P. Charpin, H. Dobbertin, Weight divisibility of cyclic codes, highly nonlinear functions on F_{2^m}, and crosscorrelation of maximum-length sequences, *SIAM J. Disc. Math.*, **13** (2000) 105–138.
3. A. R. Calderbank, G. McGuire, B. Poonen, M. Rubinstein, "On a conjecture of Helleseth regarding pairs of binary m-sequences," *IEEE Transactions on Information Theory*, **42** (1996), 988–990.
4. P. Delsarte, Four fundamental parameters of a code and their combinatorial significance, *Information and Control*, **23** (1973) 407–438.
5. S. Golomb, G. Gong, Periodic Binary Sequences with the "Trinomial Property" *IEEE Transactions on Information Theory*, **45** (1999), 1276–1279.
6. S. Golomb, Theory of transformation groups of polynomials over $GF(2)$ with applications to linear shift register sequences, *Inform. Sci.*, **1** (1968) 87–109.
7. T. Helleseth, "Some results about the crosscorrelation function between two maximal linear sequences," *Discrete Math.*, **16** (1976), 209–232.
8. H. Janwa and R. M. Wilson, Hyperplane sections of Fermat varieties in P^3 in char. 2 and some applications to cyclic codes, in: Applied Algebra, Algebraic Algorithms and Error-Correcting Codes (Proceedings AAECC-10), Gérard Cohen, Teo Mora, Oscar Moreno (Eds.), Lecture Notes in Computer Science **673**, Springer-Verlag, 1993.

The Quantum Entanglement of Binary and Bipolar Sequences

Matthew G. Parker[1] and V. Rijmen[2]

[1] Inst. for Informatikk, University of Bergen, Norway,
matthew@ii.uib.no, http://www.ii.uib.no/~matthew/mattweb.html
[2] Katholieke Universiteit Leuven, Dept. Elektrotechniek-ESAT,
K. Mercierlaan 94,B-3001 Heverlee, Belgium

Summary. We examine multi-spectra of quantum states under local unitary action, and show that optimal error-correcting codes and certain sequences represent states with high multiparticle entanglement.

1 Introduction

This paper highlights two partial entanglement measures, namely 'Linear Entanglement' (LE) (Section 3, Definition 9), and 'Stubborness of Entanglement' (SE) (Section 4, Definition 14), which is a sequence of parameters. We show, by interpreting the length 2^n indicator for $\mathbf{C}$ as an n-particle quantum state that, for those states representing binary linear error-correcting codes (ECCs), the ECCs with optimal weight hierarchy also have optimal LE and optimal SE (Theorems 9 and 14). By action of local unitary transform on the indicator of $\mathbf{C}$, we also view the quantum state as a bipolar sequence. In this context a sequence is partially described by its nonlinear order, N, and correlation immunity order, CI (Definitions 17, 18). We show that N and CI give a lower bound on LE (Theorem 15). LE is the $n - \log_2$ of a spectral 'peak' measure of Peak-to-Average Power Ratio (PAR_l (Section 3, Definition 8)). This paper refers both to PAR_l and to LE (Definition 9). We emphasise quantum states $\boldsymbol{s}$ from the set $\ell_{\mathbf{p}}$, where $\ell_{\mathbf{p}}$ is equivalent to the set of binary linear ECCs.

Main Results:

- LE and SE of $\boldsymbol{s}$ can be found from the HI multispectra (the multispectra resulting from a Local Unitary (LU) transform which is any tensor combination of 2×2 Hadamard and Identity transforms of $\boldsymbol{s}$) (Theorems 6 and 13).
- Quantum states from $\ell_{\mathbf{p}}$ which are equivalent under LU transform to binary linear ECCs with optimum or near-optimum Weight Hierarchy also have optimum or near-optimum LE and SE over the set $\ell_{\mathbf{p}}$ (Theorem 14, Corollary 3).
- LE($\boldsymbol{s}$) is lower-bounded by an additive combination of the Nonlinear Order and Correlation-Immunity of $\boldsymbol{s}$ (Theorem 15).

A qubit is a two-state particle, (s_0, s_1), meaning it is in state 0 with complex probability s_0, and state 1 with complex probability s_1, such that $|s_0|^2 + |s_1|^2 = 1$.

Definition 1 *Let* $\mathbf{l_n}$ *be the infinite set of normalised linear vectors which can be written in the form* $(a_0, b_0) \otimes (a_1, b_1) \otimes \ldots \otimes (a_{n-1}, b_{n-1})$

Entanglement exists between two or more particles if their joint probability state cannot be factorised using the tensor product. More formally,

Definition 2 *Let* s *be an* n*-qubit state.* s *is a pure entangled state if* $s \notin \mathbf{l_n}$. s *is not entangled if* $s \in \mathbf{l_n}$.

Partial Quantification of Entanglement: Currently, in the literature, there is no agreement on a standard measure or measures of entanglement for more than two qubits [5]. We therefore propose the following partial entanglement measures:

- Linear Entanglement (LE) is used to quantify distance to the nearest tensor-product state (Section 3, Definitions 8, 9).
- Stubborness of Entanglement (SE), is a series of k' parameters, β_j, specifying the residual order of entanglement after j most-destructive single-qubit measurements on $\mathbf{s}$ (Section 4, Definition 14).

Measurement

Definition 3 *Let* s *be an* n*-qubit state. The vector* $s = (s_0, s_1, \ldots, s_{2^n-1})$ *implies a measurement basis (s-basis) such that qubit* i *is measured as* η *with probability* $\sum_{\substack{k=0 \\ k_i = \eta}}^{2^n-1} |s_k|^2$, *where* k *is decomposed as* $k = \sum_{i=0}^{n-1} k_i 2^i$, *and* $\eta, k_i \in \{0, 1\}$. *After measurement,* s *becomes* $s_{x_i=\eta}$.

For the rest of this paper normalisation of $\mathbf{s}$ is omitted for clarity. Normalisation ensures $\sum_{i=0}^{2^n-1} |s_i|^2 = 1$.

We consider pure state entanglement in a closed system - we assume little or no entanglement with environment.

Proposition 1 *A binary linear* $[n, k]$ *error-correcting code (ECC),* $\mathbf{C}$*, can be represented by a length* 2^n *binary indicator,* $\mathbf{s}$*, where* $\mathbf{C} = \{(i_0, i_1, \ldots, i_{n-1}) | s_i \neq 0\}$, *where* $i = \sum_{j=0}^{n-1} i_j 2^j$, $i_j \in \{0, 1\}$. *More specifically, codeword* $(i_0, i_1, \ldots, i_{n-1})$ *in* $\mathbf{C}$ *occurs with probability* $|s_i|^2$.

A unitary transform $\mathbf{U}$ satisfies $\mathbf{U}\mathbf{U}^\dagger = \mathbf{I}$ where $\dagger$ means conjugate transpose and $\mathbf{I}$ is identity. Let $\mathbf{U}$ be a $2^n \times 2^n$ unitary transform. Let s be the joint-state vector of n qubits.

Fact 1 *If* $\mathbf{U}$ *is tensor decomposable into* 2×2 *local unitary (LU) transforms, then entanglement of* s *is the same as the entanglement of* $\mathbf{U}s$. *[8]*

Definition 4 *If $\exists\mathbf{U}$, where $\mathbf{U}$ is decomposable into 2×2 local unitaries, such that $s' = \mathbf{U}s$, then we write $s' \equiv_{LU} s$ (LU Equivalence).*

The $2^n \times 2^n$ Walsh-Hadamard Transform (WHT) is $\bigotimes_{i=0}^{n-1} \begin{pmatrix} 1 & 1 \\ 1 & -1 \end{pmatrix}$:

Proposition 2 *Let $\mathbf{C}$ and $\mathbf{C}^\perp$ be binary linear ECCs described by the indicators s and s', respectively. Then $s \equiv_{LU} s'$, where the LU transform is the WHT [6,10].*

2 *HI* Multispectra and the Set, $\ell_\mathbf{p}$

We consider spectra of a subset of quantum states after transform by LU tensor products of $\mathbf{H}$ and $\mathbf{I}$, where $\mathbf{H} = \begin{pmatrix} 1 & 1 \\ 1 & -1 \end{pmatrix}$, and $\mathbf{I} = \begin{pmatrix} 1 & 0 \\ 0 & 1 \end{pmatrix}$. We call this set of transforms the '*HI* Transform', and the associated spectra the '*HI* multispectra'. We identify a set of states, $\ell_\mathbf{p}$, which are LU equivalent to binary linear ECCs, via *HI* transforms (Theorem 1), and propose Algebraic Polar Form (APF) to separate magnitude (binary), and phase (bipolar) properties of s,

$$s = s(\mathbf{x}) = m(\mathbf{x})(-1)^{p(\mathbf{x})}$$

where $m(\mathbf{x})$ and $p(\mathbf{x})$ are both boolean Algebraic Normal Forms (ANFs) [6] in n binary variables $\mathbf{x} = \{x_0, x_1, x_2, \ldots, x_{n-1}\}$.
Example: $s(\mathbf{x}) = (x_0 + x_1 + x_2 + 1)(x_1 + x_3)(-1)^{(x_1 x_2 + x_0 + 1)} =$
$0,0,0,1,0,0,1,0,-1,0,0,0,0,1,0,0$
$m(\mathbf{x})$ is decomposed as $m(\mathbf{x}) = \prod_k h(\mathbf{x})_k$. We are interested in $m(\mathbf{x})$ comprising $h(\mathbf{x})_k$ of degree 1, and $p(\mathbf{x})$ of degree ≤ 2.

Definition 5 *"H acting on i" means the action of the transform, $\mathbf{I} \otimes \ldots \otimes \mathbf{I} \otimes \mathbf{H} \otimes \mathbf{I} \otimes \ldots \otimes \mathbf{I}$ on s, where $\mathbf{H}$ is preceded by i $\mathbf{I}$ matrices, and followed by $n-i-1$ $\mathbf{I}$ matrices. We write this as $H(i)$, or $H(i)[s]$.*

$\prod_{i\in\mathbf{T}} H(i)$ is the action of H on the subset of qubits i represented by the integers in $\mathbf{T}$.

Definition 6 *Let $\mathbf{T_C}, \mathbf{T_{C^\perp}}$ be integer sets chosen so that $\mathbf{T_C} \cap \mathbf{T_{C^\perp}} = \emptyset$, and $\mathbf{T_C} \cup \mathbf{T_{C^\perp}} = \{0,1,\ldots,n-1\}$. This is a bipartite splitting of $\{0,1,\ldots,n-1\}$. Let $\mathbf{x} = \mathbf{x_C} \cup \mathbf{x_{C^\perp}}$, where $\mathbf{x_C} = \{x_i | i \in \mathbf{T_C}\}$, and $\mathbf{x_{C^\perp}} = \{x_i | i \in \mathbf{T_{C^\perp}}\}$.*

Definition 7 *Define $s \in \ell_\mathbf{p}$, such that $s(\mathbf{x}) = (-1)^{p(\mathbf{x})}$, where $p(\mathbf{x}) = \sum_k q_k(\mathbf{x_C}) r_k(\mathbf{x_{C^\perp}})$, where $\deg(q_k(\mathbf{x_C})) = \deg(r_k(\mathbf{x_{C^\perp}})) = 1\ \forall k$, and where $x_i \in p(\mathbf{x})$, $\forall\ i \in \{0,1,\ldots,n-1\}$. $\ell_\mathbf{p}$ is the set of 'bipartite quadratic bipolar' states.*

Theorem 1 *(proof omitted) If $s \in \ell_\mathbf{p}$, then the action of $\prod_{i\in\mathbf{T}} H(i)$ on s gives $s'(\mathbf{x}) = m(\mathbf{x})$, for $\mathbf{T} = \mathbf{T_C}$, or $\mathbf{T} = \mathbf{T_{C^\perp}}$. s' is the binary indicator for a binary linear $[n, n-|\mathbf{T}|, d]$ error correcting code, $\mathbf{C}$, (or $\mathbf{C}^\perp$), if $\mathbf{T_C}$, (or $\mathbf{T_{C^\perp}}$) is used.*

3 PAR$_l$ and Linear Entanglement (LE)

3.1 Definitions

Partial quantification of entanglement for $\boldsymbol{s}$ of length 2^n (n qubits) is achieved by measuring maximum correlation of $\boldsymbol{s}$ with **any** length 2^n 'linear' sequence, $\boldsymbol{l_n}$. This correlation maximum can be expressed as a 'Peak-to-Average-Power-Ratio' (PAR$_l$),

Definition 8

$$PAR_l(\boldsymbol{s}) = 2^n max_l(|\boldsymbol{s} \cdot \boldsymbol{l}|^2)$$

where $\boldsymbol{l}$ is any normalised linear sequence from the set, $\boldsymbol{l_n} = \{(a_0, b_0) \otimes (a_1, b_1) \otimes \ldots \otimes (a_{n-1}, b_{n-1})\}$, and $\cdot$ means 'inner product' [7].

Linear Entanglement (LE) is then defined as,

Definition 9

$$LE(\boldsymbol{s}) = n - \log_2(PAR_l(\boldsymbol{s}))$$

We can alternatively express PAR$_l$ as,

Definition 10 *Let $\boldsymbol{s}' = (s'_0, s'_1, \ldots, s'_{2^n-1}) = (\bigotimes_{j \in \mathbf{T}} \mathbf{U}(j))\boldsymbol{s}$, where $\mathbf{U}(j) = \begin{pmatrix} \cos\theta_j & \sin\theta_j e^{iw_j} \\ \sin\theta_j e^{-iw_j} & -\cos\theta_j \end{pmatrix}$, $i^2 = -1$, and θ_j and w_j can take any real values, $\forall j$. Then,*

$$PAR_l(\boldsymbol{s}) = 2^n max_t(|s'_t|^2), \qquad \forall \mathbf{T} \subset \{0, 1, \ldots, n-1\}$$

3.2 PAR$_l$ for States from ℓ_p

Definition 11

$$PAR(\boldsymbol{s}) = 2^n max_i(|s_i|^2)$$

Theorem 2 *If $\boldsymbol{s} \in \ell_\mathrm{p}$, then $\boldsymbol{s}$ is LU equivalent to the indicator for an $[n, k, d]$ binary linear code, and,*

$$PAR_l(\boldsymbol{s}) \geq 2^r, \quad where\ r = max(k, n-k)$$

Proof. Let $k \leq n - k$. The indicator for the $[n, k, d]$ code has 2^k equal magnitude non-zero coefficients. By Parseval's Theorem and normalisation, the PAR of such an indicator must be 2^{n-k}. ∎

Theorem 2 implies that states, s, from $\ell_{\mathbf{p}}$ have a minimum lower bound on PAR_l (upper bound on LE) when the associated $[n,k,d]$ code, $\mathbf{C}$, satisfies $k = \lfloor \frac{n}{2} \rfloor$, with $\mathrm{PAR}_l \geq 2^{\lceil \frac{n}{2} \rceil}$. In this case LE coincides with the 'Schmidt Measure', $\mathbf{P}$ [3,1], defined as $\log_2$ of the minimal number of non-zero entries in any state, s', where $s' \equiv_{\mathrm{LU}} s$.

Theorem 3 *If $s \in \ell_{\mathbf{p}}$ then s is LU equivalent to the indicator for an $[n,k,d]$ code, and the Schmidt Measure, P, satisfies, $P(s) \leq min(k, n-k)$.*

Theorem 4 *Let $s(\mathbf{x}) = x_{\pi(0)}x_{\pi(1)} + x_{\pi(1)}x_{\pi(2)} + \ldots + x_{\pi(n-2)}x_{\pi(n-1)}$, where π is any permutation of the indeces $\{0,1,\ldots,n-1\}$. Then $PAR_l = 2^{\lceil \frac{n}{2} \rceil}$.*

Proof. Corollary 6 of Section 7 of [7]. ∎

In [1,9] Persistency of Entanglement (Section 4) of such states is proven to be $\lfloor \frac{n}{2} \rfloor$. This is the same as proof of PAR_l, because Persistency, LE, and Schmidt Measure all coincide for these states. These states also have PAR ≤ 2.0 under the one-dimension complex Discrete Fourier Transform [2,7]

Let $\mathbf{s} \in \ell_{\mathbf{p}}$. Recalling Definition 6, let $k = |\mathbf{T}_{\mathbf{C}^\perp}|$, $k^\perp = |\mathbf{T}_{\mathbf{C}}|$, and $k + k^\perp = n$. Without loss of generality we renumber integer sets $\mathbf{T}_{\mathbf{C}^\perp}$ and $\mathbf{T}_{\mathbf{C}}$ so that $\mathbf{T}_{\mathbf{C}^\perp} = \{0,1,\ldots,k-1\}$ and $\mathbf{T}_{\mathbf{C}} = \{k, k+1, \ldots, n-1\}$. Let $\mathbf{t}_{\mathbf{C}^\perp} \subset \mathbf{T}_{\mathbf{C}^\perp}$ and $\mathbf{t}_{\mathbf{C}} \subset \mathbf{T}_{\mathbf{C}}$, where $h = |\mathbf{t}_{\mathbf{C}^\perp}|$ and $h^\perp = |\mathbf{t}_{\mathbf{C}}|$. Let $\mathbf{x}_{\mathbf{t}^\perp} = \{x_i | i \in \mathbf{t}_{\mathbf{C}^\perp}\}$, $\mathbf{x}_{\mathbf{t}} = \{x_i | i \in \mathbf{t}_{\mathbf{C}}\}$, and $\mathbf{x}_* = \mathbf{x}_{\mathbf{t}^\perp} \cup \mathbf{x}_{\mathbf{t}}$. We define $\mathbf{M}$ to be a $k \times k^\perp$ binary matrix where $M_{i,j-k} = 1$ iff $x_i x_j \in p(\mathbf{x})$, and $M_{i,j-k} = 0$ otherwise. Thus $p(\mathbf{x}) = \sum_{i \in \mathbf{T}_{\mathbf{C}^\perp}} x_i (\sum_{j \in \mathbf{T}_{\mathbf{C}}} M_{i,j-k} x_j)$. Then we define a submatrix, $\mathbf{M}_{\mathbf{t}}$, of $\mathbf{M}$, which comprises only the rows and columns of $\mathbf{M}$ specified by $\mathbf{t}_{\mathbf{C}^\perp}$ and $\mathbf{t}_{\mathbf{C}}$. Let χ_t be the rank of $\mathbf{M}_{\mathbf{t}}$.

Theorem 5 *Let s' be the result of the action of $\prod_{i \in \mathbf{t}_{\mathbf{C}^\perp} \cup \mathbf{t}_{\mathbf{C}}} H(i)$ on $s \in \ell_{\mathbf{p}}$.*

$$PAR(s') = 2^{h + h^\perp - 2\chi_t}$$

Proof. We observe that $\mathrm{PAR}(s') = \mathrm{PAR}(\mathrm{WHT}((-1)^{p_t(\mathbf{x}_*)}))$, where $p_t(\mathbf{x}_*) = \sum_{i \in \mathbf{t}_{\mathbf{C}^\perp}} x_i (\sum_{j \in \mathbf{t}_{\mathbf{C}}} M_{i,j-k} x_j)$. By affine transformation of the variables in $\mathbf{x}_*$ we can rewrite $p_t(\mathbf{x}_*)$ as,

$$p_t(\mathbf{x}_*) = \sum_{i=0}^{\chi_t - 1} f_i(\mathbf{x}_{\mathbf{t}^\perp}) g_i(\mathbf{x}_{\mathbf{t}})$$

where f_i and g_i are linearly independent linear combinations of the variables in $\mathbf{x}_{\mathbf{t}^\perp}$ and $\mathbf{x}_{\mathbf{t}}$, respectively, $\forall i$. For clarity we further rewrite p_t as,

$$p_t(\mathbf{x}_*) = p_t(\mathbf{y}, \mathbf{z}) = \sum_{i=0}^{\chi_t - 1} y_i z_i$$

where the y_i and z_i are linearly independent binary variables over the spaces $\mathbf{x_{t^\perp}}$ and $\mathbf{x_t}$, respectively. It is known that the WHT over the $2\chi_t$ binary variables $\mathbf{y} \cup \mathbf{z}$ of $(-1)^{p_t(\mathbf{y},\mathbf{z})}$ has PAR $= 1$, (i.e. p_t is bent, [6]). Therefore the WHT of $(-1)^{p_t(\mathbf{x}_*)}$ over the $h + h^\perp$ binary variables in $\mathbf{x}_*$ has PAR $= 2^{h+h^\perp-2\chi_t}$. ∎

Corollary 1 *As* $0 \leq \chi_t \leq min(h, h^\perp)$, *it follows that* $PAR(s') \geq 2^{|h-h^\perp|}$

Theorem 2 also follows trivially from Theorem 5.

Lemma 1 *Let* $s \in \ell_\mathbf{p}$, *and let* s' *be a* HI *transform of* s. *Let* $s'' = H(i)[s']$).

$$\frac{PAR(s'')}{PAR(s')} \in \{\frac{1}{2}, 2\}$$

Proof. From Theorem 5. ∎

Theorem 6 PAR_l *of* $s \in \ell_\mathbf{p}$ *is found in the* HI *multispectra of* s.

Proof. Section 8. ∎

4 Weight Hierarchy and Stubborness of Entanglement (SE)

4.1 Weight Hierarchy

Definition 12 *[4] The Weight Hierarchy of* $\mathbf{C}$, *is a series of parameters,* d_j, $0 \leq j \leq k$, *representing the smallest blocklength of a linear sub-code of* $\mathbf{C}$ *of dimension* j, *where* $d_k = n$, $d_1 = d$, *and* $d_0 = 0$.

Theorem 7 *Let* s_c *be the indicator of an* $[n, k, d]$ *binary linear error-correcting code,* $\mathbf{C}$. *Let* $\mathbf{Q} \subset \{0, 1, \ldots, n-1\}$. *Let,*

$$m_\mathbf{Q} = \frac{|\mathbf{Q}| + \log_2(\mu) - n + k}{2}, \quad \text{where } \mu = PAR(s'_c) \tag{1}$$

and $s'_c = \prod_{t \in \mathbf{Q}} H(t)[s_c]$. *Then the Weight Hierarchy of* $\mathbf{C}$ *is found from the* HI *multispectra of* s_c, *where,*

$$d_j = min_{\mathbf{Q}|m_Q=j}(|\mathbf{Q}|)$$

Proof. Section 9 ∎

4.2 Stubborness of Entanglement

Theorem 8 *Let $s \in \ell_p$. A single qubit measurement on s gives s', and has one of two results:*

- *Destructive Measurement, where $|\{s'_i | s'_i \neq 0\}| = \frac{1}{2}|\{s_i | s_i \neq 0\}|$, and $PAR(s') = 2PAR(s)$.*
- *Redundant Measurement, where $|\{s'_i | s'_i \neq 0\}| = ||\{s_i | s_i \neq 0\}|$, and $PAR(s') = PAR(s)$.*

where $|s_i|, |s'_i| \in \{0, 1\}$.

Proof. Section 9. ∎

The entanglement order is the size (in qubits) of the largest entangled subsystem of the system.

Definition 13 *A most-destructive series of j single-qubit measurement over some set of measurements on s produces a final state s' such that entanglement order(s) – entanglement order(s') is maximised.*

Definition 14 *Stubborness of Entanglement (SE) is a series of parameters, β_j, $0 \leq j \leq k'$, representing the smallest possible entanglement order, β_j, after $k' - j$ most-destructive measurements of an n-qubit system, where $\beta_{k'} = n$, $\beta_0 = 0$.*

'Persistency of Entanglement', as defined in [1], is k', as it is the minimum number of measurements necessary to achieve entanglement order 0.

Definition 15 *The* C*-basis for s is the basis in which the vector becomes an indicator for the code* C*. This indicator has non-negative, multi-valued output and, for $s \in \ell_p$, the indicator has binary values. Without loss of generality, we always assume $|\mathbf{C}| \leq |\mathbf{C}^{\perp}|$. There is not a* C*-basis for every s but, for $s \in \ell_p$, there exists a* C*-basis and* $\mathbf{C}^{\perp}$*-basis.*

Lemma 2 *Let $s \in \ell_p$. Let s' be a HI transform of s. Then $n - \log_2(PAR(s'))$ destructive measurements in the s'-basis are sufficient to completely destroy entanglement in s.*

Proof. Follows from Theorem 8. ∎

Theorem 9 *Let $s \in \ell_p$. Then $LE(s)$ measurements is the minimum number of measurements required to completely destroy the entanglement of s.*

Proof. From Definition 9, Theorem 6, and Lemma 2. ∎

Theorem 10 *$k' \leq k$. In words, k destructive measurements in the* C*-basis suffice to destroy all entanglement in s, where $s \in \ell_p$ is equivalent to the binary linear $[n, k, d]$ error-correcting code,* C.

Proof. From Theorem 2 and Lemma 2. ∎

Theorem 11 *$k-j$ destructive measurements in the* **C**-*basis suffice to reduce the entanglement order of s to d_j qubits, where $s \in \ell_{\mathbf{p}}$ is equivalent to the binary linear $[n,k,d]$ error-correcting code,* **C**.

Proof. Recursive application of the proof for Theorem 10. ∎

Corollary 2

$$k' < k \quad or \quad \beta_j \leq d_j \quad if\ k' = k$$

In words, for states from $\ell_{\mathbf{p}}$, the Weight Hierarchy is an upper bound on Stubborness of Entanglement.

Theorem 12 *The series of residual entanglement orders, β'_j, resulting from j most-destructive measurements of s_c in the s_c-basis, satisfies $\beta'_j = d_j$. In words, the Stubborness of Entanglement confined to the s_c-basis is equivalent to the weight hierarchy of* **C**.

Proof. Recursive application of Theorems 8, 7, Lemma 2, and Theorem 9. ∎

Theorem 13 *Let $s \in \ell_{\mathbf{p}}$. Then the Stubborness of Entanglement of s can be computed from a set of most-destructive measurements in the HI-basis of s.*

Proof. A single-qubit measurement reduces (classical) entropy of the system by a maximum of 1 bit per measurement. This corresponds to measuring the qubit as 0 or 1 with equal likelihood $(-(\frac{1}{2}\log_2(\frac{1}{2}) + \frac{1}{2}\log_2(\frac{1}{2})) = 1)$. Destructive measurements in the HI-basis of a state s from $\ell_{\mathbf{p}}$ are always of this form (Theorem 8). For states from $\ell_{\mathbf{p}}$, entropy is LE. The theorem follows from Theorem 9. ∎

Definition 16 *A binary linear $[n,k,d]$ code is here considered optimal for a given n if $d_1, d_2, \ldots, d_{k-1}$ are maximised for each k, $k \leq \lfloor \frac{n}{2} \rfloor$, with priority to the weights, d_i, with lowest i.*

Theorem 14 *Let $s \in \ell_{\mathbf{p}}$ where s is LU equivalent to an optimal or near-optimal binary linear code of dimension $\leq \frac{n}{2}$. Then Stubborness of Entanglement is equal to the Weight Hierarchy of the code.*

Proof. We require the following Lemma.

Lemma 3 *Let s be LU equivalent to an optimal or near-optimal binary linear code, C. If C has dimension $k = \frac{n}{2}$, the HI multispectra of s only has PAR as high as 2^{n-k} for two HI transforms, one the WHT of the other. These two transform spectra, s_c and $s_{c^\perp}$ are binary indicators for C and $C^\perp$, respectively. Similarly, when C has dimension $k < \frac{n}{2}$, the HI multispectra of s only has PAR as high as 2^{n-k} for one HI transform. This transform spectra, s_c, is the binary indicator for C.*

Proof. (Lemma 3). By Theorem 7, an optimal or near-optimal binary linear code should have PAR as low as possible in the HI multispectra. It is easy to validate Lemma 3 for $1 \leq n \leq 4$ by exhaustive computation. Consider a state, s, that satisfies Lemma 3 for n even. There is (at least) one HI transform of s which gives a PAR of 2^{n-k}. Let us call this transform R. Let $x_i x_j$ be an arbitrary term in s such that R contains $H(i)$. Let us increase n by one to $n' = n + 1$ by appending $x_l x_i$, to create $s^+(\mathbf{x}) = s(\mathbf{x}) + x_l x_i$, where x_l is a new qubit added to the system. Then $R \otimes I_l$ is the only HI transform of s^+ which achieves PAR as high as 2^{n-k+1} so Lemma 3 is satisfied for n odd, as long as it is satisfied for n even. Let us increase n to $n'' = n+2$, by appending $x_l x_i + x_m x_j + x_l x_m$, to create $s^{++}(\mathbf{x}) = s(\mathbf{x}) + x_l x_i + x_m x_j + x_l x_m$, where x_l and x_m are new qubits added to the system. Then $R \otimes I_l \otimes H_m$, $R \otimes H_l \otimes H_m$, and $R \otimes H_l \otimes I_m$ are HI transforms of s^{++} which achieve a PAR of 2^{n-k+1}, 2^{n-k}, and 2^{n-k-1}, respectively. Therefore Lemma 3 is satisfied for n even as no other HI transform reaches a PAR of 2^{n-k+1}. By induction on n it follows that we can construct states s satisfying the conditions of Lemma 3 for all n and $k \leq \frac{n}{2}$. ∎

The most-destructive series of measurements on s is performed in the basis which is the HI transform, s', of s that achieves highest PAR (Lemma 2, Theorem 9). But, from Lemma 3, $s' = s_{\mathbf{c}}$ and, for $k = \frac{n}{2}$, $s' = s_{\mathbf{c}^\perp}$ also. The Theorem follows from Theorem 12. ∎

Corollary 3 *Quantum states from $\ell_{\mathbf{p}}$ which have optimum Linear Entanglement and optimum Stubborness of Entanglement are LU equivalent to binary linear codes with optimum Weight Hierarchy.*

5 Cryptographic Measures of Entanglement

Let s_H be the WHT of s.

Definition 17 *The Nonlinear Order (N) of s is given by,*

$$N(s) = n - \log_2(PAR(s_H))$$

which, by Definition 9, is a trivial upper bound on LE.

Definition 18 *[11] s has Correlation-Immunity Order t (CI-t) if s_H is zero at all indeces with binary weight $\leq t$.*

s_H is an indicator for a code, $\mathbf{C_H}$, such that $\mathbf{C_H}$ has minimum Hamming Distance $d = t+1$, (ignoring phase). Nonlinear Order and Correlation-Immunity of a state from $\ell_{\mathbf{p}}$ provide upper and lower bounds on PAR_l.

Theorem 15 *Let $s(\mathbf{x}) = (-1)^{p(\mathbf{x})} \in \ell_{\mathbf{p}}$ and have nonlinear order, N. Then,*

$$n - N \leq \log_2(PAR_l(s)) \leq n - \frac{N}{2}$$

Furthermore, if the Correlation-Immunity of s is CI-t,

$$n - N \leq \log_2(PAR_l(s)) \leq max(n - t - 1 - \frac{N}{2}, n - N)$$

where $0 \leq t + 1 \leq n - N$.

Proof. The lower bound follows from Definitions 9 and 17. For the upper bound, the application of $H(i)$ on a member of $\ell_{\mathbf{p}}$ either decreases or increases PAR by a factor of 2 (Lemma 1). Therefore, $\text{PAR}(\prod_{i \in \mathbf{T}} H(i)[s])$ is maximised for some subset, $\mathbf{T}$, if $\exists \mathbf{T}'$ such that,

$$\text{PAR}(\prod_{i \in \mathbf{T}'} H(i)[s]) = 2^{|\mathbf{T}| - |\ |\mathbf{T}| - |\mathbf{T}'|\ |} \tag{2}$$

for at least one subset, $\mathbf{T}'$, of each size i, $0 \leq i \leq n$. But, for $|\mathbf{T}'| = n$, $\prod_{i \in \mathbf{T}'} H(i)[s] = s_H$, where $s_H = \text{WHT}(s)$. We have,

$$\text{PAR}(s_H) = 2^{n-N} \tag{3}$$

Combining (2) and (3) gives,

$$2^{2|\mathbf{T}| - n} = 2^{n-N} \Rightarrow \text{PAR}_l(s) \leq 2^{|\mathbf{T}|} = 2^{n - \frac{N}{2}}$$

thereby satisfying the first part of the Theorem. Now, if s is also known to be CI-t, then,

$$\text{PAR}(\prod_{i \in \mathbf{T}'} H(i)[s]) = 2^{n-N-t-1} \tag{4}$$

$\forall \mathbf{T}'$ where $|\mathbf{T}'| = n - t - 1$. Combining (2) and (4) gives,

$$\begin{array}{ll} \text{PAR}_l(s) \leq 2^{|\mathbf{T}|} = 2^{n-t-1-\frac{N}{2}}, & 0 \leq |\mathbf{T}| \leq n - t - 1 \\ \text{PAR}_l(s) \leq 2^{n-N}, & n - t \leq |\mathbf{T}| \leq n \end{array}$$

The Theorem follows. ∎

6 Non-Bipartite Quadratic Bipolar States

We now examine quadratic bipolar states which do not have bipartite form. A 'Fully-Connected Quadratic' is a bipolar quantum state of the form,

$$s(\mathbf{x}) = (-1)^{\sum_{j=0}^{n-2} \sum_{i=j+1}^{n-1} x_j x_i}$$

These states have a PAR of 2.0 under HI multispectra, but PAR_l is 2^{n-1}. Fully-Connected States satisfy the following:

$$\mathbf{s}' = (10\ldots01) = \begin{pmatrix} w^7 & 0 \\ 0 & w \end{pmatrix}_f \prod_{j=0}^{n-1} \mathbf{NH}(j)\boldsymbol{s}$$

where $w = e^{\frac{2\pi i}{8}}$, $\mathbf{NH} = \begin{pmatrix} 1 & i \\ 1 & -i \end{pmatrix}$ (the NegaHadamard Transform [7]), $i^2 = -1$, and $\mathbf{s}'$ is the binary indicator for the $[n,1,n]$ repetition code. $\begin{pmatrix} w^7 & 0 \\ 0 & w \end{pmatrix}_f$ acts on qubit f, where f is chosen arbitrarily from $\{0,1,\ldots,n-1\}$. So Fully-Connected quadratic bipolar states are LU equivalent to $\ell_{\mathbf{p}}$. Moreover, as $k = 1$ for a repetition code, only **one** measurement is necessary (of any qubit in the **NH**-basis) to completely disentangle the state. Let $s(\mathbf{x}) = (-1)^{x_0x_1+x_0x_2+x_0x_3+x_0x_4+x_1x_2+x_1x_4+x_2x_3+x_3x_4}$. By computation, $\text{PAR}_l(s) = 8.0$ (e.g. under the WHT), suggesting that only $\log_2(\frac{32}{8}) = 2$ measurements are necessary to completely disentangle the state, for instance, by measuring qubits 0 and 1 in the H-basis. Computations suggest that $s(\mathbf{x}) = (-1)^{x_0(x_1+x_2+x_3+x_4+x_5)+x_1x_2+x_2x_3+x_3x_4+x_4x_5+x_5x_1}$ (or any of its 72 qubit permutations) has $\text{PAR}_l = 4.486$. The lowest possible PAR_l for a 6-qubit state from $\ell_{\mathbf{p}}$ is $2^3 = 8$. No other 6-qubit bipolar quadratics have PAR_l this low. No more than 4 measurements are required to completely destroy entanglement in the state. For instance, by measuring qubits 0,1,3,5 in the bipolar basis.

7 Conclusion

This paper has identified PAR, code dimension, weight hierarchy, nonlinear order, and correlation immunity, as having useful meanings in a quantum context. Most importantly, we have shown that, for quantum states which are equivalent under local unitary transform to indicators for binary linear error-correcting codes, the partial entanglement measures of Linear Entanglement and Stubborness of Entanglement can be computed by only examining the multispectra arising from tensor products of 2×2 Walsh-Hadamard and Identity matrices. The paper has focussed on bipolar quadratic entanglement, in particular entanglement that can be described using binary linear block codes. Future work will consider higher-degree and higher-alphabet states, and we will need to search outside the HI multispectra to find the desired entanglement parameters.

8 Appendix: PAR_l and Linear Entanglement (LE)

Proof. (**Theorem 6**) Let H_{e_i}, U_{e_i}, and I_{e_i} be the PARs of the resultant spectrum, $\boldsymbol{s}'$, after application of $H(e_i)$, $U(e_i)$, or $I(e_i)$, respectively, on $\boldsymbol{s}$, where $U(e_i)$ is any 2×2 unitary matrix acting on qubit x_{e_i}, and $I(e_i)$ is the 2×2 identity matrix acting on x_{e_i}. The maximum PAR scaling of $\boldsymbol{s}'$ relative to s after application of one 2×2 unitary matrix is by a factor of 2, either up or down. But from Lemma 1, when $s \in \ell_{\mathbf{p}}$, the application of H_{e_i} on s' scales PAR by a factor of 2. Therefore,

$$\begin{array}{ll} \text{either } U_{e_i} \geq H_{e_i} & \text{if } H_{e_i} = \frac{1}{2} I_{e_i} \\ \text{or } U_{e_i} \leq H_{e_i} & \text{if } H_{e_i} = 2 I_{e_i} \end{array} \tag{5}$$

Moreover,

$$\frac{1}{2} I_{e_i} \leq U_{e_i} \leq 2 I_{e_i} \tag{6}$$

Combining (5) and (6) gives,

$$\begin{array}{l} \text{if } H_{e_i} = \frac{1}{2} I_{e_i} \text{ then } H_{e_i} \leq U_{e_i} \leq 2 I_{e_i} \\ \text{if } H_{e_i} = 2 I_{e_i} \text{ then } \frac{1}{2} I_{e_i} \leq U_{e_i} \leq H_{e_i} \end{array} \tag{7}$$

Let us write

$$U(e_i) = [U(e_i) H(e_i)]\, H(e_i) = V(e_i) H(e_i)$$

for some 2×2 unitary matrix, $V(e_i)$. Then,

$$\frac{1}{2} H_{e_i} \leq U_{e_i} \leq 2 H_{e_i} \tag{8}$$

Combining (7) and (8) gives,

$$\begin{array}{l} \text{if } H_{e_i} = \frac{1}{2} I_{e_i} \text{ then } H_{e_i} \leq U_{e_i} \leq I_{e_i} \\ \text{if } H_{e_i} = 2 I_{e_i} \text{ then } I_{e_i} \leq U_{e_i} \leq H_{e_i} \end{array} \tag{9}$$

Let any product of the H_{e_i}, U_{e_j}, and I_{e_k} define the resultant PAR after application of the same tensor product of $H(e_i)$, $U(e_j)$, and $I(e_k)$.

$$\begin{array}{l} \text{either } H_{e_j} U_{e_i} \leq U_{e_j} U_{e_i} \leq I_{e_j} U_{e_i}, \\ \text{or } I_{e_j} U_{e_i} \leq U_{e_j} U_{e_i} \leq H_{e_j} U_{e_i}, \quad i \neq j \end{array} \tag{10}$$

Combining (9) and (10) guarantees that at least one of the four following conditions is satisfied:

$$\begin{array}{ll} \text{either} & H_{e_j} H_{e_i} \leq U_{e_j} U_{e_i} \leq I_{e_j} I_{e_i} \\ \text{or} & H_{e_j} I_{e_i} \leq U_{e_j} U_{e_i} \leq I_{e_j} H_{e_i} \\ \text{or} & I_{e_j} H_{e_i} \leq U_{e_j} U_{e_i} \leq H_{e_j} I_{e_i} \\ \text{or} & I_{e_j} I_{e_i} \leq U_{e_j} U_{e_i} \leq H_{e_j} H_{e_i} \end{array}$$

Continuing in this fashion, it follows that,

$$\prod_{i \in \mathbf{T} \setminus \mathbf{t}_*} H_{e_i} \prod_{j \in \mathbf{t}_*} I_{e_i} \leq \prod_{i \in \mathbf{T}} U_{e_i} \leq \prod_{i \in \mathbf{T} \setminus \mathbf{t}_*} I_{e_i} \prod_{j \in \mathbf{t}_*} H_{e_i}$$

for any integer set $\mathbf{T}$ where $\mathbf{t}_* \subset \mathbf{T}$. In words, for $s \in \ell_{\mathbf{p}}$, the PAR of any LU transform of s is always upper and lower bounded by two points in the HI multispectra. ■

9 Appendix: Weight Hierarchy and Stubborness of Entanglement (SE)

Proof. (**Theorem 7**) Decompose s_c as,

$$s_c(\mathbf{x}) = f_0(x_t = 0, g_0(\mathbf{x}\backslash x_t)) + f_1(x_t = 1, g_1(\mathbf{x}\backslash x_t))$$

for boolean functions f_i and g_i. From Lemma 1, let,

$$\gamma_t = \frac{\mathrm{PAR}(H(t)[s_c])}{\mathrm{PAR}(s_c)} \in \{\frac{1}{2}, 2\}$$

Then $\gamma_t = 2$ iff $g_0 = g_1$, which implies $\mathbf{C}$ has Hamming Distance $d = 1$. Otherwise $\gamma_t = \frac{1}{2}$. More generally we can find a subset, $\mathbf{x_w}$, of $\mathbf{x}$ such that, if $\mathbf{y_w} \subset \mathbf{x_w}$, then

$$\frac{\mathrm{PAR}(\prod_{t \in \mathbf{y_w}} H(t)[s_c])}{\mathrm{PAR}(s_c)} = \begin{array}{ll} 2^{-|\mathbf{y_w}|}, & \mathbf{y_w} \neq \mathbf{x_w} \\ 2^{2-|\mathbf{y_w}|} = 2^{2-d_w}, & \mathbf{y_w} = \mathbf{x_w} \end{array}$$

where d_w is any member of the weight distribution of $\mathbf{C}$. Indeed, for a smallest-size non-empty subset, $\mathbf{x_w}$, we have $d_w = d_1 = d$, which is the Hamming Distance of $\mathbf{C}$. Similar arguments hold for the higher weights, $d_2, d_3, \ldots, d_k$ where, at each stage, a smallest unused non-empty subset, $\mathbf{x_{w_i}}$ is added. Let $\mathbf{Q} = \mathbf{y_{w_j}} \cup \bigcup_{i=0}^{j-1} \mathbf{x_{w_i}}$. After further manipulation we arrive at the Theorem. ∎

Proof. (**Theorem 8**) Let s_c be the binary indicator for a binary linear ECC. s_c takes one of two forms:

Form 1. $s_c(\mathbf{x}) = f_0(x_i = 0, g_0(\mathbf{x}\backslash x_i)) + f_1(x_i = 1, g_1(\mathbf{x}\backslash x_i))$
Form 2. $s_c(\mathbf{x}) = f_0(x_i = \gamma, g(\mathbf{x}\backslash x_i))$

where the f_i, g_i are boolean functions, and γ is fixed at 0 or 1. After measurement of x_i we obtain $s'_c(\mathbf{x}) = s_c(\mathbf{x})_{x_i=\gamma}$ which is,

- Destructive Measurement - when s_c is of Form 1.
- Redundant Measurement - when s_c is of Form 2.

The PAR relationships follow by Parseval's Theorem. Consider measuring $s'(\mathbf{x}) = H(i)[s_c(\mathbf{x})]$. If s_c is of Form 1. then s' is of Form 1. or 2. If s_c is of Form 2. then s' is of Form 1. In either case s' is still one of the same two forms and we have the same two measurement scenarios for s' as for s_c. The proof follows recursively for the whole set, $\ell_\mathbf{p}$. ∎

References

1. Briegel H.J.,Raussendorf R. (2000) Persistent Entanglement in Arrays of Interacting Particles. quant-ph/0004051 v2 28 Aug

2. Davis J.A.,Jedwab J. (1999) Peak-to-mean Power Control in OFDM, Golay Complementary Sequences and Reed-Muller Codes. IEEE Trans. Inform. Theory **45.** No 7, 2397–2417, Nov
3. Eisert J.,Briegel H.J. (2000) Quantification of Multi-Particle Entanglement. quant-ph/0007081 v2 29 Aug
4. Kløve T.,Chen W. (1997) Bounds on the Weight Hierarchies of Linear Codes of Dimension 4 IEEE Trans. Inform. Theory **43.** No 6, 2047-2054
5. Linden N.,Popescu S. (1998) On Multi-Particle Entanglement Fortsch. Phys.. **46.** 567–578,
6. MacWilliams F.J.,Sloane N.J.A. (1977) **The Theory of Error-Correcting Codes** Amsterdam: North-Holland.
7. Parker M.G.,Tellambura C. (2000) Golay-Davis-Jedwab Complementary Sequences and Rudin-Shapiro Constructions. Submitted to IEEE Trans. Inform. Theory, Preprint available on /~matthew/mattweb.html September
8. Popescu S.,Rohrlich D. (1997) On the Measure of Entanglement for Pure States. Phys. Rev. A **56.** R3319
9. Raussendorf R.,Briegel H.J. (2000) Quantum Computing via Measurements Only. arXiv:quant-ph/0010033 7 Oct
10. Roychowdhury V.P.,Vatan F. (2001) Bounds for Weight Distribution of Weakly Self-Dual Codes. arXiv:math.CO/0104016 2 Apr
11. Xiao G.,Massey J.L. (1988) A Spectral Characterization of Correlation-Immune Combining Functions. IEEE Trans. Inform. Theory **34.** No 3, 569–571

Characteristic Polynomials of Binary Kronecker Sequences

Satoshi Uehara[1] and Kyoki Imamura[2]

[1] Dept. of Information and Media Sciences, The University of Kitakyushu, 1-1 Hibikino Wakamatsu-ku Kitakyushu, 808-0135 JAPAN, e-mail: uehara@env.kitakyu-u.ac.jp

[2] Dept. of Computer Science and Electronics, Kyushu Institute of Technology, 680-4 Kawazu Iizuka, 820-8502 JAPAN e-mail: imamura@cse.kyutech.ac.jp

Summary. Kronecker sequences constructed by the Kronecker product method from a few sequences, as well as combination sequences can be synchronized considerably faster than a conventional PN sequence of equal length. Recently interesting conjectures on the linear complexity of binary Kronecker sequences were given by Kärkkäinen. In this paper we give the exact formula of the characteristic polynomial and the linear complexity for binary Kronecker sequences obtained from two sequences called inner sequence and outer sequence. We also show that the Kronecker sequence has a prominent autocorrelation property.

Keywords – Kronecker sequence, Barker sequence, linear complexity, characteristic polynomial, autocorrelation

1 Introduction

Pseudonoise(PN) sequences used in stream ciphers and spread spectrum communications are required to be more random for keeping security and to have an impulse-like autocorrelation and a low crosscorrelation for a rapid code synchronization. Kronecker sequences [1] are one class of such sequences and constructed by combining two or more periodic sequences. Recently Kärkkäinen [2], by numerical experiments, gave interesting conjectures on the linear complexities of some Kronecker sequences.

In this paper we give the exact formula for the characteristic polynomial and the linear complexity of the binary Kronecker sequence obtained from two sequences, called inner sequence and outer sequence. We also show that the Kronecker sequence has a prominent autocorrelation property when its outer sequence is a Barker sequence. Moreover, we show an autocorrelation property when its inner sequence is an m-sequence.

2 Preparation

In this section we firstly introduce the relation between a periodic sequence and its characteristic polynomial, and show a method for calculating its linear complexity. Secondly we give some properties of Barker sequences since we will consider excellent autocorrelation properties of Kronecker sequences obtained from a Barker sequence as the outer sequence.

The linear complexity of a sequence $\{a_i\}_{i\geq 0}$ over $GF(2)$ with period N, denoted as LC($\{a_i\}$), can be determined by finding the generating function of the infinite power series $\sum_{i\geq 0} a_i x^{-(i+1)}$[3]. If we have

$$\sum_{i\geq 0} a_i x^{-(i+1)} = \frac{g(x)}{f(x)}, \qquad (f(x), g(x)) = 1, \tag{1}$$

where $f(x)$ is a monic polynomial, then we have

$$\mathrm{LC}(\{a_i\}) = \deg f(x).$$

The polynomial $f(x)$ is called the characteristic polynomial of $\{a_i\}_{i\geq 0}$.

Since the left-hand side of (1) is equal to $\frac{a(x)}{x^N-1}$ with $a(x) = a_0 x^{N-1} + a_1 x^{N-2} + \cdots + a_{N-1}$, we have

$$\frac{a(x)}{x^N - 1} = \frac{g(x)}{f(x)}, \qquad (g(x), f(x)) = 1. \tag{2}$$

For a binary sequence $\{b_i\}_{0\leq i\leq \ell-1}$ with a finite length ℓ its aperiodic autocorrelation function $\bar{R}_{bb}(\tau)$ is defined by

$$\bar{R}_{bb}(\tau) = \begin{cases} \sum_{i=0}^{\ell-|\tau|-1} (-1)^{b_i + b_{i+|\tau|}} & \text{if } |\tau| < \ell, \\ 0 & \text{if } |\tau| \geq \ell. \end{cases} \tag{3}$$

The definition of the binary Barker sequence is as follows.

Definition 1 *Let $\{b_i\}_{0\leq i\leq \ell-1}$ be a binary sequence with a finite length ℓ. If $\{b_i\}_{0\leq i\leq \ell-1}$ satisfies*

$$|\bar{R}_{bb}(\tau)| \leq 1, \quad \textit{for } \tau \neq 0, \tag{4}$$

then it is called a Barker sequence. □

Property 1 *Let $\{b_i\}_{0\leq i\leq \ell-1}$ be a Barker sequence with length ℓ. Then each of $\{\bar{b}_i\}$, $\{\hat{b}_i\}$ and $\{\hat{\bar{b}}_i\}$ is also a Barker sequence, where $\bar{b}_i = b_i + 1 \pmod 2$ and $\hat{b}_i = b_{\ell-1-i}$.*

Proof of Property 1 : From (3) we have $\bar{R}_{\bar{b}\bar{b}}$ and $\bar{R}_{\hat{b}\hat{b}}$ for $|\tau| < \ell$.

$$\bar{R}_{\bar{b}\bar{b}}(\tau) = \sum_{i=0}^{\ell-|\tau|-1} (-1)^{\bar{b}_i + \bar{b}_{i+|\tau|}} = \sum_{i=0}^{\ell-|\tau|-1} (-1)^{(b_i+1)+(b_{i+|\tau|}+1)}$$

$$= \sum_{i=0}^{\ell-|\tau|-1} (-1)^{b_i + b_{i+|\tau|}} = \bar{R}_{bb}(\tau) \tag{5}$$

Table 1. The characteristic polynomial and the linear complexity of Barker sequences.

ℓ	Barker sequence	LC		Characteristic Polynomial (Periodic)
		Aperi.	Peri.	
2	00	0	0	-
	11	1	1	$x+1$
2	01	2	2	$x^2+1=(x+1)^2$
	10	2	2	$x^2+1=(x+1)^2$
3	001	3	3	$x^3+1=(x+1)(x^2+x+1)$
	110	2	2	x^2+x+1
4	0001	4	4	$x^4+1=(x+1)^4$
	1110	3	4	$x^4+1=(x+1)^4$
4	0010	3	4	$x^4+1=(x+1)^4$
	1101	2	4	$x^4+1=(x+1)^4$
5	00010	4	5	$x^5+1=(x+1)(x^4+x^3+x^2+x+1)$
	11101	3	4	$x^4+x^3+x^2+x+1$
7	0001101	4	4	$x^4+x^3+x^2+1$
	1110010	3	3	x^3+x+1
11	00011101101	6	10	$x^{10}+x^9+\cdots+x^2+x+1$
	11100010010	7	11	$x^{11}+1=(x+1)(x^{10}+x^9+\cdots+x^2+x+1)$
13	0000011001010	7	12	$x^{12}+x^{11}+\cdots+x^2+x+1$
	1111100110101	8	13	$x^{13}+1=(x+1)(x^{12}+x^{11}+\cdots+x^2+x+1)$

$$\begin{aligned}\bar{R}_{\hat{b}\hat{b}}(\tau) &= \sum_{i=0}^{\ell-|\tau|-1}(-1)^{\hat{b}_i+\hat{b}_{i+|\tau|}} = \sum_{i=0}^{\ell-|\tau|-1}(-1)^{b_{\ell-1-i}+b_{\ell-1-i-|\tau|}} \\ &= \sum_{j=0}^{\ell-1-|\tau|}(-1)^{b_{j+|\tau|}+b_j} = \bar{R}_{bb}(\tau) \end{aligned} \tag{6}$$

From (5), (6) both of $\{\bar{b}_i\}_{0\le i\le \ell-1}$ and $\{\hat{b}_i\}_{0\le i\le \ell-1}$ are Barker sequences. Therefore it is trivial that $\{\hat{\bar{b}}_i\}$ is also a Barker sequence. □

It is known that there exist 32 binary Barker sequences and their lengths are less than or equal to 13. Table 1 shows all the previously known binary Barker sequences and their LC values and characteristic polynomials.

Let $R_{bb}(\tau)$ be the periodic autocorrelation function of a sequence $\{b_i\}_{i\ge 0}$ with period ℓ defined as

$$R_{bb}(\tau) = \sum_{i=0}^{\ell-1}(-1)^{b_i+b_{i+\tau}}. \tag{7}$$

We assume that a Barker sequence is a periodic sequence, then we have the following theorem.

Table 2. The periodic autocorrelation values of Barker sequences. ($\tau \neq 0$)

ℓ	Barker Seq.	C	ℓ	Barker Seq.	C
2	00	2	5	00010	1
	11	2		11101	1
2	01	2	7	0001101	-1
	10	2		1110010	-1
3	001	-1	11	00011101101	-1
	110	-1		11100010010	-1
4	0001	0	13	0000011001010	1
	1110	0		1111100110101	1
4	0010	0			
	1101	0			

Theorem 1 *Assume that the Barker sequence with length ℓ ($\ell \neq 2$) is a periodic sequence with period ℓ. Its periodic autocorrelation values for $\tau \neq 0$ take a fixed value, i.e., we have*

$$R_{bb}(\tau) = C \in \{0, \pm 1\}, \qquad (\tau \neq 0). \tag{8}$$

For $\ell = 2$ we have $R_{bb}(\tau)|_{\tau \neq 0} = 2$. □

In Table 2, we show the list of periodic autocorrelation values given by numerical experiments.

3 Kronecker Sequences

Let $\{a_i\}_{i\geq 0}$, $\{b_i\}_{i\geq 0}$ be periodic sequences over $GF(2)$ with period N_a, N_b and characteristic polynomials $f_a(x)$, $f_b(x)$, respectively. We have

$$\frac{a(x)}{x^{N_a}+1} = \frac{g_a(x)}{f_a(x)}, \quad (g_a(x), f_a(x)) = 1, \tag{9}$$

$$\frac{b(x)}{x^{N_b}+1} = \frac{g_b(x)}{f_b(x)}, \quad (g_b(x), f_b(x)) = 1, \tag{10}$$

$$\bar{a}(x) = \frac{x^{N_a}+1}{x+1} + a(x), \tag{11}$$

$$\bar{b}(x) = \frac{x^{N_b}+1}{x+1} + b(x), \tag{12}$$

where $\bar{a}(x)$ and $\bar{b}(x)$ correspond to sequences $\{\bar{a}_i\}_{i\geq 0}$ and $\{\bar{b}_i\}_{i\geq 0}$, respectively. The definition of the Kronecker sequence constructed from inner sequence $\{a_i\}_{i\geq 0}$ and outer $\{b_i\}_{i\geq 0}$ is as follows.

Definition 2 *Let* $\{a_i\}_{i\geq 0}$ *be the inner sequence and* $\{b_i\}_{i\geq 0}$ *the outer sequence. A binary Kronecker sequence* $\{A_i\}_{i\geq 0}$ *with period* N_aN_b *is denoted as*

$$A_i = a_{i \,(\mathrm{mod}\ N_a)} + b_{\lfloor \frac{i}{N_a} \rfloor}, \tag{13}$$

i.e., it means

$$a_0+b_0, a_1+b_0, \cdots, a_{N_a-1}+b_0, a_0+b_1, a_1+b_1,$$
$$\cdots, a_0+b_{N_b-1}, \cdots, a_{N_a-1}+b_{N_b-1}$$
□

For $i = l + mN_a$, we have

$$\begin{aligned} A_i &= a_l + b_m \\ &= a_l + 2a_lb_m + b_m = a_l(1+b_m) + b_m(1+a_l) \\ &= a_l\bar{b}_m + \bar{a}_lb_m. \end{aligned}$$

Therefore, we have

$$\begin{aligned} A(x) &= \bar{b}(x^{N_a})\cdot a(x) + b(x^{N_a})\cdot \bar{a}(x) \\ &= \left(\frac{x^{N_aN_b}+1}{x^{N_a}+1} + b(x^{N_a})\right)\cdot a(x) + b(x^{N_a})\cdot\left(\frac{x^{N_a}+1}{x+1} + a(x)\right) \\ &= \frac{x^{N_aN_b}+1}{x^{N_a}+1}\cdot a(x) + a(x)\cdot b(x^{N_a}) \\ &\qquad + \frac{x^{N_a}+1}{x+1}\cdot b(x^{N_a}) + a(x)\cdot b(x^{N_a}) \\ &= \frac{x^{N_aN_b}+1}{x^{N_a}+1}\cdot a(x) + \frac{x^{N_a}+1}{x+1}\cdot b(x^{N_a}). \end{aligned}$$

From (9)-(12), (1) we have

$$\begin{aligned} A(x) &= \frac{x^{N_aN_b}+1}{x^{N_a}+1}\cdot\frac{(x^{N_a}+1)g_a(x)}{f_a(x)} + \frac{x^{N_a}+1}{x+1}\cdot\frac{(x^{N_aN_b}+1)g_b(x^{N_a})}{f_b(x^{N_a})} \\ &= (x^{N_aN_b}+1)\cdot\left(\frac{g_a(x)}{f_a(x)} + \frac{(x^{N_a}+1)g_b(x^{N_a})}{(x+1)f_b(x^{N_a})}\right) \end{aligned} \tag{14}$$

and

$$\frac{A(x)}{x^{N_aN_b}+1} = \frac{g_a(x)}{f_a(x)} + \frac{(x^{N_a}+1)g_b(x^{N_a})}{(x+1)f_b(x^{N_a})}. \tag{15}$$

We give the relation between $g_b(x^{N_a})$ and $f_b(x^{N_a})$.

Lemma 1 $(g(x), f(x)) = 1$ *if and only if* $(g(x^m), f(x^m)) = 1$.

Proof of Lemma 1 : We can denote $(g(x^m), f(x^m)) = h(x^m)$. Let $X = x^m$. From $(g(x), f(x)) = 1$ we have $(g(X), f(X)) = h(X) = 1$. Therefore we have $(g(x^m), f(x^m)) = 1$. □

From Lemma 1 we can derive the characteristic polynomial $F(x)$ of a Kronecker sequence $\{A_i\}_{i\geq 0}$.

We assume that a pair of polynomials $\{G_b(x), F_b(x)\}$ satisfy

$$\frac{(x^{N_a}+1)g_b(x^{N_a})}{(x+1)f_b(x^{N_a})} = \frac{G_b(x)}{F_b(x)}, \quad (G_b(x), F_b(x)) = 1. \tag{16}$$

From Lemma 1 we have

$$F_b(x) = \begin{cases} (x+1)\dfrac{f_b(x^{N_a})}{x^{N_a}+1} & \text{if } x+1 | f_b(x) \\ f_b(x^{N_a}) & \text{if } x+1 \not| f_b(x), \end{cases} \tag{17}$$

and we get the following theorem from (15)-(17).

Theorem 2 *The characteristic polynomial, $F(x)$, of the Kronecker sequence $\{A_i\}_{i\geq 0}$ in Definition 2 is given by*

$$\begin{aligned} F(x) &= lcm(f_a(x), F_b(x)) \\ &= \begin{cases} f_a(x) f_b(x^{N_a}) & \text{if } x+1 \not| f_b(x) \\ (x+1) f_a(x) \dfrac{f_b(x^{N_a})}{x^{N_a}+1} & \text{if } x+1 | f_b(x),\ x+1 \not| f_a(x) \\ f_a(x) \dfrac{f_b(x^{N_a})}{x^{N_a}+1} & \text{if } x+1 | f_b(x), f_a(x). \end{cases} \end{aligned}$$

□

Let $\mathrm{LC}_a = \deg f_a(x)$ and $\mathrm{LC}_b = \deg f_b(x)$. From Theorem 2 we have the following corollary.

Corollary 1 *The linear complexity $LC(\{A_i\})$ of the Kronecker sequence $\{A_i\}_{i\geq 0}$ is given by*

$$\begin{aligned} &LC(\{A_i\}) \\ &= \begin{cases} LC_b N_a + LC_a & \text{if } x+1 \not| f_b(x) \\ (LC_b - 1)N_a + LC_a + 1 & \text{if } x+1 | f_b(x),\ x+1 \not| f_a(x) \\ (LC_b - 1)N_a + LC_a & \text{if } x+1 | f_b(x), f_a(x). \end{cases} \end{aligned} \tag{18}$$

□

4 Autocorrelation of Kronecker Sequences

We consider a periodic autocorrelation of a Kronecker sequence $\{A_i\}_{i\geq 0}$. We can denote a periodic autocorrelation function $R_{AA}(\tau)$ of $\{A_i\}_{i\geq 0}$ as

$$R_{AA}(\tau) = \sum_{\ell=0}^{N_a N_b - 1} (-1)^{A_\ell + A_{\ell+\tau}}$$

$$= \sum_{\ell=0}^{N_aN_b-1} (-1)^{(a_{\ell \pmod{N_a}}+b_{\lfloor \frac{\ell}{N_a} \rfloor})+(a_{\ell+\tau \pmod{N_a}}+b_{\lfloor \frac{\ell+\tau}{N_a} \rfloor})}$$

$$= \sum_{\ell=0}^{N_aN_b-1} (-1)^{a_{\ell \pmod{N_a}}+a_{\ell+\tau \pmod{N_a}}} (-1)^{b_{\lfloor \frac{\ell}{N_a} \rfloor}+b_{\lfloor \frac{\ell+\tau}{N_a} \rfloor}}.$$

We will consider the case where the outer sequence $\{b_i\}_{i\geq 0}$ is a binary Barker sequence. Let

$$\tau_a = \tau \pmod{N_a}, \qquad \tau_b = \left\lfloor \frac{\tau}{N_a} \right\rfloor.$$

Since a periodic autocorrelation value $R_{bb}(\tau)$ has a fixed value C for $\tau_b \neq 0$ from Theorem 1, the Kronecker sequence with a Barker sequence as the outer sequence has the following autocorrelation function,

in case of $\tau_a = 0$

$$\begin{aligned} R_{AA}(\tau) &= \sum_{j=0}^{N_b-1} \sum_{i=0}^{N_a-1} (-1)^{a_i+a_i} (-1)^{b_j+b_{j+\tau_b}} \\ &= N_a R_{bb}(\tau_b), \end{aligned} \tag{19}$$

in case of $\tau_a \neq 0$

$$\begin{aligned} R_{AA}(\tau) &= \sum_{j=0}^{N_b-1} \left(\sum_{i=0}^{N_a-\tau_a-1} (-1)^{a_i+a_{i+\tau_a}} (-1)^{b_j+b_{j+\tau_b}} \right. \\ &\quad \left. + \sum_{i=N_a-\tau_a}^{N_a-1} (-1)^{a_i+a_{i+\tau_a}} (-1)^{b_j+b_{j+\tau_b+1}} \right) \\ &= \sum_{i=0}^{N_a-\tau_a-1} (-1)^{a_i+a_{i+\tau_a}} \sum_{j=0}^{N_b-1} (-1)^{b_j+b_{j+\tau_b}} \\ &\quad + \sum_{i=N_a-\tau_a}^{N_a-1} (-1)^{a_i+a_{i+\tau_a}} \sum_{j=0}^{N_b-1} (-1)^{b_j+b_{j+\tau_b+1}} \\ &= \begin{cases} N_b \displaystyle\sum_{i=0}^{N_a-\tau_a-1} (-1)^{a_i+a_{i+\tau_a}} \\ \quad +C \displaystyle\sum_{i=N_a-\tau_a}^{N_a-1} (-1)^{a_i+a_{i+\tau_a}} & \text{if } \tau_b = 0 \\ C \displaystyle\sum_{i=0}^{N_a-\tau_a-1} (-1)^{a_i+a_{i+\tau_a}} \\ \quad +N_b \displaystyle\sum_{i=N_a-\tau_a}^{N_a-1} (-1)^{a_i+a_{i+\tau_a}} & \text{if } \tau_b = N_b - 1 \\ C R_{aa}(\tau_a) & \text{otherwise.} \end{cases} \end{aligned} \tag{20}$$

Therefore, we have the following theorem.

Theorem 3 *Let $\{b_i\}_{i\geq 0}$ be a Barker sequence with period N_b and $\{A_i\}_{i\geq 0}$ a Kronecker sequence with period N_aN_b whose outer sequence is $\{b_i\}_{i\geq 0}$. Then we can get a periodic autocorrelation value $R_{AA}(\tau)$ from (19), (20).* □

If the inner sequence $\{a_i\}_{i\geq 0}$ is an m-sequence, then we have the following corollary.

Corollary 2 *Let the inner sequence $\{a_i\}_{i\geq 0}$ be an m-sequence and the outer sequence $\{b_i\}_{i\geq 0}$ a Barker sequence. For $N_a \leq \tau \pmod{N_aN_b} \leq N_a(N_b - 1)$ a periodic autocorrelation value $R_{AA}(\tau)$ of a Kronecker sequence $\{A_i\}_{i\geq 0}$ is denoted as*
in case of $N_b = 2$

$$R_{AA}(\tau) = \begin{cases} 2N_a & \text{if } N_a|\tau, \\ 2 & \text{otherwise,} \end{cases} \tag{21}$$

in case of $N_b = 4$

$$R_{AA}(\tau) = 0, \tag{22}$$

in case of $N_b \neq 2, 4$

$$R_{AA}(\tau) = \begin{cases} \pm N_a & \text{if } N_a|\tau, \\ \pm 1 & \text{otherwise.} \end{cases} \tag{23}$$

□

5 Conclusion

For Kronecker sequences obtained from two sequences we gave the exact formula of their characteristic polynomials and their linear complexities.

Corollary 1 shows that we can get a Kronecker sequence with large linear complexity when its inner sequence has the longer period and its outer sequence has the larger linear complexity.

Corollary 2 shows that a Kronecker sequence has small out-of-phase autocorrelation values when its inner sequence is an m-sequence and its outer sequence is the Barker sequence. Corollary 2 also suggests that such Kronecker sequences with the Barker sequence with $N_b \neq 2, 4$ are useful for synchronization in spread spectrum communications.

References

1. Stark W.E. and Sarwate D.V. (1981) Kronecker sequences for spread-spectrum communication. IEE Proc. Part F. **128**, 104–109
2. Kärkkäinen K.H.A. (2001) Linear complexity of Kronecker sequences. IEICE Trans. Fundamentals. **E84-A**, 1348–1351
3. Mills W.H. (1975) Continued fractions and linear recurrences. Math. of Computation, **29**, 173–180
4. Uehara S. and Imamura K. (1997) Characteristic polynomials of binary complementary sequences. IEICE Trans. Fundamentals, **E80-A** 193–196

The Generation of PseudoRandom Numbers for the Simulation of White Gaussian Noise

Lloyd R. Welch

Department of Electrical Engineering-Systems,
University of Southern California, Los Angeles, CA 90089-2565 USA

Summary. This paper describes and analyses some methods of generating pseudorandom sequences suitable for use in the simulation of white Gaussian noise. The criteria are somewhat different than those customary in pseudorandom number generation in other applications such as Monte Carlo methods. The methods described in this paper are computationally efficient and provide independence between X_n and $X_{n+\tau}$ for a very large range of τ.

1 Introduction

Linear Congruential generators (LCG) and multiple recursive generators (MRG) are the most popular pseudo-random number generators for diverse applications. They can be designed so that the generated sequence, $\{X(t)\}$, has k-tuples,

$$(X_{t+j_1}, X_{t+j_2}, \cdots, X_{t+j_k})$$

which are almost uniformly distributed for specific k-tuple, $(j_1, j_2, \cdots, j_k)$ or for a small collection of k-tuples.

The probability model is obtained by counting occurrences of an event within one period of the sequence and dividing by the period. Uniform distribution of k-tuples even for small values of k requires making the period very large.

However, for the simulation of white Gaussian noise, different criteria are required. The usual procedure for simulating white Gaussian noise is to generate a sequence of uniformly distributed pseudo-random numbers and apply a function to yield a normal distribution. The necessary criteria is that $(X_t, X_{t+\tau})$ are independent pairs for $1 \le \tau \le T$ for large T. And more generally, that $(X_t, X_{t+\tau_1}, \cdots, X_{t+\tau_k})$ are mutually independent for a large range of τ's.

This paper presents some linear recursions over finite fields which satisfy the pairwise criterion for very large values of T and can be efficiently generated. The question of generating sequences for which the independent triple criterion is satisfied is also considered.

2 A Class of Linear Recursions

The classical method of generating pseudorandom numbers (PRN) is by a linear congruential (LCG) recursion:

$$X_{n+1} = (aX_n + b) \bmod M$$

where M is at least as large as the desired sample range. The parameters a and b are then chosen to satisfy various structural criteria such as maximizing the period and various statistical criteria such as the almost uniformly distributed k-tuples criterion. A generalization of this is the multiple recursive generator (MRG):

$$X_n = \left(\sum_{j=1}^{r} a_j X_{n-j} \right) \bmod M.$$

Here the parameters, $(r, a_1, \dots, a_r)$ are chosen to satisfy various criteria, the main one being that the period be close to M^r. If this can be achieved, the r-tuples are close to uniformly distributed. It is possible to achieve a period of $M^r - 1$ only when M is prime.

Notice that in both these cases the arithmetic is ordinary integer arithmetic modulo an integer, M. However it has long been known [1] that pseudo random sequences can be generated using finite fields. Indeed, when M is a prime number, the above arithmetic is that of a finite field. A recent description of this idea when $M = p^w$, a power of a prime with exponent greater than 1, is in ref [2, Chapter 4]. In this case, M is the size of a Galois Field and there is a conversion from field elements to integers as follows. The field elements are expressed as polynomials in a generator element and the coefficients are treated as the terms in the radix p representation of integers. A MRG recursion generates field elements and the above conversion produced the pseudorandom integer samples. When $M = p^w$ one implementation is to select a polynomial, $Q(z)$, over GF(p) which has degree w and is irreducible. The coefficients of the MRG are polynomials over GF(p) of degree less than w, and the recursion is

$$X_n(z) = \left(\sum_{j=1}^{r} m_j(z) X_{n-j}(z) \right) \bmod Q(z). \quad (1)$$

The main criterion for $Q(z)$ is that it is irreducible and the main criterion for the recursion is that r-tuples so generated are almost uniformly distributed. This is obtained by selecting $(m_1, \dots, m_r)$ to produce a sequence with a period of $\{X_n\}$ equalling $p^{rw} - 1$.

3 The Main Point of This Paper

The point of this paper is to show that when a recursion of degree r over a finite field with M elements has period M^r-1, then X_t and $X_{t+\tau}$ are 'almost' independent for τ in the range $1 \leq \tau < M^{r-1}$.

More precisely,

$$\text{Prob}(X_t = a) = \frac{1}{M} + O(M^{-r})$$

and

$$\text{Prob}(X_t = a, X_{t+\tau} = b) = \text{Prob}(X_t = a)\text{Prob}(X_{t+\tau} = b) + O(M^{-r})$$

for $1 \leq \tau < M^{r-1}$.

To show this we must count the number of t for which

$$(X_t, X_{t+\tau}) = (a, b)$$

We begin with an observation: For every τ, $X_{t+\tau}$ is a linear combination of $(X_{t-1}, \cdots, X_{t-r})$.

Proof: The recursion can be written in vector form. (See [1, Chapter III, section 2.9].)

$$(X_n, X_{n-1}, \cdots, X_{n-r+1}) = (X_{n-1}, X_{n-1}, \cdots, X_{n-r}) \begin{pmatrix} m_1 & 1\,0 \cdots 0 \\ m_2 & 0\,1 \cdots 0 \\ \vdots & \vdots\,\vdots \ddots \vdots \\ m_{r-1} & 0\,0 \cdots 1 \\ m_r & 0\,0 \cdots 0 \end{pmatrix}$$

Iterating this equation for n replaced in turn by $t, t+1, \cdots, t+\tau$ gives

$$(X_{t+\tau}, \cdots, X_{t+\tau-r+1}) = (X_{t-1}, \cdots, X_{t-r})R^{(\tau+1)}$$

where R is the above matrix. The first column of $R^{(\tau+1)}$ is the desired linear combination.

Let the (i, j)th component of R^k be r_{ij}^k. Then the problem of computing $\text{Prob}(X_t = a, X_{t+\tau} = b)$ is the problem of counting the number of non-zero r-tuples, $(X_{n-1}, \cdots, X_{n-r})$ for which

$$a = \sum_{i=1}^{r} X_{n-i} r_{i1}^1$$

$$b = \sum_{i=1}^{r} X_{n-i} r_{i1}^{(\tau+1)}$$

If the two equations are linearly independent the number of solutions is M^{r-2} provided $(a,b) \neq (0,0)$. For $(0,0)$, the number of non-zero r-tuples is $M^{r-2}-1$. From this, the 'almost' independence follows.

The remaining question is "For what range of τ are the two equations linearly independent or, equivalently, what is the smallest positive τ for which X_t and $X_{t+\tau}$ are not 'almost' independent?"

If the two equations for X_t and X_{t+T} are linearly dependent, they are related by an element of the field:

$$(r_{11}^{(T+1)}, \cdots, r_{r1}^{(T+1)}) = \gamma(r_{11}^{1}, \cdots, r_{r1}^{1})$$

where γ is some field element. It follows from the recursion that

$$X_{t+T} = \gamma X_t \text{ for } 0 \leq t < M^r - 1.$$

Therefore r-tuples with distance T apart along the recursive sequence differ by a multiplicative factor of γ. Since all non-zero r-tuples occur someplace in the recursion, R^T must be γ times the identity matrix.

Now, since γ is in GF(p^w), $\gamma^{p^w-1} = 1$. It follows that $R^{T(p^w-1)} = I$ and $T(p^w-1)$ must be a multiple of the period, $p^{wr}-1$. Therefore

$$T \geq \frac{p^{wr}-1}{p^w-1} \approx M^{r-1}$$

In conclusion, X_t and $X_{t+\tau}$ are 'almost' independent for all positive τ less than M^{r-1}.

Of course, the independence of X_t and $X_{t+\tau}$ for a large range of values of τ is only a first step in designing a generator for white gaussian noise. We would like to pick generators such that $X_t, X_{t+\tau_1}$, and $X_{t+\tau_2}$ are mutually 'almost' independent for a large range of (τ_1, τ_2). I have no results on how to achieve this but there is a 'hand wave' argument which suggests there are generators with triples 'almost' mutually independent for $1 < \tau_1 < \tau_2 < T$ where $T \approx \sqrt{2M^{(r-2)}}$.

The argument goes as follows:

The three variables will fail to be linearly independent if the three vectors,

$$\begin{matrix} (& r_{11}^{1} & , \cdots , & r_{r1}^{1} &) \\ (& r_{11}^{(\tau_1+1)} & , \cdots , & r_{r1}^{(\tau_1+1)} &) \\ (& r_{11}^{(\tau_2+1)} & , \cdots , & r_{r1}^{(\tau_2+1)} &) \end{matrix}$$

have rank less than 3.

If we think of these three vectors as selected from a uniform distribution, the probability of rank less than 3 is approximately $\frac{M^2}{M^r} \equiv \frac{1}{M^{(r-2)}}$. If we test all cases with $1 < \tau_1 < \tau_2 < T$, we would expect $\binom{T}{2} \frac{1}{M^{r-2}}$ occurrences of deficient rank. Setting T so that this number is less than 1 gives the result.

4 Implementation and A Proposed Subfamily

The easiest case to implement is when $p = 2$, since radix 2 is the dominant usage for the machine representation and the coefficients of a polynomial over GF(2) can be packed in one or more machine words. Let $Q(z)$ be a polynomial over GF(2) with degree w which is irreducible and let α be a root of $Q(z)$ in $GF(2^w)$. The recursion can be expressed as

$$X_n = \sum_{j=1}^{r} m_j X_{n-j}. \tag{2}$$

where all of the quantities are polynomials in α and the arithmetic is in $GF(2^w)$.

The difficulty of implementing fast multiplication in finite fields has limited the use of this family and the recursions recommended in [2, p 90] have trinomial polynomials defining the fields. However in our proposed subfamily, there is no restriction on the polynomials defining the field but rather a restriction on the coefficients of the recursion to be from the set

$$\{0, 1, \alpha, \alpha^{-1}\}$$

Define sets of indices, $S_0 = \{j : m_j = 1\}$, $S_1 = \{j : m_j = \alpha\}$, and $S_2 = \{j : m_j = \alpha^{-1}\}$. The recursion can then be written:

$$X_t = \sum_{j \in S_0} X_{t-j} + \alpha \sum_{j \in S_1} X_{t-j} + \alpha^{-1} \sum_{j \in S_2} X_{t-j}$$

The desired pseudorandom numbers are then the X_t viewed as integers instead of packed polynomials.

Software implementation of multiplication by α and α^{-1} when the coefficients of elements are packed into machine words is quite simple.

To multiply by α, left shift the word by 1 and test bit position w. If it is 1 then XOR with Q.

To multiply by α^{-1}, test LSB. If it is 1, then XOR with Q. Then right shift by 1.

In the case where the machine word has exactly w bits the XOR's use modified Q's and the test must precede the left shift.

With the Intel Pentium processors, multiplication is even simpler in assembly language since the shifts generate flag bits which can be used for the XOR decision.

5 Tables[1]

Some Sample Fields with Primitive Recursions					
GF degree	Recursion degree	$Q(z)$ (hexidecimal)	S_0	S_1	S_2
16	12	0x10197	$3, 4, 12$	2	$0, 1$
24	8	$z^{24} + \text{0x17}f$	$2, 3, 4, 8$	1	0
32	6	$z^{32} + \text{0x173}$	$1, 2, 3, 4, 6$		0
40	4	$z^{40} + \text{0x13}b$	$2, 4$	$1, 3$	0
48	4	$z^{48} + \text{0x17}f$	$1, 3, 4$		$0, 2$
56	3	$z^{56} + \text{0x19}b$	3		$0, 1, 2$
64	3	$z^{64} + \text{0x1}df$	$1, 2, 3$		0

References

1. Solomon W. Golomb. *Shift Register Sequences*. Holden-Day, Inc, 1967.
2. Shu Tezuka. *Uniform Random Numbers: Theory and Practice*. Kluwer Academic Publishers, 1995.

[1] The hexidecimal representation of a polynomial over $GF(2)$ is obtained by writing the sequence of coefficients, blocking the sequence in groups of 4 and replacing each group by the appropriate hexidecimal digit. For the polynomial, $z^{16} + z^8 + z^7 + z^4 + z^2 + z + 1$, the coefficient sequence is 1 0000 0001 1001 0111. The hexidecimal result is 0x10197.

Author Index

Other titles in the DMTCS series:

Combinatorics, Complexity, Logic: Proceedings of DMTCS '96
D. S. Bridges, C. S. Calude, J. Gibbons, S. Reeves, I. Witten (Eds)
981-3083-14-X

Formal Methods Pacific '97: Proceedings of FMP '97
L. Groves and S. Reeves (Eds)
981-3083-31-X

The Limits of Mathematics: A Course on Information Theory and the Limits of Formal Reasoning
Gregory J. Chaitin
981-3083-59-X

Unconventional Models of Computation
C. S. Calude, J. Casti and M. J. Dinneen (Eds)
981-3083-69-7

Quantum Logic
K. Svozil
981-4021-07-5

International Refinement Workshop and Formal Methods Pacific '98
J. Grundy, M. Schwenke and T. Vickers (Eds)
981-4021-16-4

Computing with Biomolecules: Theory and Experiments
Gheorghe Paun (Ed)
981-4021-05-9

People and Ideas in Theoretical Computer Science
C. S. Calude (Ed)
981-4021-13-X

Combinatorics, Computation and Logic: Proceedings of DMTCS'99 and CATS'99
C. S. Calude and M. J. Dinneen (Eds)
981-4021-56-3

Polynomials: An Algorithmic Approach
M. Mignotte and D. Stefanescu
981-4021-51-2

The Unknowable
Gregory J. Chaitin
981-4021-72-5

Sequences and Their Applications: Proceedings of SETA '98
C. Ding, T. Helleseth and H. Niederreiter (Eds)
1-85233-196-8

Finite versus Infinite: Contributions to an Eternal Dilemma
Cristian S. Calude and Gheorghe Paun (Eds)
1-85233-251-4

Network Algebra
Gheorge Stefanescu
1-85233-195-X

Exploring Randomness
Gregory J. Chaitin
1-85233-417-7

Unconventional Models of Computation (UMC2K)
I. Antoniou, C.S. Calude and M.J. Dineen (Eds)
1-85233-415-0

S. Rudeanu
Lattice Functions and Equations
1-85233-266-2

GPSR Compliance

The European Union's (EU) General Product Safety Regulation (GPSR) is a set of rules that requires consumer products to be safe and our obligations to ensure this.

If you have any concerns about our products, you can contact us on ProductSafety@springernature.com

In case Publisher is established outside the EU, the EU authorized representative is:

Springer Nature Customer Service Center GmbH
Europaplatz 3
69115 Heidelberg, Germany

Batch number: 10682323

Printed by Printforce, the Netherlands